T0215058

WHAT WE OWE TO NONHUMAN ANIMALS

This book strongly challenges the Western philosophical tradition's assertion that humans are superior to nonhuman animals. It makes a case for the full and direct moral status of nonhuman animals.

The book provides the basis for a radical critique of the entire trajectory of animal studies over the past fifteen years. The key idea explored is that of 'felt kinship'—a sense of shared fate with and obligations to all sentient life. It will help to inspire some deep rethinking on the part of leading exponents of animal studies. The book's strong outlook is expressed through an appeal for radical humility on the side of humans rather than a constant reference to the 'human-animal divide'. Historical figures examined in depth include Aristotle, Seneca, and Kant; contemporary figures examined include Christine Korsgaard and Martha Nussbaum. This book presents an account according to which the tradition has not proceeded on the basis of impartial motivations at all, but instead has made a set of pointedly self-serving assumptions about the proper criteria for assessing moral worth.

Readers of this book will gain exposure to a wide variety of thinkers in the Western philosophical tradition, historical as well as contemporary. This book is suitable for professionals working in nonhuman animal studies, students, advanced undergraduates, and practitioners working in the fields of philosophy, environmental studies, law, literature, anthropology, and related fields.

Gary Steiner is Professor of Philosophy Emeritus at Bucknell University.

Routledge Human-Animal Studies Series

Series edited by Henry Buller

Professor of Geography, University of Exeter, UK

The new *Routledge Human-Animal Studies Series* offers a much-needed forum for original, innovative and cutting-edge research and analysis to explore human–animal relations across the social sciences and humanities. Titles within the series are empirically and/or theoretically informed and explore a range of dynamic, captivating and highly relevant topics, drawing across the humanities and social sciences in an avowedly interdisciplinary perspective. This series will encourage new theoretical perspectives and highlight ground-breaking research that reflects the dynamism and vibrancy of current animal studies. The series is aimed at upper-level undergraduates, researchers and research students as well as academics and policy-makers across a wide range of social science and humanities disciplines.

Winged Worlds
Common Spaces of Avian-Human Lives
Edited by Olga Petri and Michael Guida

Methods in Human-Animal Studies
Engaging with animals through the social sciences
Edited by Annalisa Colombino and Heide K. Bruckner

What We Owe to Nonhuman Animals
The Historical Pretensions of Reason and the Ideal of Felt Kinship
Gary Steiner

For more information about this series, please visit: www.routledge.com/Routledge-Human-Animal-Studies-Series/book-series/RASS

WHAT WE OWE TO NONHUMAN ANIMALS

The Historical Pretensions of Reason and the Ideal of Felt Kinship

Gary Steiner

Routledge
Taylor & Francis Group

LONDON AND NEW YORK

Designed cover image: Michael Sowa, Abschied
In Anlehnung an: Irene Dische, Hans Magnus Enzensberger, Michael Sowa,
Esterhazy
© 2009 Carl Hanser Verlag GmbH & Co. KG, München

First published 2024
by Routledge
4 Park Square, Milton Park, Abingdon, Oxon OX14 4RN

and by Routledge
605 Third Avenue, New York, NY 10158

Routledge is an imprint of the Taylor & Francis Group, an informa business

British Library Cataloguing-in-Publication Data
A catalogue record for this book is available from the British Library

ISBN: 978-1-032-54584-4 (hbk)
ISBN: 978-1-032-54585-1 (pbk)
ISBN: 978-1-003-42559-5 (ebk)

DOI: 10.4324/9781003425595

Typeset in Times New Roman
by Apex CoVantage, LLC

Dedicated to Joseph P. Fell III (1931–2017)
Colleague, Mentor, Friend

CONTENTS

ACKNOWLEDGMENTS

I had the benefit of a great many sources of inspiration in the course of writing this book. Bucknell University provided sustained generous support, including the award of a Presidential Professorship that made it possible for me to take two year-long sabbaticals in close succession. Additional significant support was provided by the John William Miller Fellowship Fund at Williams College; in addition to the Fund's substantial financial support, I had the benefit of input and guidance from the Fund's chief administrator, Michael McGandy, as well as from Miller Fellows Peter Fosl, Ryan Johnson, and Katie Terezakis. I am particularly indebted to Professor Terezakis for her careful attention to and reflections on my work over a period of years, and to Dr. McGandy for providing the initial inspiration to undertake a serious study of Miller's important contributions to American philosophy.

My many discussions with some key colleagues at Bucknell University helped to shape my thinking about the historical ideal of reason in the Western philosophical tradition. Several reading groups with Linden Lewis, John Rickard, and Adam Burgos opened some very fruitful avenues for my thinking. The reading groups that I organized in collaboration with Professor Burgos on Charles Mills and José Medina proved particularly stimulating for my thinking about problems surrounding the phenomenon of alterity. I have learned, and continue to learn, a great deal about political theory from Michael James.

Several colleagues in Europe provided additional stimulus to my thinking about the issues discussed in this book. My long association with John Baldwin has afforded me some very trenchant insights into the nature and limitations of the approach to reason that I refer to as "Eleatic" in this book. In Athens, Evangelos Protopapadakis has been both a gracious host and a thoughtful conversation partner about matters pertaining to nonhuman animal ethics; it has been my good fortune, thanks to Professor Protopapadakis, to have become associated with the

Panhellenic Animal Welfare and Environmental Federation. My long-time colleague in Vienna, Erwin Lengauer, has devoted a great deal of energy to supporting my work, and most recently engineered an epic driving trip through central and eastern Europe that afforded me the opportunity to share my ideas about nonhuman animal ethics in Sofia, Bucharest, Odessa, Kiev, and Budapest. I have also benefited greatly from a number of in-depth discussions with Herwig Grimm, who has hosted me on several occasions at the Veterinary University in Vienna and whose powerful notion of the "animal-in-itself" has informed my thinking for this book at a fundamental level.

This book is dedicated to my dear departed colleague, Joseph P. Fell III, who was a student of Miller and had the breadth and depth of philosophical knowledge as well as the philosophical instincts to see the power and importance of Miller's thought. His widow, Judith Sigler Fell, has continued to promote the study of Miller in Joe's stead; Judith has a keener sense of Miller's thought and its importance than most academic philosophers with whom I am acquainted.

As always, my deepest debt of gratitude goes to Paula Davis, whose unremitting confidence in me and my work means more to me than I can express in words.

INTRODUCTION

> The very nature of philosophical truth, if it be possible to assert it, destroys every vestige of egoism, because it can be won only by the conquest of egoism, only by the hard labor of self-discipline in which every accidental prejudice is brought under the impersonal law of a self-declared truth common to all. And if it is not common to all, one reverts to arbitrary violence in order to safeguard those arbitrary values that express one's own chance for satisfaction.
>
> John William Miller, "The Quality of Philosophy is Not Strained"

What are the prospects for arriving at a truly impartial sense of the nature and moral worth of nonhuman animals, if it should turn out that the major exponents of the Western philosophical tradition have proceeded on the basis of some highly dubious assumptions? And what would motivate so many philosophers, up to this very day, to adhere to these assumptions? Fifty years have passed since Richard Ryder coined the term 'speciesism' to capture the essence of these dubious assumptions, and yet in the intervening time I do not believe that a great deal of progress has been made in the endeavor to debunk the self-serving and ultimately arbitrary assumption that members of the human species are categorically superior to members of all other species in the cosmic scheme.

When I refer to progress in this connection, I am thinking specifically of efforts to broaden the scope of moral concern so as to be more rather than less inclusive. Even people who have no concern for the fate of nonhuman animals are familiar with this effort within the strictly human community: Efforts to redress sexual violence, racial injustices, and comparable harms proceed on the basis of an insight into the historical practice of making arbitrary and unjust distinctions between human beings on the basis of characteristics such as gender and race. Those committed to gender and racial justice are appealing implicitly to a generic conception

DOI: 10.4324/9781003425595-1

of the human individual, one that sees equal inherent worth or dignity in all human agents. Progress consists in militating against selfish egoism and abandoning what John William Miller aptly refers to as "arbitrary violence" intended to "safeguard . . . arbitrary values."[1]

The arbitrariness of privileging members of one race or gender within the human community finds a parallel in the tradition's arbitrary privileging of human beings over all other beings in the natural world—not simply over nonhuman animals, but over absolutely everything in creation. Western thought has become so deeply inured to this way of seeing the place of human beings in the world that it is difficult to find the critical distance needed to call the tradition's grounding assumptions into question. One of those assumptions is that *logos* (linguistic rationality) is the ground-floor criterion of moral status, such that beings possessing *logos* are categorically superior to (i.e., matter more than) beings that are *aloga* (bereft of linguistic rationality). Another grounding assumption is that the possession of *logos* is exclusive to human beings; this latter assumption is the impetus behind derogatory terminology such as 'brutes' and 'dumb animals', language whose usage persists to this day in spite of the growing sense in many people that we are actually moving toward a more just and responsible sensibility about the fortunes of nonhuman animals.[2]

If these twin assumptions—that *logos* is the definitive criterion for superior moral status and that human beings are the exclusive possessors of *logos* in the natural world—are mistaken, then what accounts for the tradition's (and, by extension, our entire culture's) overwhelming adherence to them over such a long period of time? The answer, I hope to show, is that in reality we do not assess ideas and arguments with nearly as much impartiality as we like to think we do; instead, the very process of rational reflection and assessment is ultimately impossible to disentangle from the *affective* dimension of our lives, i.e., from *our embodied, felt encounter* with the world and with others. Exponents of the philosophical tradition have been quick to treat our affective (emotional) encounter with things as subsidiary to our putatively more authoritative mode of access to the truth (namely, through reason and language), and in a number of instances they have treated emotional states as potential threats to our moral integrity. What complicates the effort to debunk the self-serving assumptions of the tradition is the fact that the tradition has not been entirely wrong to attribute certain capacities to human beings that no nonhuman animals appear to possess. As I have argued in previous work, I believe that the tradition has been right to see special capacities in human beings that appear not to be shared by nonhuman animals. In particular, it does appear that human beings are unique in being able to formulate general rules for conduct and for holding (mature) members of the human community formally accountable for their actions. Different conclusions can be drawn from this insight. One is that human beings are burdened with *obligations* that nonhuman animals are incapable of confronting. (This would be why, for example, the fact that lions kill gazelles does *not* mean that it is fine for human beings to kill nonhuman animals.) The tradition, however, has drawn a very different conclusion: that human beings enjoy special *prerogatives* to use nonhuman animals as tools or instrumentalities for the satisfaction of human needs and desires. In other words, rather than acknowledging a certain *difference* between human and nonhuman animals, the tradition has proclaimed *the categorical superiority of human beings* over the entirety of the nonhuman.

Where major exponents of the tradition have purported to offer impartial arguments in support of human superiority, my central thesis in this book is that the human proclamation of superiority over the nonhuman is ultimately based not on arguments but on precisely the sorts of selfish tendencies that are decried by virtually all leading moral theories. These selfish tendencies are rooted in our embodied, affective constitution rather than in any sort of detached rationality, and in this regard we are much more like than unlike the nonhuman animals we have treated for millennia as what Peter Singer calls "replaceable resources."[3] Stated bluntly, there is a certain arrogance in the tradition's confident proclamation of human superiority, not only in virtue (as I hope to show) of the weakness of its supposed arguments, but also in virtue of the rather atrocious historical record humanity has left in matters ranging from racial violence in the human community to large-scale environmental devastation.

Humanity's selfish tendencies reflect not only arrogance but also an abiding pathos of *fear*. We, just like our nonhuman sentient relatives, face epic challenges in the struggle for survival. Just like nonhuman animals, we are vulnerable, embodied beings who confront contingency, uncertainty, and ultimately death. The many philosophical, literary, and artistic works devoted to our preoccupation with death have reinforced a sense in humanity that death is somehow of greater significance, or is somehow more dire, for human beings than for nonhuman animals, simply in virtue of the fact that we can contemplate death in the abstract (which is to say, in moments of reflection not directed exclusively at the present instant, such as when we lament the loss of a long-deceased loved one or anticipate our own death). I simply do not see how being able to contemplate an event such as death in abstract terms entails that death is somehow more meaningful or more important in the case of humans than in the case of nonhuman animals. The abiding sense that death matters more in the human case has been reinforced by two millennia of anthropocentric thinking, the kind of thinking that arbitrarily privileges human beings over everyone and everything nonhuman.[4]

What I hope to show in this book is that the fear that has motivated the tradition has given rise, as if in a Freudian reaction formation, to an inflated sense of our own nature and worth. Specifically, we have elevated reason and denigrated affect or emotion, in an effort to proclaim our own uniqueness and superiority in the face of a deep awareness of our own vulnerability and mortality. This effort has led to a distorted conception of the nature and reach of our rational powers, as well as of the role of our affective dispositions in the formation of moral commitments. Here I deliberately refer to moral *commitments* rather than to moral *judgments*, because

I believe that judgment is not the canonical form that moral commitment tends to take. The virtually exclusive focus on moral judgment stressed by the rationalist tradition is the product of efforts to find a criterion or capacity that irretrievably distinguishes human beings and elevates them above all other beings. In the later chapters of this book, I will examine considerations offered by pragmatist and phenomenological thinkers in an effort to rethink the nature of moral commitment in a manner that acknowledges *the essential role played by our embodiment and our affective dispositions in the very formation of moral commitments*. Rather than being an adjunct to detached rational judgment, our emotional or affective sensibilities are part of the basis of moral commitment. Thus, where Kant argues that moral love provides needed support for our moral commitments but maintains that that love is not part of the basis of moral commitment, I will urge the conclusion that *moral love is in fact part of the basis of moral commitment*. This in turn opens the prospect of exploring the respective roles and mutual interplay of reason and emotion in the establishment and revision of moral commitments.

The recognition that our affective life affords us with our primary sense of orientation in the world opens the additional prospect of rethinking our relationship with nonhuman animals. Just as we have attributed too much to ourselves, I believe that we have attributed all too little to nonhuman animals. To suppose, as a surprising number of philosophers do, that all the various impressive ways in which nonhuman animals navigate their environments are due to blind instinct is to assume a little too hastily that to possess rationality is to possess linguistic and conceptual ability in a specifically human sense. As a number of heterodox thinkers have noted, 'logos' or 'reason' can mean a variety of things, not all of them reducible to predicative judgment. Thinkers such as Porphyry and Schopenhauer recognize that any being able to reckon with causal relations in the world is a being that possesses rationality, regardless of whether that being can conceptualize causal relations in detached, theoretical terms.

Here, too, our affective life is of crucial importance: I have already suggested that the prevailing sense of human superiority has not been driven primarily by detached rational considerations, but rather by affects such as arrogance and fear. We have not thought about, studied, and related to nonhuman animals in a spirit of open inquiry. Instead, we have been guided by a prior commitment to human superiority—one that, if I am right, is ultimately an externalization of a deep sense of vulnerability and possibly even inferiority. (Inferiority? Just think about that historical record to which I referred a moment ago. Is there another animal species that comes anywhere near us in inflicting devastating harm on the environment, not to mention on others of its own kind?) If this is true, then what is needed is not simply another stab at thinking about and studying nonhuman animals, but one informed by a different pathos altogether—one of *humility*.

In this book I think about this shift as one from an anthropocentric to a non-anthropocentric ideal of living. The anthropocentric ideal has prevailed in the West for several millennia and is one according to which human beings are both unique and superior to all other sublunary beings. A comparatively non-anthropocentric ideal would be one guided by a pathos of modesty or humility, and it would be characterized by the kind of genuine *openness* that has proved elusive for the philosophical tradition. In this book I will associate this notion of openness and humility with Heidegger's notion of "letting beings be." This constitutes a clear departure from the traditional pathos of boldness and self-certainty, and it brings with it a burden that the traditional approach does not: The requisite sense of openness demands of us that we constantly challenge our own prejudices and ulterior motives in a manner that I believe the prevailing pathos of the tradition does not. It is an openness that holds the potential to move us toward an ethos of wholeness rather than fragmentation in the moral community, by which I mean *a more inclusive sense of community* in place of the highly *ex*clusive sense of moral community to which even some well-known advocates of nonhuman animals have continued to adhere.

But some progress has been made in traditionally minded philosophical efforts to rethink the nature of nonhuman animal experience as well as the moral status of nonhuman animals. Christine Korsgaard has recently made a significant step in this direction with her publication of *Fellow Creatures: Our Obligations to the Other Animals*, in which she seeks to revise Kant's views so as to render them congenial to the proposition that nonhuman animals are full-fledged moral subjects, i.e., direct as opposed to merely "indirect" beneficiaries of moral concern. Korsgaard argues that nonhuman animals are best conceived as being ends in themselves, a notion that Kant rejected on the grounds that nonhuman animals are non-rational. From this commitment Korsgaard draws some important conclusions about the various ways in which we have historically mistreated nonhuman animals. But as I hope to show in the course of this book, Korsgaard remains firmly established within the anthropocentric worldview, and as a result she does not draw out the full implications of what it might actually mean for a nonhuman animal to be an end in itself.[5] I contrast Korsgaard's appeal to the notion of nonhuman animals as ends in themselves with Herwig Grimm's comparatively open conception of nonhuman animals as beings who *set* ends for themselves, a capacity that Korsgaard explicitly denies to nonhuman animals. I believe that it is only this sort of standpoint, one that departs radically from the anthropocentric mindset in its willingness to entertain possibilities that could be threatening to our established ways of life, that truly holds the promise of bringing about a large-scale transformation of our relationship to nonhuman animals—and perhaps by extension to the entirety of the natural world.

In the first two chapters of this book, I examine the terms of the Western tradition's assertion of and adherence to a pointedly anthropocentric ideal of living, as well as the conception of reason that finds pride of place in that ideal. I focus primarily on the commitment to categorical human superiority expressed in the thought of Aristotle, Seneca, and Kant, as well as on the commitment of each of these thinkers to the superiority of reason over embodied states such as emotion. In examining these thinkers' central commitments, I show how each of these commitments is presented in the form of an explicit contrast between human and nonhuman animals according to which humans are assumed to be superior in the moral scheme. In other words, where major exponents of the tradition purport to be proceeding on the basis of straightforward, detached, impartial insights into the nature of things, these thinkers are actually motivated by a prior commitment to vindicating human superiority.

In Chapter 3, I examine the efforts of several recent thinkers to challenge and revise the "Eleatic" conception of reason urged on us by the tradition, a conception according to which reason is an eternal faculty unaffected by the flux of time and circumstance. Inspired by the Pragmatists, José Ortega y Gasset and John William Miller argue for a move away from the Eleatic ideal and toward a conception of what Ortega calls "historical reason," a conception according to which reason operates immanently within history rather than within some transcendent noumenal space. The inscription of reason within history—within the concreteness of time and circumstance—holds the promise not only of resolving debates such as that over moral realism, but of developing a conception of embodied rationality that shows us to be considerably closer to nonhuman animals than we have historically been willing to think. This move toward embodied rationality in turn opens the prospect of seeing an inner relationship between reason and the affective dimension of our embodied existence, a dimension that draws us even closer to nonhuman animals.

In Chapter 4 I seek to fill a lacuna left by the turn to historical reason. Even though Ortega and Miller make a key contribution to our rethinking of the nature of rationality, they are left with the problem of warrant: In the process of working through the various conflicts that inevitably arise in collaborative efforts to arrive at the truth rationally, to which criteria are we to appeal in order to resolve those conflicts? The Eleatic tradition appealed to criteria such as Kantian Transcendental Ideas of Reason. But the turn to historical reason was intended to demonstrate that no such permanent criteria are available to us. It is here that I believe an appeal to our embodied, affective nature can fill the lacuna: It is not reason, I will suggest, but rather our affective (emotional) engagement with things that provides us with our sense of orientation, and it is at that level of engagement that our commonality with nonhuman animals may prove to be most evident. I draw on the thought of Heidegger and Merleau-Ponty to show that there is a level of active, meaningful engagement with the world that is prior to the predicative rationality so vaunted by the tradition, and I conclude the chapter by suggesting that nonhuman animals are essentially identical with us in this regard.

In Chapter 5, I develop an ideal of "felt kinship," an ideal of moral community that includes all sentient beings rather than simply those beings who are rational in a specifically human sense. This ideal is one committed to propositions that are wholly incompatible with the anthropocentric ideal, in particular that justice is not simply a reciprocal relation that prevails among linguistic contractors but one that extends to the entire community of sentient creatures. Felt kinship is an ideal whose realization requires a dialectical interplay between reason and emotion, rather than the ascendancy of one of these faculties over the other. The tradition was right to express concern about the potentially pernicious influence of selfish affects as well as about the potential of rational reflection to help us escape the influence of such selfish tendencies. But the tradition largely failed to accept the idea that emotion provides us with our basic sense of orientation in the world, thereby failing to consider the possibility that our affective sensibilities make key contributions to our efforts to refine our sense of what matters. It is through a dialectical process in which reason and emotion mutually inform one another, I believe, that we might someday find the will to expand the scope of direct moral concern in a manner that amounts to more than lip service. In this connection, I conclude with a reflection on Tzvetan Todorov's ideal of "a well-tempered humanism," a humanism that would move us beyond racial enmities—and by extension, I propose, beyond the long-prevailing sense that nonhuman animals are either not really members of the moral community or that they "count less" in the moral scheme than human beings. I propose that the key to realizing such a well-tempered humanism is to be found in an unexpected place: in the *oikeiosis* doctrine of the Stoic philosophers, who recognized the power and significance of maternal love as the model for all forms of genuine community but who failed to explore its full potential.

The ideal of felt kinship is one according to which nonhuman animals merit both compassion and justice rather than one or the other, just as both emotion and reason are involved in the formation and revision of moral commitments. From the standpoint of Eleatic common sense, such an ideal must seem to have come straight out of Aristophanes's cloud-cuckoo land. And yet I believe that a truly radical rethinking of the moral status of nonhuman animals demands nothing less.[6]

Notes

1 John William Miller, "The Quality of Philosophy is Not Strained," p. 50.
2 See for example Gary Steiner, review of *Kant and Animals*, p. 518.
3 Peter Singer, *Animal Liberation*, p. 229.
4 I examine those two millennia in depth in Gary Steiner, *Anthropocentrism and Its Discontents*, a text to which I will refer throughout this book.
5 Thus I cannot agree with Nagel's assessment that "if [Korsgaard's position on nonhuman animals] prevailed it would be one of the largest moral transformations in the history of humanity." Thomas Nagel, "What We Owe a Rabbit."
6 In previous work I have followed the widespread convention of referring to 'humans' and 'animals', with a note at the outset that this is simply a convenient shorthand and that I naturally acknowledge that human beings are animals. But language matters; it quickly reifies and reinforces our sense of things. Thus, throughout this book I refer to 'human' and 'nonhuman animals', in order to stress a fact that exponents of the tradition tend to mention and then quickly forget.

1
BACKGROUND IDEALS OF LIVING

> Thou know'st 'tis common—all that lives must die,
> Passing through nature to eternity.
>
> *Hamlet* I.ii.72–3

1. A Counterintuitive Idea

I would like to start with a radical proposal, one that is likely to be threatening to most people: *that we should commit ourselves to an ideal of nonviolence toward nonhuman animals that is so strict that pursuing it might throw our lives into upheaval*. This proposal raises three fundamental questions: Who is the "we" here, on what grounds should this "we" be committed to such a strict ideal of nonviolence, and what sorts of changes in our ways of living are required by this proposal such that most people are likely to find it threatening?

The "we" is anyone who genuinely cares about nonhuman animals. Of course, a great many people insist that they "genuinely" care about nonhuman animals, and yet the historical record is crystal clear on the nature and extent of human brutality toward nonhuman animals.[1] There is clearly a contradiction here, which demonstrates that what constitutes genuine caring is a subject of some debate. This is not a matter that can be resolved through abstract reasoning; there is no dictionary definition of what it means to care "genuinely" about something. What seems clear, however, is that a great deal of self-serving human prejudice clouds the judgment of some highly sophisticated, highly "rational" human beings. Of course, as Socrates knew, self-serving prejudice poses a problem for various kinds of human discourse, as he attempts to explain to his interlocutor in the *Phaedrus*. Socrates's concern was the ways in which people employ sophistry rather than a love of the truth to persuade others. The love of truth is supposed to remind us of

DOI: 10.4324/9781003425595-2

our connection with others, whereas those who employ sophistry are in effect separating themselves from others and trying to manipulate them.

The problem of selfishness is taken to a fundamentally higher level when it comes to our reflections on the moral entitlements of nonhuman animals. We are heirs to a very long history of human valuing that has underwritten a pattern of unremitting violence perpetrated on nonhuman animals.[2] This history extends back to the ancient Greeks. A grounding commitment in this history is the assumption that human beings are superior to nonhuman animals in two interrelated senses: we are superior, we have told ourselves, in terms of our cognitive powers, and in turn we have told ourselves that this cognitive superiority places us in a position of moral superiority over nonhuman animals. This latter sense of superiority has been conceived in different ways, but it typically takes the form of supposing either that nonhuman animals have no moral status or that they possess a moral status inferior to that of human beings. This sense of superiority has ossified into a sort of common sense in our culture. It seems fair to say that the vast majority of people take it for granted that nonhuman animals are morally inferior to human beings. Moreover, very few people writing about the moral status of nonhuman animals today have taken any really effective measures to challenge this prejudice.

When I refer to moral superiority and moral inferiority, I am referring to people's general sense of who counts more and who counts less in the moral scheme of things. This "more" and "less" is often discussed by academic philosophers in terms of conflicts of interests and how best to resolve them. Philosophers have a special penchant for thinking about these conflicts in terms of life-and-death scenarios. Tom Regan, for example, argued for the equal inherent moral worth of all beings who are cognitively sophisticated enough to count as what he called "subjects-of-a-life," and yet he maintained that in an emergency you always sacrifice a nonhuman animal to save a human being—precisely because the human has more to lose by dying than a nonhuman animal (a dog, in Regan's discussion) has to lose.[3] Peter Carruthers insists that, in a scenario involving a burning house containing a human being and a dog, only one of whom can be saved, "no one would maintain that you ought to place the lives of many dogs above the life of a single human"—i.e., you always save the human, never the dog at the expense of a human.[4]

Moral superiority and inferiority need not be conceived strictly in terms of deadly conflicts. Indeed, as Gary Francione has noted, the vast majority of conflicts between human and nonhuman animal interests are *not* matters of life and death.[5] (So it is worth asking ourselves what is at stake for philosophers who are determined to frame our conflicts with nonhuman animals *as if* they were matters of life and death.) Instead, these conflicts pertain to a variety of ways in which human beings have long been accustomed to using nonhuman animals to satisfy our desires—as food and clothing, for forced labor, in experimentation, and as sources of entertainment (in zoos, circuses, sport hunting, animal fighting, etc.). Precious few human beings on the face of the earth can legitimately claim that using nonhuman animals is genuinely necessary for their survival. For the rest of us, our uses of nonhuman animals are matters of habit, convenience (including economic gain), and pleasure—really nothing more. Isn't it the case that these uses need to be *justified*?

A great many people insist that these uses require no justification. I beg to differ. When one sentient being exercises dominion over another sentient being—when one forcibly interferes with the life of another—and if the being exercising dominion possesses the rational capacity to think about her actions and the possible moral entitlements of the being on whom the force is being inflicted, then the individual inflicting the force has an obligation to confront the question whether her actions are justified. This is simply what it means to be a moral agent: to be in a position to step back from your actions and commitments, evaluate them from the standpoint of rational detachment, and tell yourself in all sobriety whether your actions and commitments honor your obligations to the community of which you are a part. To be a moral agent is not to be an isolated individual but rather a member of a community of moral agents united (at least in theory, if not always in practice) by the love of truth that was Socrates's concern.

Philosophers place tremendous confidence in the faculty of reason, which includes the capacity to reflect on one's own arguments and commitments in an effort to test whether those arguments and commitments are legitimate. In working out the basic rules for the administration of justice in society, John Rawls proposed a process he called "reflective equilibrium" to test the legitimacy of our underlying convictions. Rawls recognized that these convictions are not purely rational but involve "moral sentiments," and he saw reason as the arbiter of those sentiments—as a check on which ones truly honor our commitments to our sense of community and which do not. We start with our basic moral intuitions or beliefs, and we subject these to critical scrutiny in an effort to bring them into line with "our considered judgments."[6] The process of reflective equilibrium is intended to help us confront various forms of perspectival bias, including self-serving prejudice. Rawls's aim was to arrive at a generic conception of the citizen, one that confers on each member of society the sorts of basic rights and obligations that guarantee justice in the sense of fairness to all. In a just society, all are to be treated fairly, which is to say impartially: citizens of a certain gender or race, for example, should not enjoy special prerogatives that are denied to others.

For Rawls, the process of reflective equilibrium is designed to eliminate bias in the articulation of notions of citizenship and civic responsibility. He did not envision it as a tool for eliminating prejudice in the larger-than-human sphere, which includes our relationship to nonhuman animals. Rawls makes this clear near the end of *A Theory of Justice* when he states that "equal justice is owed to those who have the capacity to take part in and act in accordance with the public understanding" of the basic terms of justice. For Rawls, the beings who are owed duties of justice are beings who possess "moral personality." Because nonhuman animals lack "the capacity for a sense of justice . . . they are outside the scope of the theory of justice," which means, as it did for the Stoic philosophers in antiquity, that nothing we do to nonhuman animals can possibly count as an injustice. Rawls acknowledges, however, that it might be possible to establish some sorts of moral obligations toward nonhuman animals, although these obligations would be based on "compassion and humanity" rather than on any strict principles of justice.[7]

But even though Rawls does not intend the process of reflective equilibrium to be used for reflections on matters such as our moral obligations toward nonhuman animals (Rawls, at any rate, never uses the process with that aim in mind), the process is well suited to that purpose. Carruthers uses (or purports to use) the process for that purpose in arguing *against* the idea that nonhuman animals are comparable to humans in having "an equal right to life, and an equal right not to be made to suffer."[8] This raises two issues: one is whether the process of reflective equilibrium is suitable for a reflection on the question of our moral obligations toward nonhuman animals, and the other is how we can be sure when we employ the process we are not simply reproducing and rationalizing our pre-existing prejudices about the relative moral standing of human and nonhuman animals.

I believe that the first issue is less controversial than the second. Unless we are prepared to undertake a total critique of reason and consider our rational capacities to be no more authoritative than the cacophony of drives and feelings that form a substantial part of our nature, then we should be committed to an ideal of reason as a faculty that enables us to step back from our local, immediate commitments and test their validity by attempting to occupy the standpoint of what Adam Smith thought of as the impartial spectator.[9] I have argued in my previous work that the total critique of reason so vaunted in contemporary postmodern discourse is a wrong road, and that reason should be acknowledged to possess at least limited autonomy in the endeavor to reflect on and revise our moral commitments.[10] One of the key themes in this book is the nature and limits of reason in moral reflection, and the ways in which reason must be seen to be in dialogue with the moral sentiments if we are to arrive at a judicious assessment of the sorts of arguments that people typically offer under the imprimatur of reflective equilibrium. I hope to show that when philosophers proceed on the basis of what they purport are purely detached rational considerations, they are not only missing the key affective basis of moral commitments, but are often (if only unwittingly) reproducing exactly the sorts of human arrogance that they think they are challenging. When I refer to the problem of anthropocentrism in this book, I am referring to the human arrogance that has unremittingly placed human beings at the apex of the natural order and that has allowed us to feel entitled to make a great many pronouncements about what nonhuman animals are, what their experiential capacities are and are not, and how we may consider ourselves entitled to treat (and, to a great and troubling extent, use) nonhuman animals.

A little while ago I stated that the "we" is anyone who genuinely cares about the fortunes of nonhuman animals. But what I really mean to say is that the "we" is *all of us*, inasmuch as the very considerations that underwrite our commitments to our fellow human beings apply with equal force to our relationships with nonhuman animals. What I hope to show in this book is that if we become clear about the grounding considerations that inform our sense of moral obligation to ourselves, our fellow humans, and our community as a whole, then we will recognize a very serious inconsistency in our ways of valuing and acting. In urging this conclusion I am calling for something much more profound than simply the foolish consistency that is "the hobgoblin of little minds."[11] For even those of us who purport to include nonhuman animals as full members of the moral community either do not in fact include them there or do so in a grudging manner that ascribes to them a fundamentally subordinate status, thereby ascribing to ourselves the prerogative to exercise authority over the fortunes of nonhuman animals in a manner that goes far beyond guardianship.

And that is a tragic affair. For in an essential sense, human and nonhuman animals are inseparably bound in the condition of mortality. This is why I chose Gertrude's observation from *Hamlet* ("all that lives must die/Passing through nature to eternity") as the epigram to this chapter, and why I chose the following statement from Schopenhauer as the epigram to my book *Animals and the Moral Community*: "In all essential respects, *the animal* is absolutely identical with us. . . . The difference lies merely in the accident, the intellect, not in the substance which is the will. The world is not a piece of machinery and animals are not articles manufactured for our use."[12] In the concluding chapters of this book I will draw on the theory of "felt kinship" that I first presented in *Animals and the Moral Community*, with an eye toward showing how both reason and emotion are essential components in disclosing a living sense of our shared mortality with all sentient life.[13]

That in fact provides the answer to the second question I posed at the outset. We should be committed to a strict ethic of nonviolence because we are in all truly *essential* respects identical with other sentient creatures. All sentient creatures struggle consciously for survival and a meaningful existence. Of course, there are important differences between human and nonhuman animals due to the fact that (many if not all) humans possess linguistic agency whereas there do not appear to be any nonhuman animals who possess the kind or degree of linguistic ability that appears to be necessary for articulating, contemplating, and endorsing abstract principles. Philosophers have concluded from this difference that nonhuman animals are incapable of taking on any sort of formal responsibility toward others, and in this the philosophical tradition has been right: to the extent that nonhuman animals appear to be incapable of abstract reflection on things such as motives and expectations, it does not make sense, for example, to hold a domesticated dog or cat morally or legally responsible for causing property damage (although depending on the circumstances, we might hold the human owner responsible).

But a great many philosophers have gone further and maintained that nonhuman animals are incapable of negotiating meaning in their encounters with the world. Even those philosophers who have sought to get past the antiquated idea that nonhuman animals are merely biological reaction devices tend to succumb to the temptation to treat nonhuman animal behavior as strictly instinct-driven, which amounts to almost the same thing. These philosophers take care to point out that nonhuman animals have subjective states of awareness and a conscious relationship to the ends that they pursue (companionship, shelter, food, and survival generally), but too many of these thinkers fail to recognize that there is a contradiction between characterizing behavior as instinctual and considering it to be an open encounter with meaning.[14]

Two recent thinkers who make this mistake, I believe, are Mark Rowlands, who characterizes nonhuman animal behavior in entirely deterministic terms, and Christine Korsgaard, who suggests that the behavior of nonhuman animals is entirely instinctual.[15] One of the things I hope to show in this book is that this mistake is the product of a very strong desire to identify something "special" in human beings, some capacity that confers on us the prerogative and perhaps even the responsibility to exercise authority over nonhuman animals. This desire is easy to see in the work of philosophers who are very open in their assertion of a moral hierarchy in which human beings are superior to nonhuman animals. It is more difficult to see in the work of philosophers who claim to be offering more edified views of the nature and lives of nonhuman animals, but in a great many cases it is present in their thought as well. In this connection, a very simple litmus test turns out to be a philosopher's view about pet ownership, an institution that can be viewed from very different moral perspectives and about which I will have some things to say in this book. For now, I will simply offer as a working hypothesis the proposition that any philosopher who suggests that pet ownership is (or can in principle be) to the advantage of nonhuman animals is, if only unwittingly, succumbing to anthropocentric prejudice and is not genuinely embracing an ethic of strict nonviolence.

The mistake that many people make about the nature of nonhuman animal experience is compounded by a second, deeply significant factor: The human desire to see in ourselves some special characteristic that elevates us above nonhuman animals brings with it an inflated sense of our own ability to know what the lives of nonhuman animals are like, when in reality we ought to have the humility to admit that we ultimately *do not know* what it is like to live as a sentient creature whose experience is not informed by predicative language. So we ought to proceed in a spirit of modesty and caution when we find ourselves inclined to make pronouncements about the nature of nonhuman animal experience—not only in acknowledgment of our own ignorance, but in recognition of the vast injustices that we have perpetrated on nonhuman animals for thousands of years and that we have justified (or better: rationalized) by appealing to the following sorts of wildly self-serving claims: that nonhuman animals are driven purely by instinct, that they have no (or comparatively "dim") inner awareness, that they do not set ends for themselves but are strictly determined by the natural imperatives of survival, or that they do not tell stories about themselves. To all of these claims, Gertrude's verdict is worth keeping in mind.

Now what about people who insist that they really love nonhuman animals? This expression of love is probably easiest to see in the relationship that many humans have with their nonhuman companion animals. The question whether pet ownership is legitimate is highly controversial, particularly as the abolition of this institution would require a major sacrifice on the part of humans who love their nonhuman companion animals. But institutions involving nonhuman animals such as pet ownership can be viewed from very different perspectives, and not all of these perspectives grant centrality to specifically human interests. I was clear at the outset that my proposal is a radical one, one that makes it incumbent on us to confront the following question: is there a sense in which people who are generous and loving pet owners are participating (if only against their intention) in an institution that does violence to nonhuman animals?[16] As much as I dearly love my nonhuman companion animals, I have to say yes. When philosophers go about arguing for the moral permissibility of this or that practice, they typically present some grounding premises and seek to draw consistent conclusions from those premises. But what if, as I have suggested, the very operation of reason is not so straightforward but instead operates against the background of deep, global, implicit background considerations that are so pervasive that we tend to be scarcely aware of their influence on our thinking?

It is these sorts of background considerations—I think of them as manifesting ideals of living—that I will explore in the remainder of this chapter. One of the conclusions for which I will argue is that a number of practices involving nonhuman animals appear to be benign (or to be positively beneficial to nonhuman animals) only when seen in the light of a background ideal of living that is pointedly anthropocentric. To say that a regime such as pet ownership does violence to nonhuman animals does not mean that anyone deliberately inflicts physical harm—although this does sometimes occur. Violence can be perpetrated in much more subtle ways that easily elude the viewpoint of established common sense; the key to acknowledging this fact lies in finding the right frame of reference for addressing the matter.

The prospect of establishing and living in accordance with a *non*-anthropocentric ideal affords us such a frame of reference—if we have the will to embrace it. I suggested in the introduction that philosophy is born to an important extent in *fear*. There is ultimately something right in Hobbes's assessment of human beings as inherently "diffident."[17] What Hobbes proposed about relationships between human beings tells us a great deal about the way we have historically related to nonhuman animals. We have long exercised dominion over nonhuman animals, and the intensity of this dominion has increased steadily as our technological capacity has increased. The prospect of embracing a non-anthropocentric ideal of living is unsettling to many due to the threat it poses to our accustomed ways of valuing and living. It equally poses a threat to our long-standing self-congratulatory assumption that we are vastly more intelligent than nonhuman animals, so intelligent that we can state with confidence what nonhuman animals are and are not, and in particular what the inner lives of nonhuman animals are like.

After examining the views of several key exponents of the anthropocentric background ideal of living, I will return to this notion of a non-anthropocentric background ideal. For now, I will simply present a guiding thread, one intended to provide a contrast with what is about to follow. The radical proposal that I have been urging demands, as I have suggested, a certain modesty and an unprecedented acknowledgment of human ignorance. Consonant with this spirit of inquiry, the Austrian ethicist Herwig Grimm has proposed that we think of nonhuman animals in terms of what Grimm calls "the animal-in-itself." Grimm's expression recalls Kant's characterization of the thing-in-itself as the unknowable "beyond" that helps us to conceive the fundamental limits of human knowledge. For Kant, the thing-in-itself is a counter-concept that enables us to recognize that all of our knowledge claims are relative to our own constitution as knowers, i.e., relative to our faculties of sensibility and our conceptual apparatus. The thing-in-itself can never be an object of knowledge, but instead is a placeholder for whatever it is that lies irretrievably beyond our comprehension. When Grimm proposes that we think of nonhuman animals in terms of the "animal-in-itself," he is invoking Kant's recognition of the fundamental limits of human knowledge, the fact that our inherently finite nature makes it impossible for us to know things as they ultimately are—which, Kant suggests, is the way an infinite intelligence such as God would see them.[18] The animal-in-itself is an "ultimately unknowable, counterfactual ideal . . . that we must interrogate ever anew."[19] This does not mean that we have no sense whatsoever about the experiences of nonhuman animals; it means that we should resist the temptation to make definitive pronouncements about those experiences, particularly where our pronouncements amount to denying capacities to nonhuman animals that we arrogate to ourselves. In this connection, Grimm believes that we should proceed on the assumption that nonhuman animals are "beings who set ends for themselves," and that we should place "the burden of proof on those who would seek to restrict animal thriving."[20]

Consider how stark a contrast this approach poses to virtually all the leading approaches to the moral status of nonhuman animals today: Rather than assuming that the flourishing of nonhuman animals is, as Martha Nussbaum and others assume, fundamentally more limited than that of humans; rather than assuming, as Korsgaard does, that nonhuman animals do not set ends for themselves; and rather than assuming, as Donaldson and Kymlicka do in *Zoopolis*, that we are actually doing nonhuman animals a favor by subjecting them to a pointedly anthropocentric conception of citizenship—in other words, rather than rushing to impose human ways of thinking and valuing onto nonhuman animals, Grimm proposes that we take a fundamentally more open, questioning, fallibilistic approach that proceeds from the working hypothesis that nonhuman animals may well be very much more like us than unlike us, and that as a consequence we ought to proceed in a spirit of "respect and concern" for nonhuman animals rather than proceeding in a manner that threatens to "instrumentalize" them.[21]

Several years ago, Judith Butler published remarks in the *New York Times* concerning the oppression of women that prove thought-provoking for Grimm's idea of the animal-in-itself. Addressing the question whether men have a role to play in confronting the problem of violence against women, Butler suggests that "participation is not an entitlement; it is an obligation. But men who join that important fight against violence against women and trans people need to follow the leadership of women."[22] The analogy to nonhuman animals is this: just as men have an important role to play in the task of confronting violence against women and trans people, humans have an essential role to play in confronting the problem of violence against nonhuman animals. But—and again, this is radical—what if we think of nonhuman animals as somehow taking the lead in this struggle? To a certain widespread brand of common sense, this makes no sense at all: how can nonhuman animals lead humans if they lack the conceptual and linguistic apparatus to formulate and communicate plans of action to us, and if (as we have told ourselves for so very long) they need *us* to lead *them*?

My preliminary answer to this question is that we may need to broaden rather considerably our conception of what it means to "lead." It would be useful to start with an honest reflection on how well human beings have acquitted themselves in the historical record as leaders or managers of the world: Are there any other species that have come even close to humans in the capacity to despoil the environment and inflict violence on themselves and other species? Philosophers love to stress the hypothetical achievements that are made possible by human rationality, but the actual historical record tells a very different story. Perhaps nonhuman animals can provide us with inspiration in the endeavor to live up to the ideals that we have touted for ourselves. To this, philosophers tend to have recourse to some version of Tennyson's assessment of nature as "red in tooth and claw," as if this were true of nonhuman animals but not of human beings. But here again, I would suggest that we are ultimately no different than nonhuman animals, notwithstanding the efforts of many people to proclaim ideals of universal human rights and the like. I do not mean to denigrate these ideals, but simply to note how seldom real live human beings actually live up to them.

A hint to what it might mean to let nonhuman animals lead in a non-anthropocentric sense (i.e., in a sense other than that of proclaiming and seeking to enforce objective, linguistically articulable rules) is provided by the German philosopher Bernhard Taureck, who proposes that we start from the acknowledgment that we know precious little about nonhuman animals other than the fact that they are sentient, self-determining creatures who can lead their lives perfectly well without human interference. Before we try (if ever) to assert specific positive nonhuman animal rights or entitlements, we ought to *let animals be*, which is something that human beings have never really done. *Only once we refrain from interfering with the lives of nonhuman animals can we begin to understand what they are really like*; and only then, if ever, can we even begin to think about what we genuinely owe to nonhuman animals. In the face of this approach, any attempt to assert specific claims on behalf of nonhuman animals amounts to the imposition of human values and ideas on beings who may or may not be congenial to them. Thus, Taureck calls for an "imperative to discontinue use" of nonhuman animals and identifies this imperative as a core principle of veganism.[23] This imperative dovetails very seamlessly with the "vegan imperative" that I first discussed in *Animals and the Limits of Postmodernism* as well as with Grimm's call to approach nonhuman animals through the lens of the nonhuman "animal-in-itself."

2. Background Ideals of Living in the Philosophical Tradition

The sorts of claims made by philosophers, even those purportedly friendly to the cause of nonhuman animals, stand in stark contrast to the notion of the nonhuman animal-in-itself and the imperative to discontinue use. Each of these sets of commitments, the traditional ones and the radical alternative that I am proposing, is traceable to a global background ideal whose broad outlines can be inferred from a reflection on the foreground commitments on which philosophers tend to focus their attention. By undertaking a reflection on key traditional commitments about nonhuman animals, we will be able to bring the anthropocentric terms of this underlying background ideal into relatively clear relief. At the same time, it will become clear that this traditional ideal also contains certain essential *non*-anthropocentric elements. Examining this internal inconsistency, one whose influence can be seen throughout the history of Western philosophy, will help us to revise this traditional ideal so as to arrive at the basic terms of a non-anthropocentric background ideal of living. Put simply, the very terms of the anthropocentric ideal of living are in a tragic sense (i.e., as noted before, not simply in some empty logical sense) internally contradictory and stand in need of fundamental revision. As I will suggest in later chapters, the revision that is at stake here is not simply a matter of rational considerations but entails fundamental changes in our affective (felt) commitments as well.[24] This is what I mean when I suggest that it is not simply a matter of rationally grasping the non-anthropocentric ideal, but of having the will to embrace it.

Aristotle's thought provides the ideal point of departure for a reflection on traditional anthropocentric ideal of living, as this ideal takes its bearings from Aristotle's central commitment: that human but not nonhuman animals possess *logos*, the capacity for linguistic rationality. It would take a very long discussion to provide a full picture of Aristotle's position regarding the relative moral status of human and nonhuman animals, so here I will focus on three key features that will bring his background ideal into relief.[25] Those features are Aristotle's teleological conception of nature, the hierarchy of natural beings that he sketches, and his conception of community.

Aristotle's conception of the physical world proceeds from a doctrine of four "causes" (*aitiai*) designed to explain processes of natural change and growth. Each natural being possesses an internal principle (or "nature") that establishes that being's potential for development. The four causes are the factors influencing the being's development toward the realization of its potential: matter, form, the agency of the artificer, and the "end or that for the sake of which a thing is done." This last cause, the "end" or "final" cause, "tends to be what is best and the end of the things that lead up to it."[26] Each natural being has a natural purpose or point of fruition in its process of growth. According to this teleological conception of natural change, the "end" of an acorn is a mature oak tree; the "end" of a pig embryo is a mature pig. As will become clear in a moment, Aristotle thinks of the various ends in nature as organized hierarchically, with humans at the apex.

Aristotle accounts for this hierarchy through recourse to a conception of "soul" that is very much unlike the conception of soul that emerged through the Gospels and came to take hold of the Western imagination. Rather than thinking of soul in terms of a spiritual entity separate from the body, Aristotle conceives of soul as the principle governing a body's movement. He discusses several different kinds of soul, each corresponding to a particular natural function: nutritive, appetitive, sensory, locomotive, and rational. Plants possess nutritive soul, which is simply to say that they seek nutrition (e.g., by spreading their roots in the soil or growing toward sunlight). Animals possess nutritive soul as well as appetitive soul, which Aristotle considers to be present in any being that encounters its environment through the sense of touch and thus has desire; Aristotle believes that inasmuch as all animals possess at least this sense of touch, all possess appetitive soul. Sight, hearing, and smell pertain to beings possessing sensitive soul. Some animals possess in addition locomotive soul, the capacity to change location. Most important for our purposes is rational soul or "the power of thinking," which Aristotle acknowledges only in "a small minority" of natural beings who possess all the other forms of soul as well.[27]

To possess rational soul, on Aristotle's view, is distinctive of humans, who stand at the apex of natural beings. Aristotle offers a number of considerations that make it clear that he considers rationality to render human beings fundamentally superior to all other natural beings; for example, that humans but not nonhuman animals are capable of belief and that our capacity for rationality renders our sense of touch far more discriminating than in any other animals.[28] Considerations of this kind culminate in Aristotle's view that only human beings are truly suited to political community. Aristotle recognizes that there are various nonhuman animals, even insects, who exhibit high degrees of sociability. But he maintains that "man is more of a political animal than bees or any other gregarious animals."[29] To the extent that nonhuman animals are incapable of states such as belief, Aristotle treats them as incapable of political community on the grounds that they are incapable of doing things such as formulating rules for conduct, reflecting on their desires, and making deliberative choices about courses of action.[30] This in turn means that nonhuman animals are incapable of the specifically human form of flourishing that Aristotle calls *eudaimonia*, which is typically translated as 'happiness' but is best understood as a communal process of pursuing moral and political virtue. It is specifically in virtue of rational capacity (*logos*) that Aristotle considers human beings capable and nonhuman animals fundamentally incapable of pursuing moral and political virtue.[31]

Nonhuman animals do play an important role, however, in Aristotle's ideal of the political state. Their role is a purely instrumental one, which is to say that they function essentially as living tools for the satisfaction of human needs. Early in the *Politics*, Aristotle writes that

> after the birth of animals, plants exist for their sake, and that the other animals exist for the sake of man, the tame for use and food, the wild, if not all, at least the greater part of them, for food, and for the provision of clothing and various instruments. Now if nature makes nothing incomplete, and nothing in vain, the inference must be that she has made all animals for the sake of man.[32]

This is perhaps the most conspicuous single indication in Aristotle's writings of his commitment to a strict natural hierarchy in which human beings stand at the apex. Martha Nussbaum makes light of the significance of this passage when she offers the verdict that "this passage is from an introductory section of the work . . . concerned with stating the appearances," that it is simply "a preliminary *phainomenon*" expressing "an anthropocentric vantage point," and that therefore it is not to be taken as "a serious theoretical statement."[33] It must be acknowledged that there is a good reason to tread lightly when interpreting this passage (as I'll explain in a moment), but the bottom line in interpreting this statement is that Aristotle never backs away from this "anthropocentric vantage point"—he simply intensifies it.

The good reason to tread lightly is that Aristotle wrote a remarkable amount about nonhuman animals—well over a thousand pages, and the depth of his knowledge about the experiential capacities of a wide variety of nonhuman animals makes a lot of contemporary ethological research look somewhat unoriginal by comparison. These writings on nonhuman animals present a stark contrast with Aristotle's remarks in his psychological, ethical, and political writings, in which he categorically denies nonhuman animals rational capacities and (apparently on that very basis) assigns them an inferior status vis à vis human beings. In the zoological writings, Aristotle is considerably more willing to acknowledge capacities such as intelligence and foresight in nonhuman animals, going so far as to use terms such as *nous* (reason), *dianoia* (thought or understanding), *synesis* (sagacity), *technika* (skill or ingenuity), and *phronesis* (practical wisdom) to characterize the experiential capacities of certain nonhuman animals.[34]

Thus, at the very least there is a serious tension in Aristotle's writings between viewing nonhuman animals as fundamentally inferior to human beings and viewing them as on an experiential continuum with humans that makes a strict dividing line all but impossible to draw. What is decisive from the standpoint of probing the underlying background ideal of living that informs Aristotle's position is the fact that when it comes to conceiving the proper relationship between human and nonhuman animals, he very clearly privileges the former and treats the latter as mere instrumentalities. The "anthropocentric vantage point" that Nussbaum dismisses as a merely "preliminary *phainomenon*" is ultimately Aristotle's considered position when it comes to the moral status of nonhuman animals. At most, one might say that Aristotle does not consider himself in a position to assert that the *cosmic* meaning of the lives of nonhuman animals is to function as instrumentalities for the satisfaction of human needs and desires. But from the standpoint of *human* evaluation and conduct, that is *exactly* their meaning.

At the same time, Aristotle does make some pronouncements that suggest that he considers the servitude of nonhuman animals to be assigned to them not simply by humans but by nature. These remarks occur early in the *Politics*, so one might offer Nussbaum's caution as grounds for taking them as merely a provisional statement of prevailing common sense. But Aristotle himself never challenges or qualifies these remarks. Even more important, they anticipate very accurately prejudices that persist to this day and that contain clues to the background ideal of living that underlies them.

In the first book of the *Politics*, Aristotle presents a set of remarks concerning the cognitive capacities of enslaved persons, women, and nonhuman animals. His remarks about enslaved persons and women are notorious for their denigrating character: He suggests that some people are assigned the role of slave by nature, which is to say that their natural endowments (Aristotle focuses on rational capacity) destine them for servitude.[35] "Some are marked out for subjection, others for rule."[36] A natural slave "has no deliberative capacity at all" and "participates in reason enough to apprehend, but not to have" a rational principle. The natural human slave is one who possesses rational capacity just enough to follow rules or orders but cannot generate such rules from out of her own rational capacity. Women (and here Aristotle seems to have in mind women who are free rather than slaves by nature) possess deliberative capacity but "it is without authority," and this means that women are "by nature . . . inferior" and destined to be ruled by men.[37] Nonhuman animals, in contrast with slaves and women, "cannot even apprehend reason." They are moved by passion rather than reason; and inasmuch as "the rule of the soul over the body, and of the mind and the rational element over the passionate, is natural, and expedient," it is entirely appropriate that human beings rule over nonhuman animals. Nonhuman animals and slaves "both with their bodies minister to the needs of life."[38]

Whose life? Ostensibly the life of the *polis* and its human inhabitants. Here Aristotle makes a distinction between wild and tame nonhuman animals, suggesting that it is only tame ones who serve human needs. But clearly this is not strictly true: wild nonhuman animals sometimes do serve human needs, as is clear from activities that Aristotle discusses, such as hunting. What Aristotle appears to mean here is that tame nonhuman animals are better suited to serving human beings due to their compliant nature. "Tame animals," he states, "have a better nature than wild and all tame animals are better off when they are ruled by man; for then they are preserved."[39] The "better nature" of tame as opposed to wild nonhuman animals is due to their greater susceptibility to being marshaled into servitude to humans. Whether it is accurate to say that tame nonhuman animals are "preserved" due to this servitude is naturally a controversial matter. Just think of those figures I cited earlier regarding the number of nonhuman animals killed annually for human consumption; many of these individuals are products of domestication, which is to say that they are for all intents and purposes "tame." Then think of all the nonhuman animals used for invasive experimentation, field labor, and captive entertainment. Do sentient beings whose lives are strictly limited to these sorts of functions really benefit from being "preserved" in these ways? I will dwell on this question in the remainder of this book.

The essential function of nonhuman animals in the *polis* is akin to that of slaves: to help satisfy material needs and thereby liberate the rulers (all of them human, and at least arguably all of them men) for the cosmically higher activities of practicing virtue, participating in the administration of justice, and engaging in abstract contemplation. In the passages just cited from the *Politics*, Aristotle is clear that the (rational) soul is higher than the body. He further states that the pursuit of human excellence (virtue) is higher than pursuits such as the acquisition of material wealth.[40] Slaves and nonhuman animals can do no more than contribute to the latter aim; they cannot themselves participate in the pursuit of moral virtue, nor can either be involved in a genuine friendship with a fully rational being. But human slaves and nonhuman animals are not quite the same, for Aristotle believes that while one cannot have a friendship with "a slave *qua* slave . . . *qua* man one can; for there seems to be some justice between any man and any other who can share in a system of law or be a party to an agreement."[41] Aristotle never explains how a human being supposedly so impoverished in rational capacity can be a party to an agreement and thus a candidate for friendship with a fully rational being. What he presumably has in mind is the fact that even slaves, unlike nonhuman animals, participate in language and thus have a bond of meaning with other human beings, even if their grasp of meaning is inherently inferior to that of fully rational humans. What is clear here is that Aristotle believes that a certain kind of mutual intelligibility is required for inclusion in the political community, and that he considers slaves to participate in it whereas nonhuman animals do not.[42]

Like the vast majority of philosophers after him, Aristotle assumes either that human beings are unique in possessing freedom, or that human beings possess freedom in a manner or to a degree that renders them fundamentally superior to nonhuman animals. In one way or another, this commitment informs the views of virtually every philosopher who asserts a moral hierarchy that places human beings above nonhuman animals. Many philosophers rely on it, implicitly if not explicitly, in arguing that death is a greater harm for humans than for nonhuman animals. Notwithstanding the ubiquity of this commitment, it has been subjected to remarkably little critical scrutiny. The assumption is that nonhuman animals either are not truly free (whatever 'truly' is supposed to signify—I will present Kant's answer to this question shortly) or that the freedom possessed by nonhuman animals is fundamentally more limited than human freedom, and therefore inferior, due to limitations in cognitive capacity (the utilitarians in particular take this view, as I will also discuss). Grimm and Taureck give us reasons to reflect critically on this grounding assumption of the tradition, and to remain open to the possibility that it is ultimately driven by self-serving prejudice rather than by any genuinely reasonable considerations.

Quite apart from the question of the precise place of slaves and nonhuman animals in Aristotle's ideal conception of community, it is worth stressing that he very clearly privileges activities associated with thought over those associated with mere bodily functions—and in this he is followed by virtually the entire subsequent philosophical tradition. Given that nonhuman animals are bereft of *logos*, they are excluded from the "higher" activities of life and are relegated to the role of living instruments for the satisfaction of human needs and desires. Nonhuman animals are driven by their bodily states and are capable only of bodily pleasures, which on Aristotle's view are inherently inferior to the sorts of pleasures that attend activities involving thought.[43] This helps to illuminate Aristotle's view, which I noted earlier, that nonhuman animals are incapable of pursuing *eudaimonia* (happiness in the sense of moral virtue). Whatever edifying things Aristotle may say about nonhuman animals in his zoological texts, his verdict about them is clear: for all practical purposes, they exist to serve human beings.

From here it is a very short step to a variety of views offered by subsequent philosophers about nonhuman animals—from the ancient Stoic Chrysippus's view that the natural "end" of a pig is to be eaten to John Stuart Mill's lapidary conclusion that "it is better to be a human being dissatisfied than a pig satisfied."[44] Mill's reasoning is that nonhuman animals are capable only of bodily satisfactions, whereas human beings are capable of these as well as social, moral, and intellectual satisfactions; and he suggests that human beings are in a privileged position to decide that being human is better, inasmuch as humans have experienced all the forms of satisfaction whereas nonhuman animals have experienced only the bodily kinds. Peter Singer cites Mill's remark about the pig to support the conclusion that

> the more highly developed the mental life of the being, the greater the degree of self-awareness and rationality and the broader the range of possible experiences, the more one would prefer that kind of life, if one were choosing between it and being at a lower level of awareness.[45]

Singer offers this reasoning just after asking which I should want to be if I were given the choice between being a horse or a human being. For Singer, the answer is clear (although he does not state it explicitly): I ought to want to be a human being.

Singer's assumption is that there is some objective (in my language: cosmic) scale of preferability that transcends the perspectives of different individuals, be they human or nonhuman. He rejects the proposition that "the life of every being has equal value," arguing that

> we cannot defend this claim by saying that every being's life is all-important for it, because we have now accepted a comparison that takes a more objective—or

at least intersubjective—stance and thus goes beyond the life of a being considered solely from the point of view of that being.[46]

Is this anything more than sheer assertion on Singer's part? Even Aristotle, who as I have just demonstrated presents a pointedly anthropocentric view of human-nonhuman animal relations, expresses a sensitivity to the difference between cosmic and human perspectives. Singer, like Mill, proclaims access to some transhuman, absolute standpoint (it would *have* to be transhuman and absolute in order to permit a comparative evaluation of human and other kinds of life) from which it is supposed to be clear that human beings just happen to be lucky to have been born superior to all other natural beings. In Chapter 3 I will critically examine claims of this general kind, and I will argue for the conclusion that they presuppose access to a realm of absolute truth that is in principle unavailable to human beings.

Singer's claim of human superiority relies on the assumption that from a purely objective standpoint the life of a nonhuman animal (Singer assumes this to be true of virtually any nonhuman animal, no matter how cognitively sophisticated) involves "a lower level of awareness." This assumption is utterly question-begging: its internal logic is that specifically human forms of awareness (Singer, as noted, stresses a certain form of *self*-awareness) are *by their very nature* "superior" to all other forms of awareness, although neither Singer nor any other philosopher offers a truly cogent argument as to why or in what sense human forms of self-awareness are superior. The consideration typically offered is that self-awareness brings with it an expanded sense of time, and in particular the capacity to contemplate the remote past and the distant future. In the remainder of this book I will return to this idea and discuss the way it has been employed by a variety of thinkers. Both Singer and Regan rely on this reasoning, Singer relying on it to argue that (many if not all) nonhuman animals are "replaceable" and Regan using it (as I noted at the outset) to argue that the sacrifice of a million (or more) nonhuman animals would be justified to save a single human life.[47]

In arguing in this manner, thinkers such as Singer and Regan assume two things that have no basis in experience or observation: They assume that we know what nonhuman animal experience is like, and that the experience of nonhuman animals is "dim" in comparison with human experience. But as I have been suggesting from the outset, this is an extraordinarily convenient pair of assumptions for a species that has been deeply invested for thousands of years in a vision of itself as standing at the apex of creation. What makes us think we can plumb the depths of the nonhuman animal mind? And even if we could accomplish this, what makes us think that less complex or sophisticated intelligence (if indeed it is appropriate to characterize nonhuman animal experience that way, which I doubt) constitutes a basis for treating nonhuman animals as things we may kill with impunity should we elect to do so? How do we get from a supposed "is" to this supposed "ought," other than by dint of anthropocentric sleight of hand?

Thinkers who take the approach I have been describing draw on the pointedly anthropocentric aspects of the background ideal of living implicit (and to a great extent explicit) in Aristotle's thought: There is some cosmic scale or hierarchy of satisfactions; human beings, presumably in virtue of their rational capacity, have special insight into this hierarchy and enjoy the prerogative to make determinations about the fate of putatively "higher" and "lower" beings; human beings alone are capable of the "higher" forms of satisfaction, whereas nonhuman animals are capable only of the "lower" ones; and due to their inferior status, nonhuman animals are principally if not exclusively things to be consumed—or, in any event, dispensed with if that suits our wishes. These commitments, as I have suggested, are interwoven with certain non-anthropocentric commitments, and this makes it all but impossible to characterize a thinker as categorically anthropocentric. Those commitments will come into clearer relief once we have considered Kant's views regarding the relative moral status of human and nonhuman animals.

Kant, like Aristotle, refrains from making any sorts of definitive pronouncements about the cosmic significance of nonhuman animals in comparison with human beings. Instead, bearing fidelity to his acknowledgment that the thing-in-itself or inner nature of things is fundamentally hidden from finite beings such as humans, Kant presents his views about moral status strictly from the standpoint of human experience—and yet, like Aristotle, he ultimately proceeds as if the standpoint of finite human evaluation enjoyed absolute authority in determining the fate of nonhuman animals. In addition, Kant makes some fundamental mistakes about the way in which human reason operates; in particular, he treats human reason as functioning in a sort of ahistorical void rather than being situated in time and circumstance. In Chapter 3 I will return to this problem and examine some key phenomenological and pragmatist efforts to revise Kant's understanding of reason. Here it will suffice to examine Kant's central conclusions about the moral status of nonhuman animals and to note that Kant, like Aristotle, treats his rational insights as if they were absolute and incontrovertible. Although he offers some modifications and additions to Aristotle's views, Kant retains the central anthropocentric cast of Aristotle's position. But Kant's views also gesture toward certain non-anthropocentric underpinnings that he never fully reconciles with his anthropocentric commitments.

Kant acknowledges that "in the system of nature, a human being (*homo phaenomenon, animal rationale*) is a being of slight importance and shares with the rest of the animals, as offspring of the earth, an ordinary value (*pretium vulgare*)." This is the standpoint from which the life of every being has equal value, a standpoint that Singer rejects as lacking objectivity even though it would appear to be about as objective (i.e., impartial) a statement of our natural condition as one could articulate. From here Kant proceeds differently than Singer, presenting the special status of human beings not as an objective fact but rather as the product of a certain *sub*jective vantage point:

But a human being regarded as a *person*, that is, as the subject of a morally practical reason, is exalted above any price; for as a person (*homo noumenon*) he is not to be valued merely as a means to the ends of others or even to his own ends, but as an end in himself, that is, he possesses a dignity (an absolute inner worth) by which he exacts respect for himself from all other rational beings in the world.[48]

Now this is the sense in which viewing human beings as "persons" is a fundamentally subjective rather than objective matter: It is by means of our own rational reflections that *we assign to ourselves* the role of person. Even though Kant never quite puts the point this way, seeing or representing ourselves to ourselves as persons is a matter of *legislation*; we do not simply discover that we are persons, but instead we assert or take on this status.[49]

The idea that personhood is a role we take on recalls Cicero's discussion of the two *personae* or roles of which rational beings are capable. Our universal *persona* is the role we take on when we employ the rational capacity that "lifts us above the brute" and engage in matters of "morality and propriety." Our individual *persona* is what distinguishes each of us from others, as seen for example in "the diversity of characters."[50] For Kant, as for Cicero, it is in virtue of our rational capacity that humans but not nonhuman animals can take on roles characterized by morality, propriety, and individuality. There is a sense of selfhood at work here, as in Aristotle, that depends crucially on *logos*—which, for Cicero and Kant, as for Aristotle, is unique to human beings.

Thus Kant explicitly denies that any nonhuman animals can be considered persons. He draws a fundamental distinction between "persons" and "things," according to which all and only rational beings count as the former and all non-rational beings count as the latter. Kant's conception of rationality is complex and involves a number of functions that are not of direct relevance to the present discussion, so I will not attempt to give a full explanation of what Kant means by rationality. Here it will suffice to note that when Kant distinguishes between persons and things, he does so in the course of explaining one specific function of reason, namely, the activity of legislating and subjecting oneself to the moral law. Kant does not conceive of the moral law in traditional religious terms, as if it were ordained by some suprahuman agency. Instead, Kant secularizes the notion of moral law such that it is up to individual rational agents to determine the moral law from out of their own rational powers. Once an agent has adduced the moral law in this manner, she should recognize something that might seem paradoxical: that even though she herself adduced the law, she is henceforth strictly subject to it, i.e., obeying the moral law is not a matter of personal preference but is binding on all rational agents.

This conception of the way in which the moral law arises and gains authority signals an unprecedented view of human freedom. No longer is freedom conceived as the mere absence of external impediments, as Hobbes had suggested; instead freedom in its most authentic expression is voluntary subjection to a law that one has legislated for oneself.[51] In order to be capable of this kind of legislation, a being must be able to detach herself from personal desires and inclinations and establish a standpoint of impartiality, which for Kant means a standpoint from which the agent makes moral claims that have intersubjective validity. In this respect, the detached agent's (or "person's") moral claims share something in common with the claims of scientific rationality. Even though the notion of the thing-in-itself reminds us that we cannot know the inner nature of things and that our judgments are all relative to our own constitution as knowers (relative, that is, to matters such as the conceptual apparatus we employ), objectivity is still possible in scientific inquiry inasmuch as all rational agents employ the same basic cognitive apparatus for organizing and synthesizing experience.

One way to see the parallel between moral judgment and scientific judgment is to consider a remark that Kant offers near the end of the *Critique of Pure Reason*.[52] There he states that "all the interests of my reason . . . combine in the following three questions: 1. What can I know? 2. What ought I to do? 3. What may I hope?"[53] A central aim of the *Critique of Pure Reason* is to vindicate the prospect of absolute ("apodeictic") certainty in natural science; this is the thrust of the "what can I know?" question. An additional aim is to begin to sketch the full range of the claims of human reason, which for Kant include the establishment of a firm, unchanging basis for ethics. That basis is to enable us to address the latter two questions, about which Kant wrote a great deal. Kant presents an ideal of human reason according to which we ought to aspire to perfect our moral natures in spite of the fact that little if anything in the world of practical matters gives us any indication that such perfection might be achievable. Among other things, Kant envisions a cosmopolitan society in which war has been overcome and perpetual peace attained.

What is key for our purposes is the fact that Kant treats rational beings as participants in these functions of reason and that he categorically excludes nonhuman animals from them. For Kant, as for Aristotle, nonhuman animals function merely as tools or instrumentalities.

> Rational beings are called persons inasmuch as their nature already marks them out as ends in themselves, i.e., as something which is not to be used merely as means and hence there is imposed thereby a limit on all arbitrary use of such beings, which are thus objects of respect.[54]

Beings who are rational and self-aware, who can legislate and subject themselves to the moral law, merit respect not for what one can gain from them but rather in virtue of what they are and what they can do—namely, regulate their own lives and their relations with others through the employment of reason. When Kant qualifies this characterization of persons with the words "not to be used merely as means," he is simply acknowledging that in the world of practical affairs we inevitably need to rely on others in material ways—that I may sometimes need your help with a garden project, that you may occasionally need to borrow money from me, etc. But we are never to treat other persons merely as means; instead, we must subordinate these material relations to the fundamental *respect* that we owe to other persons.[55]

In contrast, "beings whose existence depends not on our will but on nature, have, nevertheless, if they are not rational beings, only a relative value as means and are therefore called things." Every being in the world that is non-rational in Kant's specific sense counts as a "thing" and hence *is* "something to be used merely as a means."[56] Kant states explicitly that nonhuman animals "are here regarded as things," and he cautions that when "humanity becomes an instrument for satisfying desires and inclinations . . . it is dishonoured and put on a par with animality."[57] For Kant, as for Aristotle, *nonhuman animals are things to be used to satisfy human needs and desires*. They are inherently inferior to rational beings inasmuch as they are driven entirely by desires and inclinations. And for Kant, as for Aristotle, our humanity is diminished if we behave (or are caused to behave) like mere nonhuman animals. We must rise above the influence of our desires and inclinations, evaluate them rationally, and act only on the basis of those that are commensurable with prudential considerations and the moral law.

Kant categorically excludes nonhuman animals from the moral community on the grounds that "they are not free" but "act according to rules" that are not of their devising. Whereas rational beings act "from motivating grounds," nonhuman animal behavior takes the form of "necessitation through sensory impulses."[58] This amounts to saying that rational beings are free in the sense of being self-determining, whereas nonhuman animals are driven strictly by instinct. Here it is worth noting that Kant presents a very limited conception of what it means to be "free" or "self-determining"—his criterion, as I have shown, is not simply the ability to select this or that end and pursue it, but rather the considerably more exacting ability to formulate a universally binding law in abstract and presumably linguistic terms. I believe it is more than mere coincidence that this criterion just happens to single out human beings among all known natural beings as "free." In effect it excludes all other possible forms of freedom from consideration by defining them out of existence. It is also worth noting that for Kant, freedom in this sense is not simply a feature of human beings that distinguishes us from nonhuman animals; it is a feature that elevates us above them, as is (or ought to be) clear from Kant's suggestion that we "dishonour" ourselves if we become reduced to mere animality.

To honor ourselves in Kant's sense is to actualize our freedom or, to use the term Kant employs in his writings on morality, our "autonomy." To attribute autonomy to humans and deny it to nonhuman animals is not simply to say that nonhuman animals cannot legislate and follow the moral law. It is to proceed on a dual assumption: that nonhuman animals cannot govern themselves but instead require the guidance of rational beings, *and* that those rational beings are fully *entitled* to treat nonhuman animals as mere instrumentalities or "things."

Some contemporary thinkers have sought to revise Kant's thinking so as to show that it really contains the basic elements of a robust nonhuman animal ethics, as if Kant had simply made some basic mistakes about matters such as the nature of agency or freedom. In the next section of this chapter I examine several contemporary efforts to rethink the terms of nonhuman animal ethics, and there I will offer some remarks about Korsgaard's efforts to locate resources for a robust nonhuman animal ethics in Kant's thought. For the moment I will simply raise a question: Just how realistic is it to seek the resources for a robust nonhuman animal ethics in the work of a thinker who places the moral status of nonhuman animals, domesticated ones at least, on a par with vegetables such as potatoes?[59] The answer to this question is tricky. As I have already noted, thinkers such as Kant exhibit an ambivalence between anthropocentric and non-anthropocentric commitments; so it is not difficult, as I myself am about to do, to locate in Kant certain commitments that are amenable to incorporation into a non-anthropocentric ethics. But as should become clear by the end of this chapter, it is one thing to tinker with specific ideas and commitments in a philosopher's outlook and quite another to evaluate those ideas and commitments in the light of the global background ideal of living that underwrites them. From the former standpoint, "fixing" Kant on nonhuman animals might look like a relatively straightforward affair; from the latter, it proves to be considerably more difficult and probably impossible.

The aspect of Kant's thinking that most conspicuously demonstrates his ambivalence about the moral status of nonhuman animals is his so-called "indirect duties" view. On Kant's view, duties are owed directly only to persons, which is to say to rational beings; on Kant's view it simply makes no sense to speak of a duty owed directly to a non-rational being. But Kant also fully recognizes that nonhuman animals are sentient, and this raises a difficulty: How can we consider it permissible to treat nonhuman animals exactly as we treat things such as potatoes? This was not a difficulty that Descartes had to face, as he had characterized nonhuman animal experience as fundamentally mechanistic and non-sentient.[60] Kant explicitly rejects Descartes's reasoning and acknowledges that nonhuman animals, like human beings, "act in accordance with representations" and to this extent "are members of the same genus with human beings."[61] To have representations rather than to function in a purely mechanistic manner is to have a conscious experiential encounter with the world; it is to see, hear, feel, and the like, and to do so in such a manner as to be *subjectively aware*. (Contrast the way you or I might see an object in the path of the garage door with the way the infrared safety device on the garage door bay "sees" the same object.)

That Kant acknowledges the capacity for feeling in nonhuman animals is clear from his admonition that we not act cruelly toward them. This admonition is based on the acknowledgment that nonhuman animals can suffer. But our duty not to inflict gratuitous suffering on nonhuman animals is not due to anything we owe directly *to them*. Instead that duty is really *a duty to humanity*. "With regard to the animate but nonrational part of creation, violent and cruel treatment of animals is . . . intimately opposed to a human being's duty to himself," indeed even more intimately opposed to that duty than would be the "wanton destruction of what is *beautiful* in inanimate nature."[62] Even though inanimate nature has no direct moral status and is essentially a set of resources for the satisfaction of human needs and desires, we nonetheless have a duty to ourselves not to destroy beautiful things in nature gratuitously inasmuch as doing so diminishes our humanity. Imagine an arsonist, for example, who sets the Sequoia National Park on fire: Such a person commits an affront to the dignity of persons, not to the dignity of sequoias—precisely because trees are not rational beings and hence have no share in dignity.

In stating that cruelty to nonhuman animals is even more intimately opposed to our duties to humanity, Kant is acknowledging that in a crucial respect, nonhuman animals are precisely *not* like potatoes but are in a very important sense "of the same genus with human beings." And yet he considers them to have no share in dignity for the same reason that sequoias have no share in it, namely, that nonhuman animals are not rational. The only reason not to be cruel to nonhuman animals is that cruelty "dulls [our] shared feeling of their suffering so weakens and gradually uproots a natural predisposition that is very serviceable to morality in one's relations with other men." Here Kant follows the reasoning offered centuries earlier by Saint Thomas Aquinas: that cruelty to nonhuman animals makes us more likely to be cruel to our fellow human beings.[63] But Kant goes a step further. He states that our indirect duties, which are in reality duties "*with regard to*" nonhuman animals rather than being owed directly to them—i.e., nonhuman animals are the *objects* to which these duties pertain, rather than being *subjects* who merit anything like respect—include considerations such as "gratitude for the long service of an old horse or dog (just as if they were members of the household)." Here Kant stresses that this duty is "always only a duty of the human being to himself."[64]

Here the essential instability of Kant's moral ideal comes into focus. Like Aristotle and Saint Augustine before him, Kant maintains that there can be no community with nonhuman animals.[65] What is lacking, on Kant's view, in addition to shared rationality, is the capacity for a crucial affective connection between humans and nonhuman animals.

> All passions . . . are always only desires directed by human beings to human beings, not to things; and while we can indeed have a great inclination toward the utilization of a fertile field or a productive cow, we can have no *affection* for them (which consists in the inclination toward community with others), much less a passion.

Moral community is shared by beings who are rational, who are capable of passion, and who have a "claim to freedom (a representation that no other animal has)."[66] How, then, are we to understand Kant's suggestion that a sense of "gratitude" can be expressed toward an old horse or dog, or his belief that "any action whereby we . . . treat [nonhuman animals] without love, is demeaning to ourselves"?[67] What can it mean to understand (or enact) gratitude and love not as genuine passions but merely as "inclinations"?

Of course, there is an answer to this question. We can conceive of love in the sense that people "love" their wardrobes or their cars, although that is clearly not what Kant is after here; he is after something more, as is clear from the fact that he considers cruelty to nonhuman animals to be a greater moral harm to ourselves than is the wanton destruction of natural beauty. Kant is implicitly aware of a certain *kinship* between human beings and nonhuman animals that does not obtain between human beings and non-sentient nature. But the terms of Kant's own thinking make it impossible for him to do complete justice to this kinship. The problem is that the strict person-thing binary that he has set up deprives him of any middle ground for making sense of the kind of love he wants us to express toward nonhuman animals. This problem is compounded by the fact that, in spite of his seemingly approving appeal to the notion of passion in texts such as the *Anthropology*, Kant follows Aristotle in seeing passion as a serious potential obstacle to morality, and like Aristotle he seeks to subordinate passion to the authority of reason. This endeavor to subjugate passion under the yoke of reason, while not universally endorsed, is a sort of resounding bass note in the history of Western philosophy. I offer a critique of this endeavor in Chapter 4.

A moment ago, I noted that Kant calls on us to demonstrate love for nonhuman animals. The difficulty I have identified with this call pertains equally to his suggestion that it would be appropriate to show gratitude to a nonhuman animal who has served us well. There really are only two ways to understand this suggestion. One is to take it literally, and to suppose that one can show genuine gratitude toward a mere thing. But that makes no more sense than, say, showing gratitude toward my car for starting—unless we acknowledge that, in virtue of their capacity for representations, nonhuman animals are precisely like human beings and unlike cars, and that nonhuman animals therefore really merit inclusion as *direct* beneficiaries in the moral community. To acknowledge this would be to derive considerably more significance than Kant does from his observation, noted already, that "in the system of nature a human being (*homo phaenomenon, animal rationale*) is a being of slight importance and shares with the rest of the animals, as offspring of the earth, an ordinary value (*pretium vulgare*)." Kant is simply unable—and this provides a key insight into the background ideal of living to which he, like Aristotle, subscribes—to see the natural world as anything more than the object of human domination. Kant's ultimate verdict regarding the relative moral status of human and nonhuman beings is that

> the human being . . . is the ultimate end of the creation here on earth, because he is the only being on earth who forms a concept of ends for himself and who by means of his reason can make a system of ends out of an aggregate of purposively formed things. . . . he is certainly the titular lord of nature.[68]

The other way one could interpret the call for gratitude toward nonhuman animals would be to see it as a call to act *as if* we felt gratitude toward them—in effect, to pretend that we felt gratitude. Given Kant's insistence that nonhuman animals are fundamentally incapable of having a representation of freedom or a concept of ends, this would appear to be the only possible meaning of Kant's call. But this naturally reduces the expression of gratitude to something akin to play-acting, and denudes the very idea of gratitude of its meaning. This, I think, is what Robert Nozick had in mind when he challenged the entire logic of the indirect duties view. Nozick asks the following question:

> If it is, in itself, perfectly all right to do anything at all to animals, for any reason whatsoever, then provided a person realizes the clear line between animals and persons as he acts, why should killing animals tend to brutalize him and make him more likely to kill or harm other persons? Do butchers commit more murders?[69]

The defender of Kant will hasten to point out that Kant doesn't really suggest that "it is perfectly all right to do anything at all to animals," that, for example, he believes that "agonizing physical experiments for the sake of mere speculation, when the end could also be achieved without these, are to be abhorred."[70] But this simply reinforces Nozick's observation about the incoherence of the indirect duty view: If nonhuman animals are things, mere resources for the satisfaction of human needs and desires, why should it be *abhorrent* to inflict gratuitous suffering on them? Kant's reasoning here, as I think Nozick shows, is flawed: If nonhuman animals have no inherent worth and are mere instrumentalities, then there should be no reason to suppose that harming *them* does anything at all to dull our sensibilities toward *persons*—unless, again, we acknowledge that nonhuman animals are *not* merely things. Nozick, for his part, does not think beyond this observation, other than to suggest that "animals count for something" and to hint that some sort of deontological approach might be necessary in coming to grips with the moral status of nonhuman animals.[71]

3. Anthropocentric Implications of Some Contemporary Approaches

Aristotle's and Kant's views about the moral status of nonhuman animals have had a decisive influence on philosophical thinking that has persisted up to this day. That influence is global and pervasive, and is evident even in the views of some highly influential contemporary thinkers. I have already noted Peter Singer's commitment to the superiority of human beings. He goes so far as to assert that beings who are self-aware count more in the moral calculus than non-self-aware beings, and in addition that even if there are nonhuman animals capable of self-awareness, none of them are capable of self-awareness to the degree that typical human beings are capable of it.[72] This has several key implications. One is that non-self-aware beings are "replaceable," which in less cloaked language means that they are expendable—that a given non-self-aware nonhuman animal counts no more in the moral scheme of things than, oh, say, a potato. Another implication is that we *probably* should not eat self-aware nonhuman animals—Singer never makes any unequivocal call for ethical vegetarianism or veganism, going only so far as to state that his view "brings us very close to a vegan way of life" and that practices such as zoos, rodeos, and pet ownership "raise . . . issues."[73] A third implication of Singer's position, as I have noted already, is that, all other things being equal, one should want to be a human being rather than *any* sort of nonhuman animal. I believe that anyone who takes these implications, particularly the third one, at face value is proceeding on the basis of a pointedly anthropocentric background ideal of living. After briefly considering Nussbaum's and Korsgaard's views, I will return to this claim and explain why I am convinced that it is true.

Nussbaum takes her lead from Aristotle in framing her views about the relative moral status of human and nonhuman animals in terms of capacities for "flourishing." This approach draws on Aristotle's notion, discussed earlier in this chapter, that each natural being has a "telos," "end," or natural point of fruition toward which it grows. Nussbaum characterizes a being's capacity for growth in terms of what she calls its "capabilities." On this view, an acorn has the capability to draw nourishment from the soil and undergo a process of generation that culminates in the form of a mature oak tree, after which high point of development it decays and dies. Aristotle characterizes the capabilities of different forms of life in terms of the kinds of soul that beings possess. Nussbaum dispenses with the talk of soul and simply focuses on capabilities. The picture that emerges is that beings such as acorns have relatively limited capabilities (the range or extent of their ability to flourish is less than that of sentient beings), nonhuman animals possess yet greater capacities for flourishing, and human beings possess the greatest range of capabilities for realizing their natural potential. Nussbaum has refined her list of central human capabilities over the years. According to the list she provides in *Frontiers of Justice*, those capabilities include life; bodily health and integrity; senses, imagination, and thought; emotions; practical reason; affiliation; concern for other species; play; and control over one's political and material environments. Nussbaum presents this list as being "important for each and every citizen" and as being akin to "the international human rights approach."[74]

In presenting her conception of central human capabilities as an ideal of human citizenship, Nussbaum explicitly follows Rawls when she states that she introduces her list "for political purposes only, and without any grounding in metaphysical ideas of the sort that divide people along lines of culture and religion."[75] And yet Nussbaum also argues that nonhuman animals are "direct subjects of the theory of justice." When she does so, she notes that including nonhuman animals as such subjects requires "a family of reasonable comprehensive doctrines."[76] This statement recalls two important ideas in Rawls's thought. One is Rawls's suggestion that his theory of justice pertains exclusively to relationships between citizens (human agents who are capable of entering into contractual relationships), and that any consideration of the moral status of nonhuman animals would require "a theory of the natural order and our place in it."[77] Whether or not such a theory would be a metaphysical theory, it is clear that articulating it would almost certainly divide people along the lines of culture and religion that Nussbaum cautions us to avoid. (Keep this in mind when I contrast anthropocentric and non-anthropocentric background ideals of living in the concluding section of this chapter.) So it is difficult to see how on an approach such as Nussbaum's we can fulfill the promise of including nonhuman animals as beneficiaries of justice: She relies on the notion of thinkers such as Rawls and Brian Barry that justice is best understood in terms of fairness, but she never comes to grips with the fact that people from different cultures and religions have wildly divergent views about whether the notion of fairness applies to nonhuman animals and, if it does, what fairness toward nonhuman animals requires of us.

This difficulty is brought into even clearer relief by the other important idea in Rawls on which Nussbaum relies: that of a "comprehensive doctrine." Rawls stresses that a theory of justice should sketch rights and duties only insofar as they secure a proper balance between liberty and equality among citizens. This means that "a conception of justice . . . does not apply directly to associations and groups within society," i.e., justice as fairness "does not presuppose accepting any particular comprehensive doctrine."[78] Here Rawls has in mind a variety of prerogatives that are reserved for the private as opposed to the public sphere, such as the choice of religion. (All citizens should have the right to vote, unless in virtue of some pre-arranged edict that applies equally to all citizens an individual has sacrificed that right. But no citizen should be required to belong to a particular religion or, for that matter, to any religion.)

Nussbaum seeks to bring together two mutually opposed goals: to include nonhuman animals as "direct subjects of the theory of justice," and to do so "without putting at risk any of the core metaphysical commitments of the major religions." Nussbaum suggests that "the principles [she is] advancing are political and not metaphysical."[79] I equivocated a short while ago as to whether "a theory of the natural order and our place in it" would be metaphysical. Here I am inclined to say that it is indeed metaphysical. If I am right about this, then I believe that Nussbaum cannot make good on her claim to include nonhuman animals as direct beneficiaries of justice without running afoul of a number of religious traditions (and, more generally, comprehensive doctrines). I will offer just one example in support of this conclusion: the 2015 encyclical *Laudato' Si*, in which the Pope, while urging a more compassionate relationship to nonhuman animals, argues for the categorical superiority of human beings over all nonhuman creatures and seeks on that basis to justify practices such as hunting and fishing.[80] The Pope's claim of categorical human superiority and his classification of nonhuman animals as resources (albeit resources we must treat with compassion) very much resemble Aristotle's and Kant's anthropocentric viewpoints but are explicitly rooted in religious commitments that seem impossible to reconcile with Nussbaum's call to treat nonhuman animals as direct beneficiaries of justice.

In other words, the kind of political liberalism from which Nussbaum proceeds cannot, by itself, make a place for nonhuman animals as direct subjects of justice.

A Rawlsian political liberalism specifies only second-order procedural rules designed to promote individual liberty and the coordination of the wills of different citizens.[81] It specifically refrains from positing any first-order (substantive) conception of the good, leaving it to the private judgment of individual citizens to determine this through their own "comprehensive" views of the good—views that can and do differ, not only across cultures and religions but often between individuals within a given cultural or religious tradition. There are, for example, Christians who follow the Pope in supposing that it is permissible to eat nonhuman animals, but there are also Christians who are strict ethical vegans.

When Nussbaum proceeds on the assumption that there is some objective conception of what justice for nonhuman animals amounts to—objective in the sense that it should be uncontroversial and have validity across different cultural and religious traditions—she is making a rather dubious assumption. Something more than a theory of political liberalism is needed in order to make room for nonhuman animals in our overall conception of justice. In *Animals and the Moral Community*, I argued that the conception of justice at work in political liberalism is best understood as *social* justice, and that in order to bring nonhuman animals into the moral community in a non-arbitrary way we would need to situate social justice within the larger context of what I called *cosmic* justice. I return to this idea in the concluding section of this chapter when comparing the basic terms of anthropocentric and non-anthropocentric background ideals of living.

This lacuna in Nussbaum's approach leads to some significant difficulties. In particular, on Nussbaum's view, including nonhuman animals as direct subjects of justice does not prevent us from using them in a variety of ways, including eating them, experimenting on them, making them work for us, and treating them as sources of entertainment in practices such as horse racing—provided, of course, that we treat them well. In fact, on Nussbaum's view we actually benefit nonhuman animals when we use them in practices such as horse racing, and the only obstacle to seeing this supposed fact is the same sort of "romantic fantasy" that leads some people to call for the abolition of the property status of nonhuman animals.[82] On Nussbaum's view, all of these practices are justified on the grounds that none of them "violate basic animal entitlements," which themselves are a function of the experiential capacities of nonhuman animals.[83] Nussbaum's list of nonhuman animal capacities includes life; bodily health; bodily integrity; senses, imagination, and thought; emotions; practical reason; affiliation; other species; play; and control over one's environment.[84] It is noteworthy that this list is virtually identical to the list of human capabilities; the only differences pertain to the human prerogative (or obligation, depending on how you look at it) to exercise "guardianship" over nonhuman animals, which on Nussbaum's view includes the human entitlement to decide what constitutes appropriate ways of using nonhuman animals in practices ranging from field labor to experimentation to nonhuman animal agriculture (using nonhuman animals as sources of food) to entertainment for human beings. With regard to this last use of nonhuman animals, and specifically with regard to racing horses, Nussbaum offers the justification that "it is condescending to animals (as it is to aging humans) to assume that lazing around the pasture is their only good."[85]

Take a moment and think about exactly what Nussbaum is implying here—that the only alternative to racing horses or engaging them in dressage is to leave them completely idle in a field. It never occurs to Nussbaum to ask the question whether these extremes are the only possibilities available to us and horses, nor does it occur to her to question her own assurance that the property status of nonhuman animals is fully compatible with respecting their rights. Domestication poses a very deep problem for anyone who would affirm duties of respect and justice toward nonhuman animals, and it proves to function as a sort of litmus test for distinguishing anthropocentric and non-anthropocentric background ideals of living. I stated at the beginning of this chapter that the realization of my proposal might throw our lives into upheaval. It is in connection with domestication that this threat comes fully into view. But even leaving that question aside for the moment, it should be clear that Nussbaum's basic orientation is on the human use of nonhuman animals, not on the question of what might truly be in a nonhuman animal's interest.

That Nussbaum reproduces many of the anthropocentric prejudices found in Aristotle and Kant is clear not only from the list of uses of nonhuman animals that she endorses, but also from the explicit human-nonhuman animal hierarchy that she asserts—and then promptly denies. Indeed, it is this very hierarchy that implicitly provides support for the idea that humans have prerogatives to determine the fates of nonhuman animals. Nussbaum states that "more complex forms of life have more and more complex (good) capabilities to be blighted, so they can suffer more and different types of harm." On this basis she argues, or at least strongly implies, that death is less of a harm to less complex forms of life than to more complex ones. The core logic is that complex = good, and that more complex = better. Human forms of life are presumably more complex, and thus more worthy of protection, than nonhuman forms of life including nonhuman animal forms of life. Nonetheless, Nussbaum maintains that "we should not follow Aristotle in saying that there is a natural ranking of forms of life, some being intrinsically more worthy of support and wonder than others."[86]

But the respective entitlements of human and nonhuman animals that Nussbaum urges on us seem very closely to resemble the kind of hierarchy first articulated by Aristotle and reinforced by Kant. Of course, there is a key difference, and that is that Nussbaum, like a variety of thinkers writing after Bentham and Mill, is sensitive to the fact that nonhuman animals can suffer and that this confers direct moral status on them. Bentham is famous for having observed that the moral status of a being should be based not on whether that being possesses reason or language, but on whether that being can suffer.[87] The French philosopher Jacques Derrida took this observation to signify that Bentham "changes everything" when it comes to our sensibilities regarding the moral status of nonhuman animals.[88] But in fact Bentham, like Derrida after him, changed very little. For in the same textual passage in which Bentham makes his famous statement about suffering, he goes on to assure his reader that nonhuman animals have nothing to lose by dying inasmuch as they have no anticipation of the future (this leads Singer to argue that nonhuman animals have no interest in continued existence), and that nonhuman animals are actually better off being killed by human beings than by being forced to die in conditions of wild nature (presumably by predation).

The lesson to be learned from this is that simply acknowledging that nonhuman animals are direct beneficiaries of moral concern and/or justice does not by itself guarantee that they will be accorded the status and the regard that they may truly merit. Aristotle, Kant, Bentham, and Nussbaum all assert hierarchies based on the supposed experiential capacities of humans and the supposed experiential limitations of nonhuman animals. None of these thinkers pays much if any regard to the question whether we are in a position to make definitive pronouncements about the experiential capacities of nonhuman animals. Instead, they all rush to assume that because we are unable to detect "complex" forms of life in nonhuman animals, that complexity must certainly be absent.[89] Nor does any of these thinkers give more than passing consideration to the question whether our judgments about the experiential capacities of nonhuman animals are distorted by a prior commitment to seeing nonhuman animals as objects of use. Using a nonhuman animal "nicely" is still using it. If Nussbaum comes closer than her historical precursors to arriving at an edified view of nonhuman animals, she nonetheless reproduces a very traditional prejudice when she proposes that human beings merit "equality" whereas nonhuman animals merit mere "adequacy" when it comes to the securing of entitlements.[90]

Where Nussbaum takes her cue from Aristotle, Korsgaard takes up some key ideas in Kant's thought and seeks to revise them so as to acknowledge nonhuman animals as direct moral subjects. This leads Korsgaard to offer a more direct and thoroughgoing challenge to certain uses of nonhuman animals than Nussbaum offers. And while Korsgaard offers some important considerations in favor of viewing many nonhuman animals as moral subjects rather than as mere objects, her approach nonetheless bears the imprint of the anthropocentric prejudice that characterizes the various philosophers I have been examining in this chapter. In this connection, it is worth considering whether one is likely to find inspiration for a robust nonhuman animal ethics in the thought of a philosopher who considered nonhuman animals to be on a par with potatoes. In the following remarks I hope to show that this is much more than a mere *ad hominem* concern.

Korsgaard's starting point is Kant's notion of an end in itself. As noted earlier, Kant privileges "persons," rational beings who can legislate, contemplate, debate, and subject themselves to the moral law. This is a conception of personhood that involves self-awareness and, importantly, the ability to set ends for oneself apart from those dictated by nature. As we have seen, Kant attributes an absolute superiority to persons over mere "things" or non-rational beings—absolute not in the sense that this superiority is some sort of cosmic endowment (Kant's idea of the thing-in-itself prevents him from making such a claim), but in the sense that human beings, as the beings who assert values, posit or recognize a special value in their own capacity to legislate the moral law. One could certainly push back here and ask exactly how or in what sense this kind of autonomy confers a greater value on human (rational) beings, and how its lack confers a lesser value (namely, a merely instrumental one) on nonhuman animals (non-rational beings). There are several possible answers to this question. One is that autonomy (or the ability to set ends for oneself) has an incomparable value because we, the possessors of that capacity, *say* it does. I believe that it is in this spirit that Mary Anne Warren suggests that human lives "have greater intrinsic value, because they are worth more *to their possessors*."[91] Another way to see what would motivate a thinker such as Kant to elevate human beings in this way is to consider the deep continuity between Kant's approach and the ancient endeavor to vindicate human beings as the only truly godlike beings in creation. This endeavor is evident both in ancient philosophical movements such as Stoicism and in the Christian tradition. Given Kant's endeavor to rethink religion "within the limits of reason alone," it will strike many readers as implausible to suppose that Kant elevated human beings over the rest of nature on the grounds that human beings alone have a spark of the divine. This, however, is what he fairly clearly has in mind when he characterizes the human being as "the titular lord of nature."

Korsgaard recognizes that this attempt to privilege human beings over nonhuman animals is dubious and in need of revision. She acknowledges that "we are not the only beings for whom things can be good or bad; the other animals are no different from us in that respect. So we are committed to regarding all animals as ends in themselves." It is in the very nature of a nonhuman animal to "[take] its own functional good as the end of action." The sense of "taking" here is what distinguishes sentient beings from nonsentient life: nonhuman animals, just like human animals, have a conscious relationship to themselves, their environments, and to their own striving. The life of a sentient being has an "essentially self-affirming nature," and this is all that is needed for a being to count as an end in itself.[92]

To the extent that nonhuman animals are ends in themselves, they

> obligate us by reminding us of what we as individuals have in common with them—that we are creatures *for* whom things can be good or bad, and that like them, although in our own special way, we each take our own good to be good absolutely when we engage in practical activity.[93]

Thus it is a mistake to suppose that human lives matter more because they matter more to their possessors; it is in the very nature of a sentient being to value its own life absolutely.[94]

Korsgaard rejects the kind of reasoning offered by thinkers such as Singer, who as I noted earlier argues for an objective standard according to which it is clearly better to be a human being than a nonhuman animal such as a horse. Korsgaard takes it as axiomatic that "all importance is tethered" to the subjective viewpoints of the beings who participate in acts of valuation—and for Korsgaard, nonhuman animals most clearly value their lives. From this she concludes that "there is no place we can stand from which we can coherently ask which creatures, or which kinds of creatures, are more important absolutely."[95] The only sense in which some things can be said to matter "absolutely" is that some goods are important to all sentient beings, all beings with a point of view on their existence.

Consider the sorts of capabilities that Nussbaum acknowledges in both human and nonhuman animals: These, in matters such as bodily health and integrity, are indisputably important to all sentient beings, quite apart from the question whether those beings can or do think in explicit, linguistically structured ways about them. Korsgaard challenges the common-sense belief that "human welfare is more important absolutely than that of the other animals," arguing on the basis of her conception of absolute value that this could be true only if "even from the point of view *of the other animals*, what is good-for human beings matters *more* than what is good-for those other animals themselves." This is what Singer accepts and Korsgaard rejects. She notes that

it is hard to imagine anything that could make [the thesis that human welfare is more important in some objective sense] even remotely plausible except some sort of teleological view, according to which human good is the purpose of the world towards which all things in some way strive.[96]

A candidate for this sort of teleological view is, as I mentioned a moment ago, the ancient Stoic view, according to which all earthly things were created to satisfy human needs, so that human beings may employ their unique endowment of rationality to contemplate the eternal *logos* of nature and thereby fulfill a special capacity granted to them by the gods.[97] Korsgaard rejects this sort of view on the grounds that "being valued by a deity could only give us a tethered value," a value that she rejects as a product of an incoherent and "antiquated" conception of values and the way in which they are generated.[98]

These considerations lead Korsgaard to argue that nonhuman animals are members of the moral community inasmuch as "we recognize them to be fellow creatures."[99] Nonetheless there are certain clear differences between human and nonhuman animals, in particular that nonhuman animals do not think about what is good for them and thus (in all likelihood) lack "practical identity," the ability to engage in self-evaluation.[100] All sentient beings are "selves" in the sense of having a point of view on themselves as distinct from other beings, but only beings who are autonomous in the specifically human (which is to say, in the Kantian-Rawlsian) sense can articulate and evaluate reasons for their actions and seek to unify their conception of the world.[101]

This does not mean that human lives matter more (or have a greater moral worth) than nonhuman animal lives, but it does have some significant implications. Korsgaard exhibits an ambivalence about the nature and extent of nonhuman animal freedom. On the one hand, she acknowledges that nonhuman animals have purposes and employ various strategies for fulfilling them. This is what it means for sentient life to be "self-affirming." Valuing is "inherent in a sentient creature's relation to herself."[102] This would suggest that nonhuman animals have some capacity for a relationship to the world that is "open" in the sense that they can encounter unanticipated contingencies and devise novel ways to confront and adapt to them. But on the other hand, Korsgaard is committed to the traditional view that nonhuman "animals are governed by the laws given by their instincts, while we rational beings are governed by laws we give to ourselves. . . . there is a sense in which rational beings choose which ends to pursue," whereas a nonhuman animal's "purpose . . . is given to her by her instincts."[103]

This latter commitment is worth questioning, and on a number of levels. More than anything else, as I have suggested from the outset, it is very much worth challenging our own pretension to know that nonhuman animal behavior is strictly dictated by instinct. Is this something we can legitimately claim to know? Do we have the cognitive powers to ascertain this, and might this conclusion be driven by prior anthropocentric commitments? It is also worth asking whether we really know what "instinct" signifies here. In *Anthropocentrism and Its Discontents*, I examined a long series of efforts in Western thought to assert a strict either-or between fully rational agency and putatively "blind" instinct, with little if any consideration of the possibility of forms of free, intelligent reckoning that are more than instinctual but less than (or different from) the linguistically structured agency of human beings. Korsgaard seems to want to have it both ways—to retain the historical commitment to the proposition that nonhuman animal behavior is caused deterministically by instinct, but also to acknowledge that nonhuman sentient beings exhibit some sort of open, free relation to themselves and their environments. How, after all, can a being be said to *participate in valuing* if its behavior is every bit as causally determined as the functioning, say, of a potato? We need to broaden considerably our conception of the forms freedom can take if we are genuinely to understand what it means for nonhuman animals to be members of the moral community and to be direct beneficiaries of justice.

I will have more to say about the notion of freedom in the concluding section of this chapter. For now, I simply wish to point out that by leaving the question of freedom uninterrogated, Korsgaard fails to get past some stock anthropocentric commitments—not simply about specific uses of nonhuman animals, but also about the global background commitments that give rise to judgments about those specific uses. Let us first consider specific uses. Korsgaard makes a clear departure from prevailing anthropocentric commitments when she suggests that "the use of animals in laboratory research and experiments that are painful, invasive, or fatal is unjustifiable, and even barbaric. It is an obvious case of treating animals as mere means to our ends, as well as being immeasurably cruel."[104] It is significant that Korsgaard takes such a clear stand on invasive experimentation, particularly given that even some of the philosophers who have expressed the greatest concern for nonhuman animals have sought to justify the practice.[105]

Her remarks about meat eating, however, are somewhat more equivocal, and this provides a first clue to the way in which Korsgaard remains devoted, if only against her own intention, to an anthropocentric background ideal of living. She notes the harms to nonhuman animals from conditions on factory farms, she notes the loss of biodiversity and the production of greenhouse gases caused by intensive nonhuman animal agriculture, and she notes that even "humane farming" leads to shortened lives and death. But instead of making a clear pronouncement of the kind that she makes about invasive experimentation, Korsgaard suggests only that "you might wonder whether [the considerations she has offered about nonhuman animal agriculture] show that you should become a vegetarian or a vegan." The closest Korsgaard comes to taking a clear stand on this question is to state that this is all a matter of "how you are related to that particular creature when you eat her, or use products that have been extracted from her in ways that are incompatible with her good. You are treating her as a mere means to your own ends, and that is wrong."[106]

Korsgaard rightly suggests that the question of eating nonhuman animals and using products that have been extracted from them—this latter consideration takes us beyond vegetarianism to veganism—needs to be resolved through a consideration of what is compatible with a nonhuman animal's good. This is what Korsgaard has in mind when she appeals to a notion of each being's functional good. Framed in these terms, the question becomes: what is the "functional good" of a nonhuman animal? Let us leave aside the obvious fact that the answer to this question varies with variations in the experiential capacities of different sentient beings—humans, horses, and ants have different functional goods, at least in some respects—and simply ask whether inflicting harm on a nonhuman animal, or causing its death, in any case other than a bona fide emergency is compatible with its functional good. The clear answer would have to be no. The only way such actions could be justified would be to argue, for example, that death is not really a harm to nonhuman animals, and perhaps to add that we do nothing wrong if we treat them extremely well in the process of "extracting" products from them.

Whether you find this sort of justification compelling ultimately depends on whether you are devoted in advance to an anthropocentric as opposed to a non-anthropocentric ideal of living. And it is here that we can finally address more directly the very idea of a background ideal of living and begin to explore how it serves as the necessary presupposition for the individual judgments we make— judgments not only about moral status, but more generally about how different parts or elements of the world fit together with one another. This is a theme that I will explore throughout this book. It requires a consideration of the respective roles and mutual interplay of reason and emotion in the process of making moral (as well as some other kinds of) judgments, and it also requires some careful reflection on the notion of world or environment that we tend to take for granted in ways that can seriously distort the conclusions at which we arrive. There is quite a lot to say about these matters. In the concluding section of this chapter, I will simply offer some preliminary reflections on the idea of a background ideal of living, and I will offer some observations about the way in which the move from an anthropocentric to a non-anthropocentric background ideal of living can lend support to the radical proposal that I presented at the beginning of this chapter. In particular, this move recommends a fundamental rethinking of the notions of freedom and justice that, if I can convince you of its legitimacy, will make you accept the proposition that we human animals need to make some profound changes not only to the ways in which we treat nonhuman animals, but more fundamentally to the ways in we view and value them.

4. Anthropocentric and Non-Anthropocentric Background Ideals of Living

That Korsgaard offers essentially all the elements of a non-anthropocentric conclusion about the permissibility of eating nonhuman animals (and of using products derived from them) but stops short of clearly stating that conclusion might be explained by the consideration that people take the food question very personally and that she wants to leave it to individuals to process her arguments and come to their own conclusions. But that seems implausible, given that a great many people support invasive experimentation to benefit human beings and yet Korsgaard does not hesitate to offer a categorical criticism of the practice. Something else seems to be going on here, and I believe that that something becomes clearer when we consider her views about the institution of pet ownership. A moment ago, I asked the question what a nonhuman animal's functional good consists in. How we answer this question has deep implications for the nature and extent of freedom that we acknowledge in nonhuman animals, and that in turn has fundamental implications for how we conceive of the duties of justice (if any) we have toward nonhuman animals. Korsgaard's own answer is that your typical nonhuman animal's functional good consists in "the activities of feeding themselves and reproducing successfully." As regards pet ownership, Korsgaard notes that we provide our nonhuman companions with food, but we generally deprive them of the opportunity to seek food on their own and we prevent them from reproducing. She asks whether we can provide our nonhuman companions with suitable substitutes for these activities, and acknowledges that "the institution of pet-keeping is morally more problematic than you might think at first glance."[107] Among other things, some nonhuman animals should not be kept as pets (she gives the example of birds and some rodents and at least raises a question about cats), but Korsgaard suggests that keeping dogs as pets is unproblematic inasmuch as "dogs evolved to live with people and their dependence on us is not an unnatural condition that is contrary to their good." She concludes her discussion of the pet question with the observation that "the company of animals is good-for us," and offers us the assurance that nonhuman (companion) animals take a "naked, unfiltered joy . . . in little things—a food treat, an uninhibited romp, a patch of sunlight, a belly rub from a friendly human."[108]

Much of this accords with established common sense about institutions such as pet ownership: After all, what could possibly be wrong with a nice food treat and a belly rub? To view an institution such as pet ownership within these narrow confines is to ignore the larger set of background considerations that give rise to individual judgments of that kind, among other things, the consideration that human beings enjoy the prerogative to decide what is and what is not in a nonhuman animal's interest—what accords with and genuinely promotes its functional good. Korsgaard offers an assessment about the functional good of your typical nonhuman animal that is worth interrogating: that that good consists principally if not exclusively in eating and reproducing. To interrogate an assessment of this kind is not necessarily to reject it as false. But it is to ask two interrelated questions: whether this tells us the full story about the functional good of nonhuman animals, and whether the functional good of human beings is really, in the end, so very different than that of nonhuman animals.

Let me address this latter question first, by excavating a little intellectual history that tells us a lot about human self-understanding and self-evaluation. The Enlightenment philosophies of thinkers such as Hegel and Comte gave rise to a modern sensibility according to which history traces out a path of inevitable progress. According to that sensibility, the functional good of human beings includes pursuing a rational path toward ever-improving conditions. One eventual goal of such striving might be Kant's ideal of perpetual peace. Here we are talking about a sensibility whose roots are to be found in an ancient background ideal of living according to which human beings ultimately have different and higher ends than nonhuman animals have—ends, indeed, that make us godlike—and according to which it is up to human beings to decide what is and is not good for virtually all other beings in the world.

But there is an entirely different assessment of the modern sensibility that is worth considering, one that radically calls into question the assumption that human beings are really so different than nonhuman animals; and this, in turn, provides grounds for stepping back from the prevailing background ideal and subjecting its basic tenets to critical evaluation. To take this "step back" is to follow a hint offered by Martin Heidegger about how best to rethink our global commitments about the proper place of human beings in the cosmic scheme. A focal point for Heidegger's concern was the ways in which the human endeavor to assert mastery over nature has culminated in modern technology. He saw in technology—not just in the sense of devices and techniques, but in the sense of an entire way of representing and relating to nature—a tragic loss of intimate connection with the world. Heidegger recognized that there is no knock-down argument that will convince technophiles that this loss has in fact occurred. Instead, he suggested that what is required is more than a simple "shift of attitude." What is needed is a "step back" that "takes up its residence in a co-responding which, appealed to in the world's being by the world's being, answers within itself to that appeal."[109]

What Heidegger is proposing here is more than a simple reevaluation of our logic. It amounts to taking up an entirely new way of thinking that involves not just logic but a commitment to undoing the influence of a form of thinking that Heidegger calls "representational" and that on his view consists in forcing a global interpretation on the world that is designed *in advance* to promote control or mastery. I will explore this idea of representational thinking in the remainder of this book and draw out more clearly the terms of Heidegger's concern. For now I will simply note that, translated into everyday language, Heidegger's statement about "co-responding" makes a lot more sense that might appear at first blush: In our haste to treat the world as an object to be understood in ways that enable us to manipulate it for our own purposes, we have lost the kind of open connection (co-respondence) with the world in which we let the world show itself to ourselves as what and how it genuinely is. To take the step back that Heidegger recommends involves two phases: first, a recognition that something has gone terribly wrong in our ways of relating to the world; and second, an endeavor based on that recognition to engage in what is sometimes called a "quietude of the will," a way of being in which we specifically refrain from willing (from seeking to alter or control the world) and maintain ourselves in a posture of open receptiveness. The endeavor to relate to the world in this way is profoundly difficult, due to our primary orientation on manipulation and a certain fear about what we might encounter.

Heidegger's contrast between the endeavor to manipulate the world and the endeavor to stand open for its mystery has implications for the question of what it means to be human—a question that, inasmuch as to be human is to be an animal, is inextricably bound up with the question of what it means to be a nonhuman animal. Consider the optimism of the Enlightenment that I addressed a short while ago. Karl Löwith offered a very clear verdict about the modern commitment to inevitable progress. He saw it as a secularized version of Christian eschatology, i.e., of the view that the unfolding of time traces out a progression (in the Vulgate it is called a *procursus* or running forward, as in "onward Christian soldiers") toward the Day of Judgment, when Christ would return to earth and take the elect to heaven. The Day of Judgment is the *eschaton* or point of teleological fruition of time itself. This is the sense in which the Christian conception of time is "eschatological": time itself is ordered toward the Day of Judgment. The modern conception of history presented by thinkers such as Hegel and Comte is a secularized version of this eschatology, where the driver is not the Holy Spirit but reason itself and the end point is described variously by different thinkers. For Hegel that end point is the realization of Absolute Spirit. For Kant it is a condition of perpetual peace. For Marx it is the overcoming of capitalist productive relations and the realization of a society of free producers.

Against this vision of history as tracing out a progression (one that the thinkers previously mentioned all treat as inevitable), Löwith offers a considerably more sober vision, one that recommends "a definite resignation" in place of the thought that we will eventually overcome the destructive power of time. Invoking the authority of Jakob Burkhardt, Löwith urges the conclusion that

> if there is anything to be learned from the study of history, it is a sober insight into our real situation: struggle and suffering, short glories and long miseries, wars and intermittent periods of peace. All are equally significant, and none reveals an ultimate meaning in a final purpose.[110]

None of this is to suggest that we should throw up our hands and do nothing about the epic injustices or environmental calamities that occur in the world. But it is to suggest that, as I have suggested from the outset, death is the great equalizer and that *in truly essential respects the human condition does not differ from the terms of nonhuman animal life*. Our most fundamental imperatives are nutrition and reproduction (although the problem of human population growth has become epic and stands in need of serious amelioration, so we hardly need to worry any more about the propagation of the species), and to that extent we are no different and no better than nonhuman animals. To suppose otherwise is to subscribe, if only implicitly, to exactly the kind of teleological conception of the world that Korsgaard rightly rejects as "antiquated."

Now how do things look if we dispense with the pretense of supposing that human beings are capable of achievements that elevate us about our natural condition? The immediate response to this, from the standpoint of deeply engrained anthropocentric prejudice, is to offer the sort of consideration that virtually all of the philosophers I have discussed so far hasten to point out—that human beings truly are different than nonhuman animals in possessing a form of freedom that no other being possesses. For Aristotle and Kant this meant that we humans can be self-regulating in a way that nonhuman animals cannot be; Korsgaard offers the slight modification that what is truly distinctive in human beings is our ability to set ends for ourselves that transcend the ends that are prescribed to us by nature. Löwith gives us reason to question whether these capacities really elevate us above our natural condition: After all, what does all this self-regulation ultimately *accomplish*? I realize that what I am about to suggest will seem cynical in the extreme to anyone who is invested in the story of inevitable progress, but it has to be considered. My suggestion is that the ability, indeed the necessity, to set ends for ourselves might actually be a sign of a tragic flaw in the human condition rather than a sign of superiority. One need only think of all the epic problems to which this capacity has given rise, in particular the unleashing of violence on an epic scale against the environment, our fellow human beings, and nonhuman animals. As I proposed earlier, the historical record does not offer much in the way of confirmation that the specifically human capacity for freedom elevates us above our natural condition.

At the same time, there are undeniable achievements that have been facilitated by human freedom, notably those in the arts and at least the recognition if not the implementation of the insight that the individual is the proper focal point for understanding the phenomenon of freedom. If there was one thing the tradition was right about, it is that human beings and we alone appear to be capable of grasping and living in accordance with explicitly articulated rights and responsibilities. But the tradition erred profoundly in asserting what for thinkers such as Rawls is a commonplace: that only those beings who can conceive of rights and responsibilities can be seen to be active participants in and full beneficiaries of them. It is here that the opposition between anthropocentric and non-anthropocentric background ideals of living begins to come into clearer relief.

There are adherents of what I take to be an essentially anthropocentric background ideal of living who are willing to talk about direct obligations toward nonhuman animals. Singer, Nussbaum, and Korsgaard all fall in this category. It is the kind and extent of obligations (and, correspondingly, the kind and extent of nonhuman animal entitlements) that these thinkers are willing to acknowledge that proves to be decisive. Here is one simple way to think about what I am getting at: Are nonhuman animals entitled to receive the benefit of the doubt from humans when it comes to thinking broadly about how to conceive of the nature and extent of nonhuman animal freedom? One possible concomitant of the step back to a more open relationship to the world is a moment of humility and generosity on the part of human beings. Only in such a spirit of openness, a spirit that is not merely a matter of logic but involves axiological commitment and thus a crucial affective dimension, would it truly be possible to consider questions such as whether pet ownership (and, more generally, domestication) genuinely benefits nonhuman animals.

I consider all the thinkers I have just named anthropocentric on the grounds that they all start from a view of nonhuman animals as a special kind of object with certain capacities that determine what sorts of entitlements they enjoy and what sorts of corresponding obligations we humans should be considered to have toward them. (This is the kind of thinking that Heidegger has in mind when he subjects "representational" thinking to critique. It is a kind of thinking according to which we are thinking subjects and everything else is an object, where some of these objects might prove to be sentient.) There is a certain indisputable logic here, but this logic is driven by some implicit background assumptions—assumptions which, if abandoned, lead to a different indisputable logic. In the remainder of this book I will develop and defend my claim that it is not logic alone that informs our deepest moral commitments. The background ideal to which we subscribe at any time will be subject to critical evaluation (which crucially involves reason), but that critical evaluation will involve dimensions other than reason. *A background ideal of living is a global, implicit, crucially (which is to say, always already) value-laden orientation on the world.* Such an ideal is implicit in the sense that it functions as the necessary presupposition for the operation of moral (and some other types of) judgment, rather than being an object of reason. An important function of reason will be to step back from the background ideal and seek to evaluate it by examining and reflecting on different aspects of the ideal; but it is in the nature of such ideals (as I will show in Chapters 3 and 4) not to be reducible all at once and in their entirety to objective specification.

Given these features of background ideals of living, it is not a simple matter of "logic" to decide whether or not nonhuman animals have this or that kind of freedom, nor to decide which exact obligations we humans should be recognized to have toward them. Nor will it be a matter of empirical observation. There is something inherently mysterious about the inner subjective lives of nonhuman animals; for all our historical self-congratulation about *logos* making us superior to nonhuman animals, it functions as an obstacle in our efforts to plumb the depths of nonhuman animal experience. We are necessarily forced to try to understand that experience through analogy to our own experiential capacities; and yet it is precisely recourse to analogy that has led so many philosophers and ethologists to conclude, *entirely speculatively*, that to the extent that nonhuman animals lack predicative linguistic ability, their behavior must necessarily be understood to be purely instinctive—as if there were no possible alternatives to these two extremes.[111]

There are of course many affinities between human and nonhuman animal experience, and in recent years philosophers and ethologists have become increasingly open to appealing to embodiment and our emotional lives as a route toward a better appreciation of the lives of nonhuman animals. In particular, much has been written about the power of empathy to open us to the lives and the predicament of our nonhuman companions in mortality. Some thinkers go a bit far in this direction, Elisabeth de Fontenay calling for an ethics based on "pathocentrism" and Coetzee's Elizabeth Costello stating that she "was hoping not to have to enunciate principles . . . open your heart and listen to what your heart says."[112] I will examine these efforts and assess the contribution made by the emotions to our moral sensibilities in Chapter 4. Here I will simply note that, like reason, our affective constitution plays a crucial role in maintaining, evaluating, and revising background ideals of living. In the end I will argue that those who, like Elizabeth Costello, would dispense with reason and principles in favor of looking into their hearts simply invert the old priority of reason over emotion and ultimately fail to grasp the crucial *interplay* of reason and emotion in the moral life. It is in the very nature of our lives as valuers that *both* play an indispensable role. The extent to which both are at work in the lives of nonhuman animals, no one can say with certainty.

This lack of certainty can be interpreted in two rather different ways, one of them anthropocentric and the other non-anthropocentric. On the anthropocentric approach, we acknowledge that the lives and capacities of nonhuman animals are a bit of a mystery, but we nonetheless exercise a certain dominion over nonhuman animals—not simply physical dominion, but a kind of cognitive dominion in the sense that we consider ourselves entitled (perhaps even obligated) to make assumptions about nonhuman animals that may or may not do justice to their free capacity to act on their own behalf. On a non-anthropocentric approach, we make a more robust effort to acknowledge both our own ignorance and our pre-existing interest in justifying various uses of nonhuman animals. The non-anthropocentric approach acknowledges a history of using nonhuman animals and recognizes the need to step back from that historical commitment and undertake measures to move beyond it.

One might think of the gulf between anthropocentric and non-anthropocentric background ideals of living along the lines of the dispute in contemporary political theory between ideal and non-ideal theory: Instead of placing the primary emphasis on an ideal outcome of some sort, perhaps we should take a very sober look at the actually existing power relations (which, let's face it, are stacked overwhelmingly in favor of human beings) and think from them toward a better future. Even if this distinction between ideal and non-ideal theory may ultimately be a sort of false dichotomy, it nonetheless is instructive for thinking about the long-standing predicament of nonhuman animals—that they are and long have been at our mercy, and that this has deprived them *and us* of the opportunity to experience what their freedom truly is.

Another way to think about the difference between the anthropocentric and non-anthropocentric background ideals of living is to consider the distinction between social and cosmic justice, to which I alluded earlier in this chapter.[113] Social justice pertains to relations between human agents who are capable of the sorts of contractual relationships on which thinkers from Epicurus to Rawls based their conceptions of justice. The sphere of social justice is one characterized by reciprocity in rights and obligations between linguistic agents who can grasp, reflect on, debate, and revise rights and duties through cooperative efforts. Contemporary political philosophers such as Jürgen Habermas and Rainer Forst make contributions to justice understood in this specifically anthropocentric sense.

Cosmic justice is a notion that appears to many anthropocentrically minded thinkers to be a contradiction in terms. Such thinkers believe, as Rawls and the Stoic philosophers maintained, that a being must comprehend the very notion of justice in order to be a beneficiary of justice. Thus for the Stoic thinkers, nothing we do to a nonhuman animal can possibly count as an injustice. To take seriously the idea that a being who is not rational and linguistic in specifically human terms can be a beneficiary of justice is to adopt a fundamentally different, non-anthropocentric perspective. It is to step back from narrowly human concerns and begin to take seriously the proposition that non-human beings can be harmed in ways that infringe on their natural entitlements. Which natural entitlements might those be? That is and will always be a very difficult question to answer, and as I have already suggested it cannot be answered merely through empirical investigation. How one answers this question will always be influenced more than anything else by the background ideal of living she embraces. The anthropocentric approach sees nonhuman animals as entities with certain capacities. The non-anthropocentric approach proceeds in an entirely different manner. It starts, as I have been suggesting, from a certain modesty and generosity, and acknowledges something that Löwith proposed: that when we think exclusively in terms of human affairs and history, "we miss that *one* world that is older and more enduring than the human world." This larger world "does not appear to be an object like other things; it encompasses everything and cannot itself be grasped."[114]

The world, taken in its most all-encompassing sense, is not an object. That it is not an object makes it very difficult but not impossible to discuss. Erazim Kohak offers a way of thinking about the world in this sense when he suggests that we humans are "integrally linked to the being of nature . . . [and] fundamentally *at home* in the cosmos, not 'contingently thrown' into it as an alien context and 'ek-sisting' from it in an act of Promethean defiance."[115] Kohak draws on some themes from Heidegger here, in particular the idea of human beings as "ek-sistent" or always maintaining a certain distance from nature and ourselves due to our ability to contemplate things and take a free stand on them. Kohak's point is one that Heidegger came to in his writings after *Being and Time*: that even though human beings struggle with the gap between themselves and the world in their endeavor to cultivate meaning, we need to recognize that all valuing and willing proceed on the basis of our being always already integrated or located in a world that is not reducible to objective grasping or insight. Kohak's point, following Heidegger, is that any thinking that fails to take our rootedness in this irreducible, mysterious sense of world as its point of departure will necessarily be impoverished—a claim that, from the standpoint of anthropocentric thinking, will naturally seem far-fetched.

Heidegger develops this notion through a reflection on the very idea of ethics. Whereas traditional thinking conceives of ethics in terms of specific rights and responsibilities on the part of beings possessing certain capacities, and where traditional thinking typically conceives of the word 'ethos' in terms of character (in the sense of an individual's moral character), Heidegger proposes that we think of 'ethos' in terms of dwelling or "the abode of the human being." This radical reflection on ethics points us toward a fairly elementary observation: that we find ourselves in the midst of a world that is not of our making, that we are not born knowing our proper place in the scheme of things, and that we face the task of finding our proper place or abode. Heidegger urges the conclusion "that thinking which thinks the truth of being as the primordial element of the human being, as one who eksists, is in itself primordial ethics."[116] To think the truth of being in this sense is to step back from our entire traditional way of valuing and thinking about our proper place, to acknowledge a certain sense of lostness, and to seek to redress that sense of lostness by "letting beings be" rather than seeking to force interpretations on them that are cut to the measure of human mastery.[117]

From the standpoint of anthropocentric values and ways of being, this ideal of letting beings be probably sounds downright kooky. It certainly could be said to contain a mystical element inasmuch as it acknowledges fundamental limits to human knowledge and draws a certain ethical orientation from that acknowledgment: one of caution and humility in our assessment of our own powers and entitlements, and a commitment to giving nonhuman beings the benefit of the doubt when it comes to assessing their powers and entitlements. Reason(ing) alone, as I have suggested from the start, is not enough by itself to move anyone to undertake the shift of perspective that is required to take on this kind of humility, although reason will play a role in coming to the commitment to make that shift. Equally important is an affective component that I have characterized in earlier work as "felt kinship" and which I will develop further in Chapter 5. There is an important insight in pathocentric approaches to the moral status of nonhuman animals, but that insight must be brought into dialectical relationship with rational considerations.

A reflection on the respective roles and mutual interplay of reason and emotion in the moral life promises to set the stage for a fundamental, non-anthropocentric rethinking of our entire sense of place in the world. This kind of wholesale rethinking will make it possible to revisit controversies such as the legitimacy of domestication (including pet ownership) and the allegation that the sort of approach I am proposing would entail "apartheid" between human and nonhuman animals.[118] The apartheid claim is that if we seek to discontinue uses of nonhuman animals altogether, we would effectively suspend all relations between human and nonhuman animals. I believe that thinkers who continue to defend domestication, and those who make the apartheid allegation, are very sincere in the concerns they express. I also believe that they are proceeding from a place of fear—specifically, fear of what the world might look like if we "let beings be" and give nonhuman animals the space for once to dictate the terms of their being and their relationships with themselves, with one another, and with human beings.

Rather than proceeding from a concern for how things would become more difficult for us (although that is most definitely a genuine concern), we ought to have the courage to abandon the traditional anthropocentric ideal while retaining the commitment to peaceful interrelationships that it at least purports to contain. Grimm's insight into the nonhuman "animal-in-itself" and Taureck's imperative to discontinue use are crucial elements in this endeavor to question our own longstanding values. The one forces us to rethink our pretensions to know what nonhuman animals are and what is good for them. The other demands of us that we act on the basis of that rethinking in a way that honors our historical pretension to be committed to consistent and thoroughgoing justice.

Notes

1 Both the numbers of nonhuman animals affected and the means we employ are staggering. According to the United Nations Food and Agriculture Organization, every year worldwide over seventy billion land animals are killed for human consumption. If we include fish and other sea creatures, the number likely exceeds a trillion per year. Jonathan Balcombe, *What a Fish Knows*, p. 7. For a detailed discussion of the conditions to which we customarily subject nonhuman animals, see David Nibert, *Animal Oppression and Human Violence*.

2 I examine this history, starting with the pre-Socratic Greeks and proceeding to the twentieth century, in *Anthropocentrism and Its Discontents*.

3 Tom Regan, *The Case for Animal Rights*, pp. 233, 243, 351. Regan states that even "a million dogs ought to be cast overboard if that is necessary to save the four normal humans." If this is true, then consistency would entail that *any* number of nonhuman animals should be sacrificed to save just one human being. Naturally I am simply

considering the question of moral entitlement here, not the obvious practical consequences that would follow from denuding the earth of potentially trillions of nonhuman
animals.

4 Peter Carruthers, *The Animals Issue*, p. 9. Here Carruthers notes that people might judge
differently if the human involved were a moral reprobate such as a murderer or a child
molester.

5 Gary L. Francione, *Introduction to Animal Rights*, pp. 49, 55, 153.

6 John Rawls, *A Theory of Justice*, pp. 44, 42.

7 *A Theory of Justice*, pp. 442, 448.

8 *The Animals Issue*, p. 99.

9 Adam Smith, *The Theory of Moral Sentiments*. I examine this notion in Chapter 5.

10 I argue at length for this conclusion in Gary Steiner, *Animals and the Limits of Postmodernism*, where I present a contrast between a "total" and an "immanent" critique of
reason. The latter acknowledges the limits of reason while maintaining a commitment to
the crucial contribution it makes to moral reflection.

11 Ralph Waldo Emerson, "Self-Reliance," p. 138.

12 Arthur Schopenhauer, "On Religion," p. 375.

13 See Gary Steiner, *Animals and the Moral Community*, Ch. 5 and 6.

14 I return to this phenomenon of meaning in Chapter 4, where I discuss Heidegger's analysis of the as-structure of intelligibility.

15 See Mark Rowlands, *Can Animals Be Moral?* and my review of Rowlands in *The Philosophers' Magazine*; Christine M. Korsgaard, *Fellow Creatures*. I present the outlines
of Korsgaard's views on nonhuman animals later in this chapter.

16 I present further thoughts on this question in Gary Steiner, "Kathy Rudy's Feel-Good
Ethics."

17 Thomas Hobbes, *Leviathan*, Ch. 13, p. 87.

18 Kant thus opposes the sensible intuition of embodied, finite beings with the "intellectual
intuition" that an infinite intelligence such as God would possess. Immanuel Kant, *Critique of Pure Reason*, pp. 88, 90 (B68, B71).

19 Herwig Grimm, "Das *Tier an sich*?," p. 69.

20 "Das *Tier an sich*?" p. 65.

21 "Das *Tier an sich*?" p. 66. I will discuss the views of Nussbaum and Korsgaard in the
remainder of this book.

22 George Yancy, "Judith Butler" (interview with George Yancy).

23 Bernhard Taureck, *Manifest des veganen Humanismus*. Taureck's call to let nonhuman
animals be recalls Heidegger's notion of "letting beings be." See my remarks in *Animals
and the Moral Community*, p. 141. I examine this notion in the concluding section of this
chapter.

24 For an earlier treatment of these themes, see Gary Steiner, "Toward a Nonanthropocentric Cosmopolitanism," Chapter 5 of *Animals and the Limits of Postmodernism*.

25 I examine Aristotle's views in much greater depth in *Anthropocentrism and Its Discontents*, Chapter 3.

26 Aristotle, *Physics*, 2.3 at 193a28–31, 194b33, 195a24–5, Aristotle, *The Complete Works
of Aristotle*, vol. 1, pp. 330, 332, 333.

27 Aristotle, *On the Soul*, book 2, Chapter 3, *The Complete Works of Aristotle*, vol. 1,
p. 659f.

28 *On the Soul*, book 3, Chapter 3 at 428a20–21 and book 2, Chapter 9 at 421a20–26, *The
Complete Works of Aristotle*, vol. 1, pp. 681, 670.

29 Aristotle, *Politics*, book 1, Chapter 2 at 1253a8–10, *The Complete Works of Aristotle*,
vol. 2, p. 1988.

30 Aristotle does not explicitly state that nonhuman animals are incapable of political community, and I can imagine a case being made for the proposition that he grants nonhuman animals the capacity to form their own political communities even though they
cannot participate in specifically human community. But the fact that he ultimately treats

nonhuman animals as resources for the satisfaction of human needs strongly suggests to me that he does not take the idea of nonhuman animal political community particularly seriously, if indeed he takes it seriously at all.

31 Aristotle, *Nicomachean Ethics*, book 10, Chapter 8 at 1178b20–29 and Aristotle, *Eudemian Ethics* book 1, Chapter 7 at 1217a24–5, *The Complete Works of Aristotle*, vol. 2, pp. 1863, 1926. Plato and Aristotle are fully aware, however, that nonhuman animals are capable of virtue in a different, limited sense—the sense in which, for example we say that the excellence or "virtue" (*arete*) of a mother canine is that she is highly effective in providing for her young.

32 *Politics*, book 1, Chapter 8 at 1256b15–22, *The Complete Works of Aristotle*, vol. 2, p. 1993f.

33 Martha Craven Nussbaum, *Aristotle's De Motu Animalium*, p. 96.

34 See my discussion of the zoological texts in *Anthropocentrism and Its Discontents*, pp. 69–76. (In attributing these capacities to nonhuman animals, Aristotle adds the qualification that nonhuman animals act simply "as if" they possessed them. Aristotle, *History of Animals*, book 7, sec. 1 at 588a17–588b4.)

35 Aristotle states that one can be a slave "by convention" or "by nature." He goes on to explain that some people, regardless of whether they are slaves by nature, may wind up in conditions of enslavement through "hunting or war." But he also states that "men [should not] study war with a view to the enslavement of those who do not deserve to be enslaved," and here he means those who are not assigned the role of servitude by nature. *Politics*, book 1, Chapter 7 at 1255b38–40 and book 7, Chapter 14 at 1333b39–40, *The Complete Works of Aristotle*, vol. 2, pp. 1992, 2116.

36 *Politics*, book 1, Chapter 5 at 1254a22–3, *The Complete Works of Aristotle*, vol. 2, p. 1990.

37 Aristotle states that "although there may be exceptions to the order of nature, the male is by nature fitter for command than the female." *Politics*, book I, Chapter 12 at 1259b2–3, *The Complete Works of Aristotle*, vol. 2, p. 1998.

38 *Politics*, book 1, Chapter 5 at 1254b6–8 and 1254b24–5, *The Complete Works of Aristotle*, vol. 2, p. 1990, Cf. book 1, Chapter 7 at 1255b18–19, p. 1992: "There is one rule exercised over subjects who are by nature free, another over subjects who are by nature slaves." I.e., differences in deliberative capacity correspond to different forms of rule.

39 *Politics*, book 1, Chapter 5 at 1254b10–12, *The Complete Works of Aristotle*, vol. 2, p. 1990.

40 *Politics*, book 1, Chapter 13 at 1259b17–21, *The Complete Works of Aristotle*, vol. 2, p. 1999.

41 *Nicomachean Ethics*, book 8, Chapter 11 at 1161b1–8, *The Complete Works of Aristotle*, vol. 2, p. 1835.

42 As I noted earlier, Rawls categorically excludes nonhuman animals from the sphere of justice; in doing so, he is implicitly following Aristotle's reasoning.

43 *Nicomachean Ethics*, book 10, Chapter 5 at 1175a26–28, *The Complete Works of Aristotle*, vol. 2, p. 1858.

44 See Porphyry, *On Abstinence from Killing Animals*, book 3, sec. 20, p. 91; John Stuart Mill, "Utilitarianism," p. 140.

45 Peter Singer, *Practical Ethics*, p. 92.

46 *Practical Ethics*, p. 91.

47 Singer argues that self-aware beings (who among other things have "a sense of their own future") are non-replaceable, whereas non-self-aware beings are replaceable, his reasoning being that there is no net utilitarian "loss" if we bring another nonhuman animal of the same kind into existence when we kill another of that kind. *Practical Ethics*, p. 120; see also *Animal Liberation*, p. 229.

48 Immanuel Kant, *The Metaphysics of Morals*, p. 186.

49 In his writings on morality, Kant discusses legislation as a rational being's activity of adducing the moral law. I won't go into a detailed defense of my claim here, but will

simply suggest that the same considerations that recommend seeing the moral law as a matter of legislation also recommend seeing our self-assertion as persons to be a matter of legislation.

50 Cicero, *On Duties*, p. 109.

51 See *Leviathan*, p. 145f.

52 For reasons that I need not go into here, on Kant's view there are differences between objective scientific judgments and the sorts of judgments that agents make in aesthetics and morality.

53 *Critique of Pure Reason*, p. 635 (A804–5/B832–3).

54 Immanuel Kant, *Grounding for the Metaphysics of Morals*, p. 36 (Ak. 428).

55 Adam Smith, who is better known for the notion of the invisible hand and widely misunderstood as having sanctioned ruthless individualism in the economic sphere, issues a similar call to subordinate selfish motives to a sense of respect for others when he cautions against "mean rapacity" in capitalist relations. See Adam Smith, *An Inquiry into the Nature and Causes of the Wealth of Nations*, vol. 1, p. 493. This passage is best interpreted against the background of Smith's remarks in *The Theory of Moral Sentiments* about the impartial spectator, which I have already mentioned and to which I return in Chapter 5.

56 *Grounding for the Metaphysics of Morals*, p. 36 (Ak. 429). At the same time, Kant makes room for a class of human beings who stand in need of rational guidance; he includes in this class of "passive citizens" women, children, apprentices, and several others. *The Metaphysics of Morals*, Doctrine of Right, sec. 46, p. 92 (Ak. 314).

57 Immanuel Kant, *Lectures on Ethics*, pp. 147, 156.

58 *Lectures on Ethics*, pp. 126, 234.

59 *The Metaphysics of Morals*, Doctrine of Right, sec. 55, p. 115 (Ak. 345). On the following page, Kant states that human beings are "co-legislating members of a state," whereas nonhuman animals are merely "property." This corresponds directly to Kant's person-thing distinction.

60 See my discussion of Descartes in *Anthropocentrism and Its Discontents*, Chapter 6.

61 Immanuel Kant, *Critique of the Power of Judgment*, p. 328 (Ak. 464).

62 *The Metaphysics of Morals*, Doctrine of Virtue, sec. 17, p. 192 (Ak. 443).

63 On Aquinas's indirect duties approach, see *Anthropocentrism and Its Discontents*, p. 131.

64 *The Metaphysics of Morals*, Doctrine of Virtue, sec. 17p. 192f. (Ak. 443). See also *Lectures on Ethics*, p. 212.

65 On Aristotle's and Augustine's denials of community with nonhuman animals, see *Anthropocentrism and Its Discontents*, pp. 62, 119. Both base this denial of community on a lack of commonality between humans and nonhuman animals.

66 Immanuel Kant, *Anthropology from a Pragmatic Point of View*, p. 369 (Ak. 268).

67 *Lectures on Ethics*, p. 434.

68 *Critique of the Power of Judgment*, sec. 82 and 83, pp. 294, 298 (Ak. 426, 431). Here Kant affirms Aristotle's verdict that nonhuman animals exist "for the human being, for the diverse uses which his understanding teaches him to make of all these creatures."

69 Robert Nozick, *Anarchy, State, and Utopia*, p. 36.

70 *The Metaphysics of Morals*, p. 193 (Ak. 443).

71 *Anarchy, State, and Utopia*, pp. 35, 42.

72 *Practical Ethics*, pp. 51–2, 103.

73 *Practical Ethics*, pp. 56, 58. See also p. 121: "Even in the case of animals with some self-awareness, killing for food will not always be wrong."

74 Martha Craven Nussbaum, *Frontiers of Justice*, pp. 76–8.

75 *Frontiers of Justice*, p. 79.

76 *Frontiers of Justice*, p. 389.

77 *A Theory of Justice*, p. 448; see also John Rawls, *Political Liberalism*, pp. 21, 246.

78 John Rawls, *Justice as Fairness*, p. 26.

79 *Frontiers of Justice*, p. 391.
80 I examine the encyclical in detail in Gary Steiner, "The Moral Schizophrenia of Catholicism."
81 See my remarks on first-order conceptions of the good and second-order procedural rules in *Animals and the Moral Community*, pp. 146–50. There I also present an overview of Brian Barry's conception of justice as fairness.
82 *Frontiers of Justice*, pp. 386 (killing a free-range nonhuman animal painlessly "is no harm at all," although on p. 393 she proposes "moving gradually toward a consensus against killing at least the more complexly sentient animals for food"), 400 ("laboring animals" are entitled to "dignified and respectful labor conditions"), 404 (at least some experimentation is permissible provided that we "[improve] the conditions of research animals"), 377 (there is nothing wrong with practices such as training horses and dogs, and involving horses in dressage and racing, nor with maintaining the property status of nonhuman animals). For a very different assessment of the property status of nonhuman animals, I refer the reader to Gary Francione's highly influential work, in particular *Animals, Property, and the Law*.
83 *Frontiers of Justice*, p. 405.
84 *Frontiers of Justice*, pp. 393–400.
85 *Frontiers of Justice*, p. 378.
86 *Frontiers of Justice*, pp. 360–1, 386. The sense of hierarchy that I have just noted in Nussbaum's thought also informs her most recent book, Martha Craven Nussbaum, *Justice for Animals*; thus, even though Nussbaum may outwardly appear to make an advance over thinkers who deny duties of justice toward nonhuman animals, the kind of justice we owe them is very much diminished in comparison with the duties that Nussbaum considers us to owe to our fellow human beings.
87 Jeremy Bentham, *An Introduction to the Principles of Morals and Legislation*, p. 311n. For a more detailed examination of Bentham, see my remarks in *Anthropocentrism and Its Discontents*, pp. 162–4.
88 Jacques Derrida, *The Animal That Therefore I Am*, p. 27.
89 A slightly different strategy is to assert that we have to go on the evidence available to us, not on the evidence we wish we had, and that this gives rise to a rebuttable presumption that nonhuman animals lack certain capacities—the rebuttal coming, if ever, in the form of indisputable evidence that they possess those capacities. See Gary E. Varner, *Personhood, Ethics, and Animal Cognition*, p. 115. As I hope will become clear in the final section of this chapter, this is a rather stingy way of proceeding, particularly given the well-known difficulties involved in ascertaining the nature of nonhuman animal experience.
90 *Frontiers of Justice*, p. 381. This call for equality for humans and adequacy for nonhuman animals recalls Nozick's consideration (which he ultimately rejects) of Kantianism for humans and utilitarianism for nonhuman animals. *Anarchy, State, and Utopia*, pp. 39, 42.
91 Mary Anne Warren, "The Rights of the Nonhuman World," p. 192. Kant explicitly rejects this claim of greater intrinsic value when he states that being able to set ends for himself gives man "only an extrinsic value for his usefulness." *The Metaphysics of Morals*, p. 186 (Ak. 434).
92 *Fellow Creatures*, p. 145f.
93 *Fellow Creatures*, p. 147.
94 *Fellow Creatures*, p. 64f.
95 *Fellow Creatures*, p. 10; see also p. 59.
96 *Fellow Creatures*, p. 11.
97 I present a detailed analysis of the Stoic view in *Anthropocentrism and Its Discontents*, pp. 77–92. I examine Seneca's views in depth in Chapter 2 below.
98 *Fellow Creatures*, p. 12.
99 *Fellow Creatures*, p. 148.

100 *Fellow Creatures*, pp. 23, 46.
101 *Fellow Creatures*, pp. 30, 33, 43–6, 80, 84.
102 *Fellow Creatures*, p. 168; see also Christine M. Korsgaard, *The Constitution of Agency*, pp. 175, 205: Nonhuman animals are governed by "mere reaction."
103 *Fellow Creatures*, p. 119.
104 *Fellow Creatures*, p. 228.
105 Conspicuous in this regard is Schopenhauer, who decried animal cruelty but nonetheless maintained that vivisection is permissible if practiced only rarely and "only in the case of very important investigations that are of direct use" to human beings. "On Religion," p. 374f. It is more than a bit ironic that Schopenhauer goes on here to offer the assessment, noted near the beginning of this chapter, that the world is not a piece of machinery made for our use.
106 *Fellow Creatures*, p. 222f.
107 *Fellow Creatures*, p. 233f.
108 *Fellow Creatures*, pp. 235, 237.
109 Martin Heidegger, "The Thing," *Poetry, Language, Thought*, p. 179; see also p. 183. Heidegger, for his part, was relatively unconcerned with nonhuman animals, famously characterizing them as "world-poor" and maintaining that they do not share in mortality but merely "perish." See my remarks in *Anthropocentrism and Its Discontents*, pp. 204–12 and *Animals and the Limits of Postmodernism*, pp. 100–12.
110 Karl Löwith, *Meaning in History*, pp. 199, 25. See also my examination of the notions of eschatology and secularization in Chapter 5 of Gary Steiner, *Descartes as a Moral Thinker*.
111 On the nature of nonhuman animal experience and the ways in which it may be like and unlike human experience, see my remarks in Chapters 1–3 of *Animals and the Moral Community*. There I make my own admittedly speculative effort to account for nonhuman animal experience in a way that avoids the strict either-or urged on us by traditional thought.
112 Elisabeth de Fontenay, "Pourquoi les animaux n'auraient-ils pas droit à un droit des animaux?" p. 153; J.M. Coetzee, *The Lives of Animals*, p. 37.
113 I first presented this distinction and discussed it in greater depth in Chapter 5 of *Animals and the Moral Community*.
114 Karl Löwith, "Welt und Menschenwelt," p. 295f.
115 Erazim Kohak, *The Embers and the Stars*, p. 8.
116 Martin Heidegger, "Letter on 'Humanism'," p. 271. Cf. pp. 257–60, where Heidegger characterizes the problem of ethics in terms of homelessness, the lack of a sense of proper place in the world.
117 Martin Heidegger, "On the Essence of Truth," p. 143.
118 *Fellow Creatures*, p. 179. Donaldson and Kymlicka make the same claim; see Sue Donaldson and Will Kymlicka, *Zoopolis*, pp. 7, 49, 62, 178, 180.

2
THE ESSENTIAL ROLE AND PITFALLS OF REASON IN MORAL JUDGMENT

> As the sole being on earth who has reason, and thus the capacity to set voluntary ends for himself, [the human being] is certainly the titular lord of nature.
>
> Kant, *Critique of the Power of Judgment*

1. Background Ideals of Living and Our Basic Understanding of Reason

The dominant tendency in Western philosophy has been to conceive of reason as preeminent among human experiential faculties. Reason has been taken to be the one experiential faculty that can attain true impartiality, and thus as the one faculty that can and should provide regulatory guidance to our other experiential faculties. Descartes, for example, sees in reason the capacity and the responsibility to assess and regulate the data and impulses transmitted to us by our sensory apparatus and our emotional states. In this respect, Descartes proves to be exemplary of a long-standing commitment in Western thought to the proposition that reason must arm itself against the emotions and the potentially distorting influence of sensory experience: Descartes, like a number of influential thinkers before and after him, frames the relationship between reason and emotion as a war in which the former needs to exercise dominance over the latter.[1] One of my aims in this book is explore this historical commitment to the idea that the emotions are dangerous and stand in need of regulatory guidance from reason. A related aim is to show that, in its efforts to fortify the human condition against the potentially pernicious influence of the emotions, the tradition has provided us with a distorted picture of both the nature of reason and its role in human moral choice.

It would be wrong, however, to suggest that the tradition has completely mis-understood the nature of reason. In many respects, the tradition has articulated a

DOI: 10.4324/9781003425595-3

perfectly unobjectionable conception of rationality. But the tradition has interwoven into its conception certain commitments about the authority and implications of human reason that bear the unmistakable imprint of global anthropocentric values. One indication of these specifically anthropocentric commitments is the tradition's repeated framing of human rational capacity as standing in contrast with the putatively "inferior" experiential capacities of nonhuman animals. Almost all the major exponents of the tradition's anthropocentric conception of reason have proceeded on the assumption that (typical mature) human beings are rational and that virtually no nonhuman animals possess rationality. The assertion that nonhuman animals are what Aristotle called *aloga* (lacking in linguistic rationality) is no mere consequence of the tradition's conception of rationality, but rather is absolutely *axiomatic* for the ideal of reason that leads to views such as the Epicurean-Rawlsian notion that justice obtains exclusively among rational beings. It is an essential feature of the tradition's entire enterprise of asserting the superiority of human beings over nonhuman animals.

I suggested in Chapter 1 that the anthropocentric background ideal of living proceeds from this assertion of categorical human superiority. Now I want to delve deeper into that ideal and show how it underwrites a conception of reason that reinforces key anthropocentric commitments. The thinkers I am going to examine in this chapter all proceed as if characterizing the faculty of rationality were a straightforward affair and as if the nature of reason were more or less transparently obvious. But observing and reflecting on human experiential capacities is not the detached objective affair that adherents of anthropocentric values appear to believe. Instead, I want to suggest that certain affective predispositions or commitments drive the endeavor to represent reason as impartial and autonomous *in toto*, when in fact reason can be (and often is) marshaled in the service of justifying hierarchical and exclusionary forms of conduct as if they were perfectly unobjectionable. In certain respects, as in the operations of formal logic, reason functions in an ideologically neutral manner. But in other respects, and most particularly in the case of moral judgment, reason is dangerously susceptible to the influence of pre-existing values and commitments—and it is most dangerously susceptible to such influence precisely when it purports to be operating in a completely impartial manner. Kant's views on the supposed essential inferiority of black Africans and Hegel's dismissive remarks about "the Oriental world" have become well-known instances of the sort of rationalization and self-deception that I have in mind here. What holds for these regrettable views about non-Europeans holds, *mutatis mutandis*, for the accounts of nonhuman animals presented by Kant and a long line of other thinkers in the Western tradition: simply put, they are ripe for debunking as transparent assertions of dominion over those who are essentially like us but who outwardly appear to be "other."[2] It is in this spirit that I noted Schopenhauer's dictum about the essential sameness of human beings and nonhuman animals in Chapter 1.

The anthropocentric background ideal of living involves two interrelated commitments about the nature of rationality: that reason provides impartial insight into its own nature and workings, and that reason is possessed either exclusively or preeminently by human beings. The first of these commitments is a commitment to the kind of "immanent" critique of reason to which thinkers such as Aristotle and Kant subscribe, the one at least implicitly, and the other quite explicitly: reason is able to occupy a standpoint of detached impartiality from which it can assess and critique a great many things, its own nature and limits included. That is the double meaning of the genitive 'of' in the title of Kant's *Critique of Pure Reason*: Reason is what is to be subjected to critique, and reason is the faculty that will perform this critique—on itself. To have faith in the power of immanent critique is to ascribe at least limited autonomy to reason in the endeavor to reflect on the state of human understanding and address the three questions that Kant considers central to the human vocation: what can I know, what ought I to do, and what may I hope?[3] To have this faith is to reject Nietzsche's verdict that reason is simply one of our many bodily affects, a verdict that corresponds to a comparatively "total" critique that deprives reason of its supposed autonomy.[4]

Along with a commitment to the power of immanent critique comes the more pointedly anthropocentric commitment to the proposition that reason is, in essence, the exclusive possession of human beings. "In the beginning was the *logos* [the word]," begins the Book of John, thereby framing the life of Christ in terms of the linguistic rationality out of which the divine act of creation arose and in virtue of which we human beings, and we alone among earthly creatures, resemble and stand in special proximity to God. This was a commitment shared by the ancient Greeks, and it is one that has persisted in Christian tradition up to this day.[5] Its correlate is either the assumption that nonhuman animals are driven entirely by their sensory and emotional states, or the more restrictive view (initiated by the Stoic philosophers in antiquity) that nonhuman animals do not really experience emotions at all but are instead driven by something like blind impulse. Taken together, these two commitments—to the power of immanent critique and to the proposition that only human beings possess rational capacity—serve the anthropocentric aim of ascribing to human beings absolute priority, authority, and prerogative over all other beings in the natural world. This is the view according to which Kant judged domesticated nonhuman animals to possess a moral status equivalent to that of potatoes, and according to which he asserted without any apparent hesitation (and in peculiar tension with his assertion that we cannot know the inner nature of things) that the human being enjoys the status of "the titular lord of nature."[6] It is also the view that led Kant to the lapidary conclusion that "as far as reason alone can judge, a human being has duties only to human beings."[7]

This ideal of reason, which follows from rather than being in any way prior to the anthropocentric background ideal of living, nonetheless has positive elements very much worth defending in any robust conception of moral judgment. Our emotional proclivities do stand in need of regulatory guidance from reason, even if reason does not ultimately enjoy complete autonomy over our affective life. My aim in this chapter is to explore the traditional ideal of reason, with an eye toward disentangling the positive elements from the starkly anthropocentric ones: In what ways, and to what extent, should the traditional ideal of reason be defended, and what are its tragic limits? The urgency of these sorts of questions should become apparent by the end of this chapter; but to answer these questions even in a preliminary way will require the remainder of this book, and in particular an extended reflection on the essential role played by our affective (embodied) constitution in the formation of moral commitments as well as on the dialectical interplay between reason and emotion that I believe lies at the core of the human vocation. It will also be necessary, in Chapter 3, to examine some recent attempts to "historicize" reason—but more about that later.

Before delving into some historical accounts of the nature and operation of reason, I would like to point out what strikes me as a fascinating peculiarity of those accounts: They all outwardly appear to be self-verifying. By this I mean that, when one takes them at face value, it is very appealing to see them as "obviously" correct in their assertion both of the authority of reason over other experiential faculties and of the proposition that reason is the exclusive possession of human beings. In this respect, the conception of reason advocated by the tradition recalls Thomas Kuhn's notion of a scientific paradigm: Like scientific paradigms as Kuhn characterizes them, anthropocentric ideals of reason involve criteria for judging their own adequacy that appear to be unassailable when taken at face value. The central idea is that, whereas instinctual impulses are "blind" and emotional states are idiosyncratic in the sense that they narrowly reflect the momentary impulses of the individual rather than taking into account the interests of others in one's group or community, reason involves stepping back from the immediacy of the moment in order to assess and pass detached judgment on the legitimacy of fleeting impulses that very often lead us to do things we simply should not do.

Up to a point, this makes perfect sense: We often do find ourselves in the thrall of momentary impulses and end up regretting having acted on the basis of those impulses when taking time to reflect on them would have led to a wiser or more prudent decision. A little later in this chapter I will draw on Seneca's views on emotional states such as anger in order to shed light on this positive role that reason can and often does play. Where the traditional ideal of reason goes awry is in supposing that emotions or affective dispositions can never serve as the basis of moral motivation. It will take much of the rest of this book for this to become completely clear; in this chapter I will give a preliminary indication of what I have in mind by examining Kant's appeal to the notion of "moral feeling," a specifically affective sense of commitment to the moral project that Kant (wrongly, on my view) insists is an important *concomitant* to moral commitment but is nonetheless not part of the *basis* of moral commitment. On the view that I seek to defend in this book, certain emotional states and predispositions do constitute part of the very basis of moral commitment, rather than being mere adjuncts to the kind of motivation that many thinkers such as Kant have treated as purely rational.

The traditional exponents of reason proceed as if there were cosmically objective criteria for the evaluation of their conception of reason (including the proposition that it is an exclusively human endowment), when in fact the ideals of reason in question are underwritten by interests to which exponents of reason turn a blind eye in their efforts to vindicate the anthropocentric view of reason they embrace. In this respect, advocates of the traditional conception of reason make claims that are very much like Peter Singer's suggestion that, given the choice, a horse (and, by implication, any living being) would prefer to be a human being—as if there were some objective *scala naturae* according to which being human ought to be preferable to being any other sort of living being.[8] This sort of thinking is exemplary of the anthropocentric background ideal that I sketched in Chapter 1: any being in its right mind (provided it had one) would "of course" prefer to be human rather than any other form of being.

It is exactly this kind of assumption that I seek to challenge by contrasting the anthropocentric ideal with the non-anthropocentric ideal for which I have drawn inspiration from thinkers such as Grimm and Taureck, who proceed from a confession of human ignorance and the recognition that we do not really know what the experiential realities of nonhuman animals are, at least not in any fine-grained way. The endeavor to see the positive aspects of the traditional ideal of reason as well as the inherent limitations of that ideal is born in important part from an interest in giving nonhuman animals the space to reveal themselves to us as they are rather than as we would have them be. It is born of the Heideggerian endeavor to "let beings be."

This sense of openness to the ultimate mystery of the way things are (including the nature of nonhuman animal experience and potential) has something in common with Thomas Kuhn's conclusion that there is "no theory-independent way to reconstruct phrases like 'really there'; the notion of a match between the ontology of a theory and its 'real' counterpart in nature now seems to me illusive in principle."[9] I leave it to philosophers of science to debate the question whether Kuhn was right to suggest that there is no cosmic yardstick on the basis of which we can legitimately claim that, say, quantum physics gives us a more adequate or "correct" interpretation of nature than Newtonian physics, or whether Newtonian physics constitutes an objective improvement over the Aristotelian model. What is important for the present discussion is that Kuhn's assertion about the theory-dependence of claims to rightness seems to me to shed light on the nature of moral claims as well as on claims about the way in which reason "objectively" functions. Whether or not "all seeing is theory-laden" when it comes to observing and assessing natural processes, it strikes me that the way in which the traditional thinkers I am about to examine characterize the faculty of reason is "theory-laden" in the sense that it is motivated and shaped by very specific anthropocentric commitments. In Chapter 3 I will return to this idea and the notion of reflective judgment in order to show how reason can contribute to the authority of moral judgment even though that kind of judgment, by its very nature, is never fully objective.[10]

If there is a "bottom line" expressed in the ideal of reason that is motivated by the anthropocentric background ideal of living, it is that reason's primary function is to assert and maintain *control*. That control can be cognitive, as when we appeal to reason to tell us about the inner nature of things, about the natural world, about the essence of moral commitment, etc. The control can also be physical, as when we employ reason instrumentally to produce devices and procedures for exercising dominion over our surroundings. That these two forms of control are naturally interrelated is most obvious in the case of science and technology, where we design technical devices on the basis of knowledge we have established about the world we seek to control. In suggesting that the entire ideal of reason to which the tradition has subscribed is driven by anthropocentric ideology, I am suggesting that both of these uses of reason presuppose the sort of ideological commitment that led Kant to assert that the human being is "the titular lord of nature"—an assertion that is, I believe, inseparable from a hierarchical view of things according to which nonhuman animals are ultimately nothing other than resources ripe for exploitation in the service of human mastery. And as I have suggested already, Kant is far from alone in being committed to such an anthropocentric conception both of reason and of human superiority.

2. Two Early Exponents of Anthropocentric Rationality: Aristotle and Seneca

A guiding thread that runs through the history of Western philosophy up to the relatively recent past is the belief that reason is essentially ahistorical and autonomous. Words such as 'autonomy' are tricky in this connection, because Kant in particular conferred a special meaning on it in the late eighteenth century, and that meaning was predicated on the emergence and consolidation of a conception of the rational individual that owes a significant debt to Luther and Descartes—Luther attributing to each Christian individual the capacity to interpret Scripture for him- or herself rather than having to rely on the tutelage of ecclesiastical authority, and Descartes secularizing this notion of moral conscience by attributing essentially equal rational capacity to all human minds.[11] It is no mere accident that Descartes, the thinker who considered reason to be a timeless rational faculty equally shared by all (typical mature) human beings, believed that the employment of reason in sciences such as physics would ultimately render human beings "the masters and possessors of nature."[12] In Descartes, as in other major exponents of reason in the tradition, the commitment to the autonomy of reason and the commitment to human superiority and control over the rest of nature (nonhuman animals included) are virtually inseparable.

This is true at least in incipient form even for Aristotle, a thinker writing two millennia before thinkers such as Descartes, Hobbes, Locke, and Kant developed and solidified the notion of individual agency in its modern sense and before the systematic mastery of nature became an explicit goal of modern science and technology. For Aristotle, it is not specifically the rational agent who is autonomous, but instead it is reason (*nous*) that has an unchanging, eternal character. Aristotle is unwavering in his assessment of "man [as] the best of the animals," but he also believes that "there are other things much more divine in their nature even than man, e.g., most conspicuously, the bodies of which the heavens are framed."[13] One of these things that stands higher than man is reason itself, which Aristotle views not simply as a human faculty but rather as "an independent substance implanted within us" that is "incapable of being destroyed." This "thought or the power of reflection" is "immortal and eternal."[14]

It is our participation in *nous* that places human beings, among all earthly creatures, in special proximity to the gods. That this is Aristotle's view is clear from the premium he places on *theoria* or pure contemplation over the kinds of thinking that are directed at practical activities.[15] *Eudaimonia* or "perfect happiness," that state we strive to achieve through the pursuit of virtue, "is a contemplative activity" much like "the activity of God," which itself "must be contemplative" inasmuch as divine beings have no material needs. Humans, the most fortunate among us at least, are capable of contemplation, whereas nonhuman animals "have no share in happiness [*eudaimonia*]" inasmuch as "they in no way share in contemplation."[16]

And even in practical matters such as securing our material needs, there is a fundamental distinction between human beings and nonhuman animals: Nonhuman animals act "by nature," whereas human beings act on the basis of reason. In Chapter 1 I noted that commentators such as Nussbaum caution us against taking too seriously remarks in early sections of the *Politics* such as the passage in which Aristotle asserts that nonhuman animals are excluded from political community with humans inasmuch as they lack *logos* or rational capacity, Nussbaum's reasoning being that those early sections contain expressions of the prevailing common sense of Aristotle's time rather than Aristotle's own considered views. I suggested that I find this interpretation of the *logos* remark implausible or at least incomplete, because Aristotle never retracts or qualifies it (although he does make some passing remarks in the zoological texts to the effect that many nonhuman animals exhibit various forms of intelligence and ingenuity). Here I will add that there are a good number of other passages in Aristotle's writings in which he draws a clear dividing line between human beings and nonhuman animals, one of the most categorical coming much later in the *Politics*: "Animals lead for the most part a life of nature, although in lesser particulars some are influenced by habit as well. Man has reason, in addition, and man only."[17]

When Aristotle states that nonhuman animals lead primarily "a life of nature," he is saying in effect that nonhuman animals behave essentially via instinct. This is clear from a passage in the *Physics* in which Aristotle states that "animals other than man . . . make things neither by art nor after inquiry or deliberation" but rather "by nature." Here Aristotle gives the examples of a swallow making its nest and a spider spinning its web.[18] Dierauer, in his magisterial study of views about nonhuman animals in antiquity, takes "by nature" in this connection to signify instinct, and I think he is absolutely right.[19] Like the Stoic philosophers after him, Aristotle views behaviors such as nest building and web spinning to be products of innate impulse, with absolutely no thought or subjective reckoning involved. Seneca, for example, cites the example of bees (hive building) and spiders (web spinning) in arguing that there is no learning in nonhuman animals but simply conformity to capacities with which they are fully endowed at birth.[20] His reasoning is that there cannot possibly be any thought involved in activities such as web spinning, inasmuch as spiders construct their webs perfectly from birth, without any instruction.

For Aristotle as for the Stoics, this notion of innateness corresponds to a difference in the sorts of ends that human beings and nonhuman animals can pursue: Nonhuman animals are able to pursue purely biological needs, whereas the possession of rational capacity enables human beings to set ends for themselves and pursue virtue for its own sake—an activity that requires thought, the capacity that renders human beings most godlike. Nonhuman animals, on Aristotle's view, are not actually capable of action in the full sense of the term. Action requires "sensation, reason, [and] desire." But nonhuman animals, being bereft of reason, are subject fundamentally to sensation and thus lack the capacity for "choice" (*proairesis*), which on Aristotle's view is "the origin of action." Nonhuman animals have no capacity for "moral excellence," which is "a state concerned with choice, and choice is deliberate desire."[21]

If nonhuman animals are incapable of deliberation, then they cannot reflect on the proper ends of action but are subject to the natural imperatives of survival, to the measure of which their innate endowments (such as nest building and web spinning) are cut. For Aristotle, this makes nonhuman animals perfect candidates for treatment as living instrumentalities for the satisfaction of human material needs, as I noted in Chapter 1. The satisfaction of these material needs affords human beings the leisure required for deliberation about the best ways in which to pursue "higher" ends such as moral virtue. Human beings can, in virtue of their linguistic rationality, "choose" in the sense that they can reflect on and make informed decisions about ends, and in turn they can deliberate about the most appropriate means to employ in pursuing those ends.[22] Nonhuman animals, on the other hand, can act in a "voluntary" manner, which means that they can pursue material ends fixed by nature; rather than being driven by rational deliberative capacity, nonhuman animals (like children, on Aristotle's view) behave on the basis of "appetite" (*epithymia*), which amounts to saying that nonhuman animals are driven by the prospect of enjoying pleasure and avoiding pain.[23]

It is on the basis of this distinction between the experiential capacities of human beings and nonhuman animals that Aristotle attributes states of character to the former and denies them to the latter. For example, nonhuman animals cannot properly be said to be capable of the virtue of courage, "because, driven by pain and passion, they rush on danger without seeing any of the perils," whereas Aristotle freely refers to the capacity for courage in human beings.[24] The same holds, *mutatis mutandis*, for the other virtues (such as temperance), inasmuch as the pursuit of virtue requires a recognition of "what reason directs."[25] And while one might seek to defend Aristotle by suggesting that in making this distinction he is not asserting or presupposing any sort of hierarchy, it is clarifying to bear in mind his statement that "the rule of the soul over the body, and of mind and the rational element over the passionate, is natural and expedient," as well as his complementary statement that the pleasures of thought are "superior" to those of sensation.[26] Again, I find the fact that the first of these statements comes early in the *Politics* not to signify that it is anything other than Aristotle's considered view, for he never retracts or significantly qualifies it; in fact, it as well as the statement about the superiority of the pleasures of thought fit hand in glove with his notion that humans, among all earthly beings, are most godlike in virtue of their capacity for *theoria* (contemplation of the eternal and unchanging).

This premium on rational operations over "merely" bodily ones such as sensation and passion has persisted for quite a long time in the history of Western philosophy, and has even exercised an influence on utilitarian thinkers who are often thought of as having fundamentally changed the terms of thinking about nonhuman animals by invoking the capacity for feeling rather than rationality as the basis for moral status. In Chapter 1, I noted that Jeremy Bentham's conviction that the capacity for feeling is the proper basis for inclusion in the moral community did not stop him from suggesting that it is perfectly fine (and in fact that we are doing nonhuman animals a favor) by killing and eating them. Similarly, John Stuart Mill considered human beings to be superior to nonhuman animals in the moral calculus on the grounds that human beings can experience moral, intellectual, and social pleasures in addition to bodily ones, whereas nonhuman animals (supposedly) can experience only bodily pleasures and pains. In this respect, as I indicated a short while ago, Aristotle anticipates Mill by attributing to nonhuman animals only the capacity to feel pleasure and pain, not any capacity to reflect on their experiences.

And while Aristotle does acknowledge that nonhuman animals can experience spirited desire or anger (*thymos*) as well as appetite, his conception of feeling in nonhuman animals is not the same as his conception of feeling in rational (which is to say, human) beings: Nonhuman animals, as I have noted, are moved by appetite (*epithymia*) rather than by reason; and appetite, on Aristotle's view, "contains no deliberative element."[27] That means that nonhuman animals are moved immediately by their volitional states and cannot step back and reflect on them in order to assess whether they *should* be so moved. This is perhaps the most straightforward sense in which Aristotle believes that nonhuman animals, as *aloga*, are not really capable of choice in the full sense of the term: Choice is undertaken deliberately, with reflection, whereas mere volition occurs comparatively blindly—not in the literal sense that a nonhuman animal fails to see the threat in its path, but in the sense that it has no capacity to assess or evaluate the threat (and select from various alternative courses of action) as a rational being could.

Again, one could object to the interpretation that I am presenting, by suggesting that Aristotle never explicitly asserts a cosmic hierarchy according to which human beings are categorically superior to nonhuman animals, and by noting that the most direct statements to this effect occur comparatively early in the *Politics* and therefore may simply reflect the prevailing common sense of Aristotle's time. People now occasionally say the same about Aristotle's dismissive remarks about women early in the *Politics*. What I have tried to show is that none of Aristotle's statements should be viewed in complete isolation from the full body of his writings, as doing so inevitably entails a loss of the guiding *ethos* from out of which those statements emerge. That *ethos*, as I argued in Chapter 1, is a pointedly anthropocentric one that sees human beings as an intermediate form of being located above nonhuman animals and below (but in special proximity to) the gods. I argued in Chapter 1 that the only way to arrive at a complete grasp of a thinker's or a culture's overall orientation on the world is to see that individual statements always emerge from an implicit, global background ideal of living that gives shape and direction to individual statements and commitments. That is as true for Aristotle as it is for virtually any other thinker. One very straightforward and clarifying way to consider whether Aristotle's statements in Book 1 of the *Politics* are really at odds with his considered views is to consider the extent to which nonhuman animals were domesticated and used as resources for the satisfaction of human needs in Athenian society.[28] There is virtually nothing in Aristotle's writings, not even his acknowledgments of ingenuity in many nonhuman animals in his zoological writings, that runs counter to the proposition that he took nonhuman animals for granted as living tools for the satisfaction of human needs. And the basis for this classification appears quite clearly to be Aristotle's conviction that nonhuman animals are fundamentally *aloga*.

In my first book on the moral status of nonhuman animals, I noted that the Stoic philosophers make a decisive step beyond Aristotle in proclaiming as an explicit commitment the idea that earthly things other than human beings were made specifically to serve as resources for the satisfaction of human needs and desires.[29] Where Aristotle asserts human superiority but never explicitly states that the gods fashioned nonhuman animals expressly to satisfy human needs, the Stoics are all too willing to make the additional step of proclaiming as a cosmic principle that material things were created to serve human beings. Epictetus, for example, argues at length that nonhuman animals exist for our sake and also that their subjection to human beings is due to their being bereft of *logos*.[30] Epictetus embraces Diogenes's ideal of striving to be a cosmopolitan, a citizen of the world, and bases the prospect of world citizenship explicitly on our capacity for *logos*.[31] In doing so, Epictetus proclaims a clear hierarchy and cautions us to avoid descending "to the level of wild beasts" by abandoning our reason.[32]

By the time Epictetus presented this notion of a cosmic hierarchy in which "dumb beasts" exist to serve rational beings, Seneca had presented a detailed account of the ideal of reason on which later Stoics such as Epictetus would rely. In his moral essays and letters, Seneca presents an ideal of reason that has much in common with Aristotle's account, although as we shall see, Seneca departs from

Aristotle's account of emotional states such as anger. Seneca argues that reason does not simply distinguish human beings from nonhuman animals, but is a clear sign of human superiority over them. In addition, he counsels the total conquest of the emotions by reason on the grounds that emotion is fleeting and untrustworthy. The ideal that Seneca sketches is one according to which "the happy life" involves calm detachment from desires for goods of fortune, as part of the endeavor to live on a plane with the gods. These aims are made possible for human beings in virtue of our rational capacity, the lack of which makes it impossible for nonhuman animals to attain happiness. As in Aristotle, the possession of *logos* renders us godlike, and its lack consigns nonhuman animals to the role of instrumentality. This is an account according to which true happiness requires a sense of the duration of time, a sense that nonhuman animals lack because their lack of *logos* imprisons them in an eternal present. The closer a being is to the eternity of the gods, the higher that being ranks in the cosmic hierarchy.

Being blessed with reason (*ratio*), mature human beings are capable of contemplating and striving to attain the good. The good, on Seneca's view, "is a matter of the understanding," not of the senses, and "we assign it to the mind." Thus the good "does not exist in dumb animals or little children."[33] Indeed, "in dumb animals there is not a trace of the happy life. . . . The dumb animal comprehends the present world about him through his senses alone. He remembers his past only by meeting with something which reminds his senses," and "the future . . . does not come within the ken of dumb beasts." As regards memory, Seneca denies to nonhuman animals the capacity for episodic memory, the ability to call up a given memory at will, and offers the example that a horse can recall a familiar road only when brought into direct view of it. As regards the future, in effect it does not exist for "dumb beasts" inasmuch as they cannot employ reason to contemplate it. "Animals perceive only . . . the present."[34] This commitment anticipates Bentham's remark that nonhuman animals have nothing to lose by dying inasmuch as they have no anticipation of the future, and it likewise anticipates Schopenhauer's suggestion that the behavior of nonhuman animals is strictly determined by present sense impressions.[35]

If we can speak of a "good" for nonhuman animals, then on Seneca's account that good consists in the satisfaction of material needs and most specifically in the provision of food. The good for a human being, on the other hand, is "not that of his belly" nor any kind of bodily pleasure, but instead consists in a contentment located in detached understanding rather than in any kind of merely sensory experience.[36] Bodily goods are certainly good for the health of the body, "but they are not absolutely good."[37] The only unqualified good is "virtue itself." If it were the case that "those things which dumb animals share equally with man are goods, then dumb animals also will lead a happy life; which is of course impossible" inasmuch as nonhuman animals are *aloga*.[38] *Of course impossible*: No creature imprisoned in an eternal present has the capacity to step out of the present moment and see that moment as one in a long line of moments stretching indefinitely into the past and the future. This in turn means that nonhuman animals cannot contemplate alternative courses of action, the implications and consequences of different actions and events, or the trajectory of their lives as a whole—all capacities requisite for the pursuit of virtue or the happy life.

The sense of a hierarchy here, one explicitly grounded in the question whether or not a given being possesses reason, is unmistakable.

> What quality is best in man? It is reason; by virtue of reason he surpasses [*antecedit*] the animals, and is surpassed only by the gods. Perfect reason is therefore the good peculiar to man; all other qualities he shares in some degree with animals and plants.[39]

In language that strikingly anticipates Descartes's much later call to render human beings "the masters and possessors of nature," Seneca asserts that it is in virtue of our possession of soul (*animus*) that we are destined to exercise "dominion over all things" and become "owner of the universe."[40] One of the "things" over which we must exercise dominion is our own body, whose emotional states pose grave threats to the specifically mental or spiritual integrity the attainment of which places us in proximity to the gods. As regards the pursuit of virtue or the happy life, "the gratifications of the body" are "unessential"; it is the responsibility of our mind (*mens*) to "be the master" of both our senses and the external things to which our senses expose us.[41]

One commitment that is telling here is that any capacities we share with nonhuman animals are "unessential," which in effect is to assign an inferior status to them. Like all anthropocentric thinkers, Seneca is at pains to identify some quality or qualities that not only are unique to human beings but that render human beings fundamentally superior to all other earthly creatures. This is a textbook definition of anthropocentric ideology—the assertion of something unique about human beings that confers on us, and us alone, the prerogative to pass judgment on and make use of nonhuman beings in any way we see fit. The endeavor to live in accordance with this anthropocentric orientation is hampered by the inescapable fact that many nonhuman animals exhibit ingenuity and other forms of subjective awareness that overlap considerably with human capacities. Aristotle openly acknowledges this overlap in his zoological texts, only to disavow it almost entirely in his psychological, ethical, and political writings.[42] What he leaves us with, if not in his own writings but in the legacy of anthropocentrism he leaves in his wake, is a reductive account of the lives of nonhuman and human animals according to which the former are driven by merely physical impulses (such as hunger) whereas the latter are driven by these but are in addition capable of reflective thought that facilitates godlike contemplative distance from our bodily imperatives.

A core commitment in this approach to living that is characteristic of the Stoic philosophers is a fundamental devaluation of bodily goods. Seneca frames this devaluation in terms of the need to eschew goods of fortune (*fortuna*), which are goods that exist in the world around us and hold the promise of giving us pleasure—but also the promise of causing us misery. Seneca counsels "indifference to fortune" on the grounds that "the paltry and trivial and fleeting sensations of the wretched body" are incapable of affording us lasting satisfaction; the "pleasures and pains" provided by our surrender to fortune, "those most capricious and tyrannical of masters, shall in turn enslave. Therefore we must make our escape to freedom."[43] Acquiescence to bodily goods is servitude, inasmuch as it constitutes a capitulation to the vicissitudes of time and thus places us at a distance from the gods, who are our proper models for conduct.

> Therefore let the highest good mount to a place from which no force can drag it down, where neither pain nor hope nor fear finds access, nor does any other thing that can lower the authority of the highest good; but virtue alone is able to mount to that height . . . she will keep in mind that old injunction, "Follow God!"[44]

To follow the example of the gods is to rise above the fluctuations of time, which remain forever outside our control, and to ascend to a state of detached rational contemplation. If there is true freedom in human life, it is to be attained by modeling our lives on divine perfection, to orient ourselves on eternity rather than on time. Instead of succumbing to the fleeting temptations of the flesh, which as we know often disappoint and in any case are never reliable, we "should stand unmoved both in the face of evil and by the enjoyment of good, to the end that—as far as is allowed—you may body forth God." This turn away from fortune and toward God, which anticipates Augustine's language of turning inward toward God in the *Confessions*, is born of a wish to be "placed beyond the reach of any desire" and thus to dispense altogether with reliance on "any outside thing."[45]

This ideal brings with it several noteworthy commitments. One is that the good is not really pleasure, but consists instead in the kind of satisfaction one enjoys in virtue of having achieved a certain kind of mastery—over one's life, over one's ultimately unfulfillable desires, and over the external world. This last aspect of mastery is not literal—we never really succeed in mastering the world around us—but rather figurative, consisting in a state of detached indifference to goods of fortune. The wise individual does "nothing for the sake of pleasure" but instead "tosses aside" all goods of fortune such as luxury, popular opinion, and social superiority.[46] The goal is an enduring state of satisfaction that transcends mere bodily pleasure.

> The happy life is to have a mind that is free, lofty, fearless and steadfast—a mind that is placed beyond the reach of fear, beyond the reach of desire, that counts virtue the only good, baseness the only evil, and all else but a worthless mass of things, which come and go without increasing or diminishing the highest good.

This highest good does not rely on fortune but instead is that which the wise individual "alone is able to bestow upon himself."[47]

This rarefied view of the good is intimately bound up with anthropocentric hierarchy: As I noted a moment ago, Seneca is at pains to discount those experiential capacities that we share with "dumb beasts," and this means that he has to find a way to discount the satisfaction of bodily needs as well as the pleasure that motivates so much of the lives of sentient beings. Nonhuman animals are governed by mere "impulse" (*impetus*), whereas human beings are governed by reason and thus are unique in being able to emulate "the universe and God."[48] A rational being has "a better and surer light [than physical eyesight], by which I may distinguish the false from the true," in comparison with which nonhuman animals possess "a mind that is all darkness."[49] Lacking reason, nonhuman animals are slaves to "the paltry and trivial and fleeting sensations of the wretched body."[50]

What Seneca ultimately makes too little of is the fact that it is precisely the possession of rationality that creates problems for human beings that, on Seneca's own account, are not possible for nonhuman animals. Like many philosophers before and after him, Seneca acknowledges that the capacity for evil is unique to human beings. Take, for example,

> the forum with its thronging multitude. . . . It is a community of wild beasts, only that beasts are gentle toward each other and refrain from tearing their own kind, while men glut themselves with rending one another. They differ from dumb animals in this alone—that animals grow gentle toward those who feed them, while men in their madness prey upon the very persons by whom they are nurtured.[51]

Along with the capacity for calm, detached, godlike contemplation, *logos* brings with it a set of capacities not only that nonhuman animals are said to lack but that facilitate forms of mischief in the human community that are unknown among nonhuman animals. Human beings are able to make the sorts of comparisons between themselves and others that Rousseau would later cite as the basis of enmity among human beings, plan and execute revenge, and seek to exercise dominance over their entire environment in a manner impossible for nonhuman animals. Thus the attribution of rational capacity to humans and its denial to nonhuman animals is a double-edged sword: It holds the promise (a dubious one, in my judgment) of enduring satisfaction, but at the same time it creates significant problems (enmity, strife, and an entire body of criminal and civil law) that appear to be unknown in nonhuman animal communities. Moreover, thinkers such as Seneca hasten to congratulate humanity on our at least theoretical capacity to ameliorate these problems, quite apart from the fact that nonhuman animals do not suffer from these difficulties in the first place, and quite apart from the fact that the historical record offers little in the way of confidence about our capacity to actualize the ideal of a truly righteous life envisioned by thinkers such as Seneca.

The way to see this is to delve a bit deeper into two interrelated features of Seneca's account of the human condition: the premium he places on satisfaction that endures (in contrast with the fleeting pleasures of the body) and his strict denial that nonhuman animals experience emotions (which on Seneca's account pose a grave threat to the prospect of spiritual integrity). Seneca maintains that emotions are inherently destructive inasmuch as they are directed exclusively at short-term aims and distort our sense of truth, which as I noted a moment ago, is a sense that transcends our physical endowments and gives us an intimation of the eternal; and the eternal, on Seneca's view, is the only proper guide for our lives. "The wise man . . . lives on a plane with the gods" and avoids being "led astray by delights which are deceptive and short-lived."[52] To be virtuous is to act uniformly, not impulsively.[53]

The value of uniformity and constancy is that they facilitate the ideal of control that guides Stoic thinking about how best to reckon with the vicissitudes of fortune: We know that goods of fortune can harm us, and that even when they benefit us their benefit is temporary. The Stoic ideal is to rise above these fluctuating goods of fortune so as to fortify oneself in a realm of unchanging, reliable reason. Reason in this sense is to function as a citadel that shields us from the pain of loss or injury— but also from the lure of pleasure.

> Nature has given to us an adequate equipment in reason; we need no other implements. This is the weapon she has bestowed; it is strong, enduring, obedient, not double-edged or capable of being turned against its owner. Reason is all-sufficient in itself . . . [it] grants a hearing to both sides, then seeks to postpone action, even its own, in order that it may gain time to sift out the truth. . . . Reason wishes the decision that it gives to be just.[54]

Noteworthy here is the strict opposition that Seneca strongly implies, and almost renders explicit, between justice and the ever-shifting parameters of material existence. Like Aristotle before him, Seneca conceives of reason as having an eternal character, in comparison with which anything in the realm of time and alteration is inherently inferior and not to be desired for its own sake.

One implication of this viewpoint is that nothing that is good for a nonhuman animal can possibly constitute a part of the good for human beings. Interestingly, this does not mean that devotees of truth must forswear material goods in the interest of pursuing virtue. The Greek Stoics had classified goods of fortune as *adiaphora* or "indifferents," just as Seneca does. This means that they are not to be desired for their own sake. But that does not prevent them from being objects of desire; some things that are *adiaphora* are also *proegmena* or "preferreds," and the Greek Stoics classified nonhuman animals and a great many other material beings as such.[55] Fully aware that human beings, as embodied, possess material needs, the Stoics employ the notion of preferreds to designate a class of material things that promote human bodily welfare even though these things are not part of the true good. Seneca's verdict on material goods is that "the wise man does not deem himself undeserving of any of the goods of Fortune. He does not love riches, but he would rather have them."[56] Seneca concludes *De vita beata* with the suggestion that there is nothing wrong with a philosopher who owns a large house or eats lavishly.[57]

This naturally places Seneca's ideal of spiritual purity in a special light, as it implicitly acknowledges the impossibility of escaping the sorts of bodily states and challenges that unite us with nonhuman animals. According to that ideal, we continue to rely on goods of the body, but we take on the attitude that deprivation of these goods can in no way harm us—"no wise man can receive injury or insult," Seneca tells us in his text on the virtue of constancy.[58] To experience true freedom is to have a mind that rises above injury and is its own source of pleasure, one "that separates itself from all external things in order that man may not have to live his life in disquietude, fearing everybody's laughter, everybody's tongue."[59] To experience true happiness, to be truly virtuous, involves withdrawing—in reflection if not in material fact—from external things, which include both goods of the body and the reception we receive from our fellow human beings. For it is only in a space of detached reflection, one in which we are the sole authors of our estimation of ourselves, that we can have any assurance of complete control over our lives.

This space of reflection affords us, as it had for Aristotle, the prospect of enduring, reliable, predictable truth. Truth in this sense is thought *sub specie aeternitatis*: namely, from a universal standpoint that transcends time and its vicissitudes. Only that which remains the same is a candidate for truth and thus for being valued. Seneca's call to "cherish humanity" at the conclusion of "On Anger" is a call to honor a specific sense of humanity, one modeled on this notion of universality and timelessness.[60] The specific ideal of humanity here has two important implications: It is predicated on the premium of mind over body that I noted earlier, and it entails—or, more accurately, it presupposes—the categorical exclusion of nonhuman animals from participation in the good.

Seneca's diagnosis of embodiment is predicated on the assumption that timeless things are "higher" than temporal ones, as well as on the assumption that temporal goods (goods of fortune) are ultimately destructive (because fleeting and unreliable) even when they have the status of "preferreds." In these respects, Seneca anticipates St. Augustine's *contemptus mundi*, the sense that our devotion to merely earthly things diverts our attention from the higher task of living in accordance with nature, which for humans amounts to living in accordance with the gods. By classifying the human mind or soul as "higher" than earthly things, and by denigrating "merely" earthly things as inessential to the good life, Seneca places himself in a position in which characterizing nonhuman animals both as inferior and as tools for the satisfaction of human needs becomes a fairly straightforward affair.

Like many thinkers in the tradition, Seneca characterizes the experience of sentient beings in one of two ways, each an extreme: Either a being is rational and can transcend the individual moments of time so as to think time as a whole (and all that that entails), or that being is enmired in an eternal present. As I noted a short while ago, Seneca conceives of nonhuman animals as the latter kind of being. For them, in effect there is no past or future, only whatever shows up in the current moment. This has a number of implications, the most immediate of which is that nonhuman animals are incapable of reflecting on their impulses so as to make actual choices about whether to follow those impulses. As in Aristotle, nonhuman animals are moved by their bodily states rather than by anything like rationally informed choice. And as in Aristotle, "rationally informed choice" for Seneca means choice that is linguistically and conceptually structured—a capacity that is categorically denied to nonhuman animals in virtue of their being *aloga*.

But unlike Aristotle, who attributes at least some emotional states (such as fear and anger) to nonhuman animals, Seneca follows the Stoic line of thinking in denying that nonhuman animals are capable of any sorts of emotional states.[61] Nonhuman animals are moved entirely by impulses toward pleasure and away from pain. Following the Stoic conviction that emotions are structured in terms of *lekta* (propositional states), Seneca maintains that nonhuman animals, lacking rationality, cannot experience states such as fear or anger. A state such as anger, "while it is the foe of reason . . . is, nevertheless, born only where reason dwells. . . . Dumb animals lack the emotions of man, but they have certain impulses similar to these emotions." Indeed "wisdom, foresight, diligence, and reflection have been granted to no creature but man," inasmuch as nonhuman animals are "incapable of speech."[62] This indicates the sense in which the minds of animals are "all darkness": There is no space whatsoever for a conscious reckoning with, and evaluation of, one's circumstances if one lacks reason. One's behavior is a strict effect of externally given causes, so there is no room for reflecting on one's responses or tendencies in a way that would contribute to the formation and growth of character and to establishing a commitment to the good rather than to indulgence in earthly things. If there is a good for nonhuman animals, it is a specifically bodily good; one oriented, as I noted above, on the consumption of food. In the final analysis, Seneca considers nonhuman animals to have more in common with inanimate objects such as rocks than with rational beings.[63] In this respect, Seneca is not far removed from the equivalence that Kant will later draw between domesticated nonhuman animals and crops such as potatoes.

If nonhuman animals are incapable of emotional states, then at first blush it seems odd for Seneca to suggest that "you become like [a dumb animal] if you get angry."[64] All Seneca means by this simile is that a person driven by passion rather than by reason has succumbed to a base impulse that is oriented on the immediacy of the present, and to this extent is being buffeted about by factors that are external in comparison with the inwardness of reason. Where in nonhuman animals the tyranny of impulse dominates, in the passionate human being it is not mere impulse but explicit *assent* to a driving passion that dominates. The essential difference is that assent to a state such as anger explicitly involves choice on the part of a rational agent who in that moment of choice is not acting rationally, whereas in a nonhuman animal there is no more actual choice than in, say, a rock beginning to roll down a slope.[65] Being "incapable of speech," nonhuman animals

> do not have fear and anxiety, sorrow and anger, but certain states similar to them. These, therefore, quickly pass and change to the exact reverse, and animals after showing the sharpest frenzy and fear, will begin to feed, and their frantic bellowing and plunging is immediately followed by repose and sleep.[66]

For Seneca, these observations suggest that there is no subjective, internal reckoning with contingencies on the part of nonhuman animals, but simply blind reactions based on the lure of pleasure and the threat of pain. If a human being behaves in the mercurial fashion that Seneca attributes to nonhuman animals (where there are precipitous shifts from fear to calm, etc.), then that person is exhibiting "a complete lack of self-control and blind love for an object" driven by the prospect of pleasure.[67] Thus nonhuman animals and human beings in the grip of passion are similar in the respect that both are driven by pleasure, which on Seneca's account is not even part of the highest good because it pertains exclusively to goods of the body.[68] Human beings differ from nonhuman animals in virtue of possessing reason, which facilitates *assent of the will* to emotional tendencies such as anger—the will, instead of hewing closely to the dictates of reason, loses control and gives itself over to the dictates of the irrational.[69]

The sense in which the experiential states of nonhuman animals are similar to those of impassioned human beings is that in both cases there is an absence of detached reflection on the appropriateness of the impulse—in the case of nonhuman animals because they are incapable of reflection, and in the case of human beings because they relinquish at least temporarily the authority of their reason. What Seneca, like many philosophers before and after him, never stops to ask is whether we can truly make sense of the subjective experiences of sentient nonhumans when we are determined to characterize those experiences as completely devoid of freedom in the sense of a capacity to engage in active linguistically-based evaluation of one's situation. Instead he depicts a strict either-or between rational beings who can become godlike through reflection on the light of truth on the one hand, and beings with inherently "dark" minds whose only tangible contact with reality is the prospect of pleasure (through the consumption of food and the like) and the avoidance of pain on the other. For Seneca, as for many exponents of anthropocentric ideology, there simply is no middle ground between these two extremes.

By framing sentient life in terms of this strict either-or, Seneca never has to confront the fact that human beings are arguably inferior to nonhuman animals in virtue of the human susceptibility to passion. Seneca does acknowledge that human beings are more capable of viciousness than nonhuman animals, in virtue of our susceptibility to let states such as anger distort our sense of what is good and desirable; his remarks about the best ways to avoid or diffuse anger reflect his awareness that our rational capacity for planning often does augment rather than ameliorate the extent, duration, and severity of our anger, thereby distorting the magnitude of perceived goods and evils.[70] Beyond his passing acknowledgment that passions such as anger make human beings capable of inflicting harm on a scale impossible for nonhuman animals, Seneca has virtually nothing to say about the fact that the respective ways in which human beings and nonhuman animals manage to live "in accordance with nature" differ vastly, indeed such that human beings arguably compare quite *un*favorably to nonhuman animals. Seneca omits to note this fact, I think, because he is so deeply invested in the anthropocentric image-of-God model according to which we more than compensate for the inadequacies of our embodiment with our rational capacity.

Seneca, like many thinkers in the tradition, takes our very possession of rational capacity as a sign of our superiority, quite apart from the question how or whether we actually live up to the promise of that capacity. Ethicists who draw hierarchical comparisons between human beings and nonhuman animals tend to point toward achievements in fields such as mathematics, the arts, and philosophy as clear signs of the categorical superiority of human beings *tout court*, when in fact a great many human beings never exhibit or act on the sorts of intellectual and creative capacities that tend to be attributed to humanity as a whole and denied to the entirety of nonhuman animal life—and this consideration is entirely independent of the challenge I have raised to the idea that there is some cosmic yardstick against which we can judge intellectual pursuits to rank "higher" than bodily ones. What does it matter if in principle I have the capacity to grasp Gödel's Theorem, or to compose symphonies, or learn Mandarin Chinese, if I never actually master any of these skills? Am I "superior" in virtue of capacities that I never actualize? Can I be sure that, because some other humans are capable of achievements of these sorts, I am capable of them as well? And on top of all this there is the hierarchy question: why seize on capacities of these specific sorts in making judgments about relative moral worth?

Instead of addressing these questions, Seneca simply takes the position that passions are destructive because they are fleeting. Even a moderate passion constitutes "nothing else than a moderate evil."[71] What Seneca leaves us with is an ideal of human agency devoid of (because elevated above) passion. Is a passionless life worth living? Is it a reasonable aspiration for a being whose very essence is bound up with embodiment? Seneca sidesteps these questions because of his unremitting devotion to an ideal of spiritual purity whose terms include a fundamental suspicion of the body. Seneca's cheerful assurance that the enjoyment of material wealth is compatible with virtue does less to militate against this suspicion than to demonstrate Seneca's implicit ambivalence about goods of the body and the access we gain to them through our emotional states. At the same time, Seneca is right to express at least some suspicion about the untoward influence of passion: unbridled passion poses a very distinct danger to the integrity of reflective judgment. In devoting an entire text to the passion of anger, Seneca endeavors to draw out the nature of this danger and to explain how contemplative reason can neutralize or at least ameliorate the threat posed by unbridled passion.

Seneca makes a very shrewd strategic choice in seizing upon anger as paradigmatic of passion. In framing his entire account of the passions in terms of such a violent and destructive emotional state, Seneca disregards the positive (not to mention, the fundamental) role played by the passions in the lives of sentient beings. Because Seneca does not recognize any viable distinction between positive and negative affects, a statement such as "behaving under the influence of blind rage makes us like dumb beasts" becomes generalized into "behaving under the influence of any passion whatsoever makes us like dumb beasts." And if anger "is contrary to nature," the implication is that acting on the basis of *any* passion is likewise "contrary to nature."[72] This reflects Seneca's anthropocentric orientation: for human beings to act in accordance with nature is to heed the call of detached reason (which, interestingly, constitutes an attempt to deny key aspects of our "nature," namely, all those associated with embodiment), whereas for nonhuman animals to act in accordance with nature is to react to impulses that are not subject to reflective evaluation but function as strict causes. That nonhuman animals arguably end up doing a considerably better job of living in accordance with nature than human beings do appears to be a consideration that had little if any influence on Seneca's estimation of the relative moral worth of human beings and nonhuman animals.

3. The Enlightenment's Chief Exponent of Anthropocentric Rationality: Kant

In Chapter 1 I provided an overview of the anthropocentric elements of Kant's philosophy. I focused in particular on his person-thing distinction, according to which only rational beings have inherent moral worth and all non-rational beings (nonhuman animals included) have a merely instrumental worth, as well as on a certain tension in Kant's thinking that is apparent in his call for "indirect duties" with regard to nonhuman animals. The tension, again, is this: Why should we have any moral regard for beings who from a moral standpoint count as mere resources— why have any more concern for, say, a horse than for a potato or a fertile field? The answer to this question is unaffected by the fact that, on Kant's view, we have no duties that we owe directly to nonhuman animals, but instead all our duties are to humanity; indeed, that observation simply enhances the tension. On the one hand, being sentient (able to suffer) is not enough to qualify a being as a direct subject of moral concern and possessor of moral worth. On the other hand, Kant recognizes that beings such as horses are precisely *not* like potatoes and fields inasmuch as they have a subjective good (one of which they are *aware*, even if not in the mode of theoretical detachment) that can be enhanced or interfered with. Even if only against his own intention, Kant implicitly acknowledges that sentient beings are significantly *un*like non-sentient beings—we simply do not tolerate treating a horse or a dog with the nonchalance with which we might toss a log on the fire.[73] I believe it is on the basis of this acknowledgment that Korsgaard has attempted to rectify Kant's thinking so as to make room for nonhuman animals as beings toward whom we can be said to have direct moral duties.

The reason I do not share Korsgaard's optimism about the prospects for rehabilitating Kant on the moral status of nonhuman animals is that entirely too much in Kant's worldview speaks against such a possibility. As I suggested in Chapter 1, this fact is difficult to see if one focuses on individual statements instead of exploring the terms of the background ideal of living that underwrites them. In this respect, my approach takes a cue from Descartes's advice to the reader of the *Meditations* to "grasp the proper order of [Descartes's] arguments" rather than simply trying "to carp at individual sentences."[74]

I find it difficult to picture Kant being persuaded by the "Kantian" account of nonhuman animal agency and moral status that Korsgaard proposes. Of course, her intent is not to show that Kant would be persuaded, but instead to show that key elements of Kant's thought, denuded of their anthropocentric prejudice, lead to a robust nonhuman animal ethic. Even so, I believe that Kant's binary person-thing distinction is so essential to his entire outlook that by removing it we would in effect be talking about another philosophical approach altogether; to remove it would make about as much sense as, say, attempting to rehabilitate Seneca so that passion actually played a positive role in the good life. In the following remarks, I will try to persuade you that this is the case by looking more closely at Kant's ideal of reason, as well as at his extended efforts to avoid the implication that feeling or emotion plays any foundational role whatsoever in moral judgment. I believe that Kant's reasons for utterly divorcing rationality and moral judgment from emotion share much in common with Seneca's: to neutralize the pernicious effects that passion can have on our judgment; to assert human uniqueness; and, in turn, to proclaim human superiority over the rest of nature.

Karl Löwith once made an observation about Kant that must have struck many readers as implausible: that Kant's thought bears the deep imprint of dualistic metaphysics, in spite of the fact that Kant sought to reframe religion "within the limits of reason alone."[75] When assessed against the background of a thinker such as Seneca, however, this interpretation gains considerably in plausibility: In spite of his open acknowledgment that human beings are embodied and subject to the potentially dangerous influence of bodily "inclination," Kant shares Seneca's interest in characterizing embodiment as an unfortunate accident of our being rather than as part of its essence. Thus Kant, like Seneca, places an absolute premium on detached reason and characterizes practically all other aspects of our constitution ("moral feeling" will be a noteworthy and revealing exception) as inimical to the interests of reason.

In Chapter 1 I noted Kant's acknowledgment of a basic distinction in our self-conception: between ourselves as natural beings (*homo phaenomenon*) possessing "an ordinary value" on a par with that of nonhuman animals, and ourselves as persons (*homo noumenon*) possessing a value "exalted above any price."[76] By now it should be clear that, within the framework of morality, Kant bases his entire outlook on the putative superiority of persons over all other beings—a textbook indication of anthropocentric ideology. Near the beginning of this chapter I linked this orientation to Kant's unqualified assertion that "as far as reason alone can judge, a human being has duties only to human beings." *As far as reason alone can judge*: What Kant, like Seneca, never seriously considers is the possibility (indeed, as I hope to show by the end of this book, the fact) that there are elements in the process of forming and revising our moral commitments that are vital albeit themselves not purely rational. In my book on Descartes as a moral thinker, I called these elements "extrarational" and argued that moral commitment inevitably has recourse to such elements even and perhaps especially in those moments when moral thinkers such as Kant most insistently deny that that is what they are doing.[77] After examining some key elements of Kant's conception of practical (i.e., moral) reason, I will try to show this by looking at his treatment of the notion of "moral feeling."

Kant's motivations for subordinating felt states such as passion to our rational capacity are very much in line with Seneca's. In particular, Kant warns that "the great mutability of things, and the storms of my passions" pose a threat to our pursuit of the good and thus to our endeavor to live up to the "supreme goodness" symbolized by the divine.[78] Nonetheless he conceives of ethics as a set of relationships exclusively between rational beings, which—even though Kant sometimes speculates about the prospect of intelligent life on other planets—amounts to viewing ethical right and duty as holding exclusively between human beings. It is in this sense that Kant believes that, from the standpoint of our rationality, we can have duties only toward human beings. This excludes not only nonhuman animals (who, in virtue of lacking "self-consciousness," cannot possibly count as direct beneficiaries of moral consideration) but God as well. "The question of what moral relation holds between God and human beings goes completely beyond the bounds of ethics and is altogether incomprehensible to us," although it is not difficult to imagine what Kant has in mind here.[79] Like Rawls after him, Kant conceives of rights and duties as complementary notions, and of duties as holding reciprocally between agents possessing equal moral worth and equivalent experiential capacities. As regards experiential capacities, Kant is clear in the *Critique of Pure Reason* that our basic experiential apparatus is of a different kind than that of a being such as God: Where our access to reality outside us is mediated through "sensible intuition" (the givenness of things subject to the limitations of space and time) and subject to *a priori* forms of conceptualization, a being such as God would possess "intellectual intuition," the unmediated givenness of things in their entirety without the mediation of limitations such as space and time.[80]

Another way to see why Kant excludes God from the reciprocity of rights and duties is to consider that God, by definition, is not the kind of being that needs guidelines such as the categorical imperative in order to know the right thing to do. Inasmuch as God's perfections include moral perfection, a being such as God does not need to engage in any sort of reflection in order to know the right thing to do. (In fact, it is an obscure matter to try to understand what it would mean for a perfect, eternal being to endeavor to act in the first place: Did God create because of some desire or lack?).

And as there can be no true reciprocity between God and a finite rational agent, there also can be no such reciprocity between human beings and putatively nonrational beings such as nonhuman animals. Rawls would later follow Kant's reasoning in proposing that only those beings who can *conceptualize* notions such as right and duty can properly be said to be beneficiaries of rights and holders of duties. Telling in this connection is the ease with which thinkers such as Kant find ways to extend rights to beings such as human children, who cannot yet grasp abstract notions such as rights: Kant, as I noted in Chapter 1, makes a place for such beings by classifying them as "passive citizens," beings lacking in "civil personality" (namely, the capacity to act on their own behalf in civil matters) but participating in personhood (recall Cicero's conception of our individual and universal "personae") to a sufficient extent to merit civil protection by "active citizens."[81]

Nonhuman animals merit no such protection because, being *aloga*, they share no commonality with us humans that could serve as a basis for our recognition of them as direct beneficiaries of moral concern. This, in sum and substance, places Kant squarely in the company of Aristotle and Seneca, both of whom place human beings at the apex of a hierarchy of created beings. Kant goes one step further in characterizing rational beings as standing infinitely above nonrational beings. I say "infinitely" because Kant is clear that rational beings possess "an absolute worth," in comparison with which the value of nonrational beings is merely "conditioned" inasmuch as everything nonrational is merely "a means to be arbitrarily used by this or that will."[82] Of course, 'arbitrarily' here does not mean that we have license to inflict wanton cruelty on nonhuman animals; instead it means that we are free to set ends for ourselves, the pursuit of which may well involve a variety of uses of nonhuman animals, some of them destructive, although we must avoid the *gratuitous* infliction of harm on nonhuman sentient life.[83]

This setting of boundaries, which follows directly from Kant's person-thing distinction, tells us a great deal about the ideal of reason that is at work in thinkers such as Kant: *Everything nonhuman in the earthly realm is an instrumentality*. Exactly as Aristotle and St. Augustine before him had done, Kant asserts that *there is no bond of "community" between human beings and nonhuman animals* that could inform a robust moral outlook encompassing the two.[84] Thus practical reason, which takes up questions of moral right and duty, has no direct bearing on nonhuman animals, who are almost entirely subject to the whims of instrumental reason. The only real limit that Kant recommends on our treatment of nonhuman animals is to avoid the infliction of harm that lacks at least some tangible human benefit. In this respect, Kant is not far removed from Peter Carruthers, who maintains that virtually any uses of nonhuman animals are permissible provided that they are undertaken for more than "trivial reasons, [and not simply] to obtain sadistic pleasure."[85]

The gathering ideal of community that guides Kant presupposes a bifurcated existence that parallels Seneca's split between reason and embodiment: The mind is properly directed at "higher" concerns than the mere welfare of the body, and it pursues those higher concerns by treating nonhuman animals and the rest of nature as material for the investigation of natural processes, the appreciation of beauty, the cultivation of the moral sense, and the pursuit of Kant's ideal of perpetual peace. Kant summarizes these aspirations of reason in the three questions he poses near the end of the *Critique of Pure Reason*: "What can I know? What ought I to do? What may I hope?"[86] Kant devotes the *Critique of Pure Reason* primarily to the first of these questions, promising the attainment of knowledge with "apodeictic" (which is to say, absolute) certainty by beings who are rational in the specific way in which human beings are rational (which on Kant's view includes beings who experience space and time as we do, who employ the same basic scientific concepts as we do, who unify their experiences in one single consciousness as we do, and who think the boundary between the knowable and the unknowable as we do). The knowledge here is of two kinds, scientific and metaphysical, and neither is the affair of sentient nonhumans—at least not of those who are *aloga* and who thus exist simply as resources for the satisfaction of the needs and desires of rational beings, who on Kant's view enjoy the status of "titular lords of nature."[87] Nonhuman animals, on Kant's view, do not engage in scientific inquiry and they do not engage in metaphysical speculation.

Nor on Kant's view do the latter two questions, what I ought to do and what I may hope, have any place in the lives of nonhuman animals. As I noted in Chapter 1, Kant is quite clear in his writings on morality that a being must be rational (indeed, in a very specific sense) in order to be an active participant in moral judgment, as well as to be a full and direct beneficiary of moral concern (Kant's "passive citizen" or Tom Regan's "moral patient").[88] Moral agency is grounded in pure practical reason, which gives rise to the categorical imperative and a set of Transcendental Ideas of Reason to which nonhuman animals have no access. As a rational guideline for what counts as morally legitimate behavior, the categorical imperative requires the ability to think in universal terms. Indeed, all the major formulations of the categorical imperative involve conceptual abstraction of a kind impossible for beings who are by definition *aloga*: the universal law formulation, the means-ends formulation, and the kingdom of ends formulation each take the form of "an a priori synthetic practical proposition," and thus a being must be able to think in "pure" (*a priori*) terms that give rise to general rules for conduct in order to contemplate and make use of the categorical imperative.[89]

The categorical imperative provides us with a general rule for deciding whether certain actions count as morally permissible, where the guiding criterion is whether through our conduct we show respect for the absolute inherent worth of rational beings. Like Aristotle and Seneca before him, Kant sees the human condition in terms of a tension between personal "inclination" and higher rational considerations: each of us is subject to strong drives toward selfishness and partiality (bias toward some and against others), but each of us is also capable of employing reason to rise above our partisan affiliations and form judgments about right and wrong that are based on a recognition of the essential sameness and moral equivalence of all rational beings. When we act on the basis of inclination, we are acting "for a selfish purpose" such as personal gain, the desire to help someone we like, or the endeavor to harm someone at whom we are angry.[90] Kant recognizes the danger to morality posed by acting on the basis of such elective affinities; like Seneca, he recognizes how emotional states such as anger can distort our perception of the value of the ends we pursue, thereby placing us at odds with rational beings to whom we ought to be relating on the basis of a sense of commonality and mutual respect.

Kant contrasts inclination with the concept of duty, which on his view is the only suitable basis for making authoritative moral judgments. On Kant's view, the more an individual pursues happiness (not in Aristotle's sense of *eudaimonia*, but in the more straightforward sense of personal enjoyment or material prosperity), the more likely that person is to come into conflict with the demands of morality, which are based on the idea of respect for persons (rational beings) and the endeavor to embrace moral ideals that transcend the realm of merely material relations. On Kant's view, "existence has another and much more worthy purpose, for which, and not for happiness, reason is quite properly intended."[91] Acting on the basis of duty rather than inclination is the way to honor the demands of morality, where "duty is the necessity of an action done out of respect for the law," and where "the law" means the moral law that requires us to exhibit respect for persons by rising above personal preferences and enmities.[92]

Because nonhuman animals are, on Kant's view, not rational (at least not in the morally relevant sense), they are excluded not only as custodians but also as beneficiaries of the moral law. Not only are they incapable of contemplating what duty requires and acting accordingly, but given that they are *aloga* they are not the kind of beings to whom anything like respect (or, for that matter, any sort of direct moral concern) can be extended. That nonhuman animals cannot contemplate duty in Kant's specific sense will be, to many minds, a fairly straightforward matter: If nonhuman animals cannot reckon with any of the three major formulations of the categorical imperative, then it follows trivially that they cannot really be considered moral agents in Kant's sense. The "universal law" formulation of the imperative calls on us to "act as if the maxim of your action were to become through your will a universal law of nature."[93] To the extent that nonhuman animals cannot think in terms of universals and cannot grasp laws of nature (that is one implicit take-away from the *Critique of Pure Reason*), it would hardly make sense to expect them to be able to employ the universal law formulation of the categorical imperative. The same holds for the other two major formulations of the imperative: Whether we are called upon to "treat humanity, whether in your own person or in the person of another, always at the same time as an end and never simply as a means," or to "act as if [you] were through [your] maxim always a legislating member in the universal kingdom of ends," it is impossible on the account of rationality presented by Kant to envision a being bereft of linguistic rationality contemplating the difference between ends and means or thinking about what would constitute "a universal kingdom of ends."[94]

All three formulations of the categorical imperative converge on the ideal of perfect respect for rational beings, an ideal whose realization requires the capacity for highly abstract thought and whose terms categorically exclude nonhuman animals as beneficiaries. To be a moral agent, on Kant's view, is to possess the freedom to rise above sensory impulses and act on the basis of "motivating grounds"; nonhuman animals lack this freedom, and are subject to the "*bruta necessitas*" posed by "sensory impulses" and inclinations.[95] Kant is very clear in his discussion of morality that only rational beings count as "persons," and that persons are the only beings who even potentially merit membership in "a universal kingdom of ends," that hypothetical place where all members of the community (all of whom are rational in Kant's specific sense) exhibit perfect respect for one another. Nonhuman animals are entirely subject to laws of physical nature, i.e., "laws of efficient causes necessitated from without," whereas rational beings can merit membership in a universal kingdom of ends by *choosing* their maxims (i.e., "self-imposed rules") in accordance with the moral law. Like Aristotle and Seneca, Kant sees the behavior of nonhuman animals as being determined by factors other than choice, which is facilitated by reason. To this extent, nonhuman animals are part of "nature as a whole . . . viewed as a machine," although as I noted in Chapter 1, Kant resists the Cartesian temptation to characterize nonhuman animals purely in terms of mechanism.[96]

What Kant does not do when he rejects Descartes's characterization of nonhuman animals as biological mechanisms is explore just what sorts of "representations" he considers nonhuman animals to experience. Kant's intent in rejecting Descartes's view of nonhuman animals as lacking subjective states of awareness is to acknowledge that many nonhuman animals are capable of sensory awareness, which includes the capacity to experience pleasure and pain. This is an implicit acceptance of the criterion for consciousness that Thomas Nagel would later offer: that to call a being conscious is to accept that "there is something it is like" to be that being.[97] Nagel recognizes that the behavior of conscious beings is often determined by mental states or external causes, but he suggests that those factors do not exhaustively describe the nature of consciousness. Here I think that Nagel touches on a widely neglected aspect of the philosophical investigation into the lives of nonhuman animals, one neglected by Kant in spite of his willingness to acknowledge that nonhuman animals experience "representations." For the notion of representation has a wide array of senses, ranging from data registering on an automated detection device (say, a Geiger counter or an infrared safety device on a garage door opener) to a richly detailed encounter with meaning (say, my unforgettable encounter with Rembrandt's "Anatomy Lesson of Dr. Nicolaas Tulp" at the Mauritshuis in the Hague some years ago). When Kant states that nonhuman animals experience representations, he does not make his meaning entirely clear. What is clear is that he believes that nonhuman animals can suffer (and thus that we are not entitled to be indifferent in our treatment of them, due to the potentially pernicious effects on our fellow humans that might ensue) but that he believes that nonhuman animals cannot set ends for themselves and therefore are excluded from the spheres of moral and legal right. To deny nonhuman animals the capacity to set ends for themselves is to reject Tom Regan's belief that any sentient creature possessing a "psycho-physical identity over time" (some at least rudimentary, pre-linguistic sense of oneself as a self persisting through the successive moments of time) also possesses the capacity for "preference autonomy," the kind of autonomy exhibited, for example, when a creature decides where to hide its food, where to establish a dwelling place, or what to eat if there is more than one option on a given occasion.[98] Kant denies this capacity for autonomy in nonhuman animals when he hews closely to a definition of autonomy as a rational capacity to legislate and act in accordance with the moral law: there simply are no other kinds or degrees of autonomy, which on Kant's view is a strict either-or affair: Either a being possess the full range of cognitive abilities that underwrite moral agency, or that being is driven purely by instinct and thus categorically excluded from the sphere of beings possessing full and direct moral status.

Kant, like many other exponents of anthropocentric ideology, never acknowledges the potentially vast varieties of consciousness and autonomy that lie between these two extremes. Is the only real "autonomy" in the world the kind guided explicitly by pure practical reason and its "a priori synthetic practical propositions"? How do we know this? And could there be an ulterior motive concealed by the outward appearance of detached objectivity in Kant's (not to mention Seneca's or Rawls's) characterization of reason and the value it confers on its possessors? After all, as I have urged from the beginning, what can easily appear to be neutral, straightforward, uncontroversial characterizations of the faculty of reason begin to look less than impartial when we recognize the extent to which those characterizations proceed from an assumption of (or endeavor to vindicate) human uniqueness and superiority. Where the tradition would have us believe that it is our possession of linguistic rationality that confers uniqueness and superiority on us, I am suggesting that the relationship is really the other way around: it is our investment in seeing ourselves as special that has guided the formulation of the very ideal of reason with which we have lived at least since Aristotle. After offering some additional, concluding remarks about Kant's ideal of reason, I will return to the notion of autonomy and examine some heterodox attempts to broaden our appreciation of what it can signify.

Kant fills in his conception of the moral life by elaborating the notion of the moral law in terms of Transcendental Ideas of Reason, concepts that are formed through the illicit use of categories of the understanding—illicit in the sense that we seek to think beyond the bounds of concrete experience and gather the seemingly fragmented moments of experience together into a coherent whole. Because these Ideas are formed through an illicit use of the categories of the understanding, they enable us to think (conceptualize) things that cannot be known as objects of scientific investigation.

What "as a whole" means for Kant becomes clear from the examples he provides of Transcendental Ideas of Reason, or what he also calls "postulates of pure practical reason": He includes among them the notions of God, freedom, and immortality, all of which he believes are needed in order to conceptualize fully the tasks of morality. The idea of immortality, for example, is not meant to proclaim the immortality of the soul as a metaphysical truth—Kant learned his lesson about metaphysics from Hume, who had urged us not to extend our inquiries beyond "experience and observation"—but rather represents the possibility of perfecting our moral nature: Kant starts from the proposition that no rational being in the world "is at any time capable" of attaining the kind of holiness that signifies "the perfect fit of the will to moral law," inasmuch as we find ourselves in the thrall of inclination. From there, Kant suggests that it would take "an endless progress," i.e., an infinite amount of time, for any rational being to achieve moral perfection, and he posits the immortality of the soul as a way of thinking past the inherent temporal limitations of our mortality toward the at least hypothetical goal of endless improvement of our nature—a goal that fits together seamlessly with the Enlightenment ideal of inevitable progress that I discussed near the end of Chapter 1.[99]

Similarly, Kant posits the Idea of God as a postulate of pure practical reason as a means for conceptualizing morality in its entirety, which includes a notion of the highest good. Where Seneca characterizes that good in terms of an ideal of contemplative satisfaction and at least qualified detachment from earthly goods, and where some thinkers view the highest good entirely in terms of moral integrity, Kant proposes that the highest good would be not simply everyone doing their duty but everyone being rewarded with happiness in proportion to their worthiness to be happy—a world in which reward and punishment would be meted out in direct relation to each individual's merit would be a higher good than all simply doing right. It is in this connection that the Idea of God is relevant: In a world that gives us no reason to hope that individuals will reap their just reward, we postulate God to symbolize the possibility of a power that could guarantee the meting out of justice. On Kant's view, we cannot conceive the highest good "except by presupposing the Highest Intelligence." Not surprisingly, given Löwith's observation about the imprint of Christianity on Kant's thought, Kant sees a special affinity between this approach to the highest good and "the Christian principle of morality."[100] In this connection, the remarks I offered in Chapter 1 about the pointedly anthropocentric cast of Christianity are very much worth bearing in mind.

What is revealing about Kant's appeal to the postulates of pure practical reason—in particular, God, freedom, and immortality—is that they all follow from and reinforce an anthropocentric background ideal of living. The system of transcendental Ideas presented by Kant presupposes an anthropocentric hierarchy that places human beings in special proximity to God; it is impossible, on Kant's view, to imagine a nonhuman animal having any direct participation in "holiness," which requires an inwardness that Kant denies to nonhuman animals when he asserts that they are driven by the *bruta necessitas* of blind impulse. By the same token, the very notion of freedom or autonomy that Kant places at the center of his theory of moral commitment is one predicated on a categorical distinction between beings who can set ends for themselves and think in highly abstract (linguistically informed) terms and beings who are bereft of these capacities.

A final consideration in connection with Kant's ideal of reason is his unremitting insistence that feeling can never be the basis of moral commitment. I have already noted Kant's concern about the ways in which inclination can and often does distort our judgment, making us do things that exhibit disrespect for persons (either ourselves or other rational beings). He makes a compelling case for the proposition that inclination tends to make us treat rational beings as mere means or instrumentalities, when in fact we must always remain mindful of their moral status as persons. Just as Seneca notes that inflamed passion can lead us to a distorted sense of the value of things, Kant recognizes that the conditions of our embodiment make us all too willing to treat persons as instrumentalities. Kant also recognizes that in the world of embodiment it is frequently impossible to avoid some sort of use-relationship with our fellow persons, which is why he presents the second main formulation of the categorical imperative as follows: "Act in such a way that you treat humanity, whether in your own person or in the person of another, always at the same time as an end and never simply as a means."[101] While we may not be able to avoid material reliance on our fellow persons altogether, we can avoid doing them wrong by always being mindful that we are relying on a fellow person who possesses inherent moral worth and thus merits absolute respect.

Kant's suspicions about inclination have much in common with Seneca's misgivings about passion: Inclination, like passion, possesses a fundamental irrationality, although the nature of the irrationality is not quite the same in the two cases. Passion, on Seneca's view, is always incommensurable with reason. Passions are "blind and violent impulses over which [reason] will itself have no control"; anyone who surrenders to passion is subject to the "tyranny" of the irrational.[102] If the mind "plunges into anger, love, or the other passions, it has no power to check its impetus" but completely loses control of itself.[103] Thus, on Seneca's view, it is better never to let oneself fall into a state such as anger in the first place.

Kant, on the other hand, recognizes a basic distinction between instrumental and practical (moral) reason, the one facilitating the acquisition of goods that one desires and the other prescribing a system of universally shared rights and duties. Acting on inclination typically involves some sort of instrumental reason: I may need to tell you a really good lie in order to get you to lend me money that I (secretly) know I cannot repay you, or I may find it useful to devise some sort of deception in order to manipulate you into doing something you would not otherwise elect to do. Many objectives that can be achieved through deception and other forms of instrumental rationality are "rational" in the sense that they express an awareness of how to obtain desired outcomes (this is what Kant has in mind when he discusses "hypothetical" as opposed to "categorical" imperatives), but they are often precisely not rational in a morally relevant sense. This is why Kant, both famously and controversially, offers as one of his relatively few concrete examples of the categorical imperative that it is never morally worthwhile to lie to a person, inasmuch as doing so constitutes a failure to respect that person's inherent worth by treating the person as a mere source of gratification (e.g., as an enfleshed ATM to provide me with the money I want).[104] Where Seneca acknowledges no prospect of a coordination of reason and passion, Kant recognizes the ways in which instrumental rationality can serve selfish or partisan ends, but he remains consistent with Seneca in seeing emotional states as threatening our moral integrity.

Like Seneca, Kant characterizes all emotional states as potential threats to morality. Seneca sees emotional states ranging from anger to love as posing the same sort of threat: Love, after all, can move us to do things that cannot be morally justified, as in cases of nepotism in which the motivating factor for hiring an individual is not who is the best candidate for the position but instead who is the person toward whom I have the fondest feeling or the closest personal bond. Love, in this sense, is on a par with hatred: the one causes us to exaggerate the value of certain individuals, while the other causes us to underestimate people's worth.

In Chapter 4 I will return to the question of emotion, and I will suggest that emotions are not best viewed as all of a piece in the way urged by Seneca and Kant—indeed, as I will show in a moment, Kant himself does not really see all emotional states as posing a threat to our moral integrity. While certain emotions do appear to be incommensurable with reason, others play a more fundamental role in the moral life and are not only commensurable with reason but in fact function as indispensable grounds for the operation of reason in moral judgment. Love will be a prime example of a foundational emotional capacity. Even though love can take very different forms, ranging from the acquisitive (selfish) to the generous (other-embracing), we make a fundamental mistake if we suppose that every instance of love is the distorting, acquisitive kind. The distinction I have in mind is gestured toward by St. Augustine in the *Confessions*, in which he contrasts the sinful love of the pears he stole in his youth with the love of something greater than oneself represented by *caritas* or love of one's community (a community that for St. Augustine is governed by the Christian God).[105] It is that sense of a deep bond with one's community (a notion that naturally can be understood as encompassing different groups, as the distinction between anthropocentric and non-anthropocentric amply illustrates) that I believe forms an essential part of the basis of moral commitment—it constitutes that "extra-rational" dimension of moral commitment that, if I am right, *must* accompany and underwrite the putatively "detached" rationality so vaunted by the tradition.

While one might think that some sort of love, perhaps the "higher" love esteemed by St. Augustine, would have a place in the moral outlook of a thinker whose entire philosophical outlook is influenced by dualistic metaphysics (even if he himself disavows any such metaphysical commitments), Kant nonetheless categorically rejects the proposition that any kind of love forms part of the basis of moral commitment. In quite a few places in his writings, Kant discusses "moral feeling" or "practical love," arguing that such feeling is an important *supplement* to moral motivation but that it does not constitute part of the *basis* of moral commitment. Early in the *Grounding for the Metaphysics of Morals*, Kant makes a distinction between "practical" and "pathological" love, stating that the former "resides in the will and not in the propensities of feeling, in principles of action and not in tender sympathy; and only this practical love can be commanded."[106] To this extent, Kant recognizes a distinction between the kind of love whose sole motivating consideration is the equal inherent worth of all rational beings, and the kind of love that expresses partisan preferences. To say that the latter cannot be commanded is to dismiss it as fully incommensurable with practical reason. To see practical (moral) love as amenable to being commanded is to acknowledge, if only implicitly, some hierarchy of feeling according to which at least one emotional state (that of "practical love") is commensurable with reason. Indeed, moral feeling is the only emotional state that Kant acknowledges to be commensurable with reason, all others being dismissed as inherently "pathological" in their tendency to draw us away from fulfilling our moral obligations. What Kant does not fully acknowledge is that *moral feeling or practical love is indeed a species of feeling*; this is exactly the conclusion he seeks to avoid when he states that practical love "resides in the will and not in the propensities of feeling."

To say that practical love or moral feeling resides "in the will" or "in principles of action" is not to say that such feeling is part of the basis of morality. Instead, on Kant's view, duty and the moral law are fundamentally *prior* to any such feeling. Kant states that respect for the law is the only feeling that can be known *a priori*, but he also states that "the cause that determines this feeling lies in pure practical reason."[107] He further states that "respect for the moral law is . . . the sole and undoubted moral drive."[108] To this extent, Kant's account of the precise place of moral feeling in the edifice of moral judgment exhibits a certain, telling ambivalence: On the one hand, he characterizes the feeling of respect as the one fundamental moral drive, while on the other hand he maintains that pure practical reason is the cause of the feeling of moral approval. This ambivalence raises many questions, one of which is exactly how best to understand the notion of respect that lies at the core of Kant's moral thought; a related question, and a crucial one for the view of moral judgment I am developing in this book, is whether a rational insight (or mental "representation" in Kant's specific sense) is capable *by itself* of giving rise to a motivation.

Kant is unequivocal in *The Metaphysics of Morals* that moral feeling forms no part of the basis of practical laws. "Moral feeling[s] . . . have nothing to do with the basis of practical laws but only with the subjective effect in the mind when our choice is determined by them, which can differ from one subject to another."[109]

This verdict applies even to the feeling of respect, which Kant characterizes as "merely subjective, a feeling of a special kind, not a judgment."[110] Being "subjective," moral feelings are "not dictated by reason but would be taken to be a duty only instinctively, hence blindly."[111] And yet, of all our feelings, "respect . . . is the only one which we can know completely a priori and the necessity of which we can discern."[112]

4. Questioning the Traditional Commitment to the Primacy of Reason

Here, I think, we find the crux of Kant's ambivalence about the human relationship to nonhuman animals, and about the precise extent to which reason proves authoritative in moral judgment. Kant openly acknowledges the significance of a *felt* dimension to moral commitment: if he didn't, he would not have devoted so much attention, in so many texts, to articulating his view of the nature and significance of *moral feeling*, in one place going so far as to classify the feeling of respect as knowable "completely a priori" even though it is "merely subjective." What it would mean to have *a priori* knowledge of a feeling is something Kant never explains. He proceeds as if he wants to acknowledge as fairly uncontroversial the proposition that moral commitment has an irreducibly (*a priori*?) affective (felt) dimension, but his guiding commitment to elevating human beings above nonhuman animals leads him to insist that the entire basis of moral commitment is rational insight or detached cognitive "representation" rather than any factor on the order of "inclination" or impulse. As I noted earlier, Kant acknowledges (criticizing Descartes) that nonhuman animals have representations, but he believes that their representations are not of the detached cognitive variety. Instead, nonhuman animals are driven by impulse or the immediacy of instrumental causes; and to the extent that human inclinations (including emotions) function as a kind of instrumental cause (we simply act on the basis of them, without reflection), Kant is at pains to divorce all feeling from the basis of moral commitment. That we are rational is enough, on Kant's view, to enable us to engage in moral judgment, as well as to confer on us the status of "titular lord of nature."

These statements go right to the core of Kant's anthropocentrism. Recall his distinction between *homo phaenomenon* and *homo noumenon*, the one referring to the natural, embodied condition of subjection to external necessity (*bruta necessitas*) that we share with nonhuman animals, and the other referring to the "higher" plane of existence reserved for beings who are rational in ways that permit us to address the questions "What can I know? What ought I to do? and What may I hope?" Kant's interest in excluding even moral feeling from the basis of morality is driven by his aim of arriving at a foundation for moral choice that is objective and thus is the domain exclusively of human beings. Kant himself is unclear as to whether objectivity in morality is really possible. As I have noted, he characterizes the categorical imperative as "an a priori synthetic practical proposition," which suggests that apodeictic certainty is possible in moral judgment. But on the other hand, much of what Kant writes about morality demonstrates an awareness that moral judgment is, in crucial respects, *un*like matters of logic and mathematics. That Kant offers precious few specific examples of duties required by the categorical imperative is, I think, a sign of this awareness.

Whether or not true objectivity is really attainable in moral judgment—in Chapter 5, I will present a view according to which moral judgment is more like Kant's notion of aesthetic or teleological judgment than like logical judgment— Kant's reasons for seeking to exclude all feeling from the basis of moral judgment are clear. As Kant observes, feeling is notoriously susceptible to variations among different individuals. This is arguably the case even for practical love: what you find morally compelling might leave me cold, and what I find morally redeeming you might find pernicious. Kant appeals to a cause prior to feeling in an effort to secure moral judgment against these sorts of vicissitudes. Kant is right that these pose a serious threat to moral judgment. The question is whether Kant is right to characterize all feelings as equally susceptible to this drawback.

Kant is confident that he is right to characterize *all* feelings as posing a threat to the integrity of moral judgment. "Feeling, whatever may arouse it, always belongs to the order of nature." To include any elements from the realm of mere nature in the basis of morality would render "the doctrine of virtue . . . corrupted at its source."[113] This is a core commitment of Kant's that I have been challenging: that anything that recalls our continuity with nonhuman animals poses a threat to a very specific, anthropocentric ideal of humanity. "A human being has a duty to raise himself from the crude state of his nature, from his animality (*quoad actum*), more and more toward humanity, by which he alone is capable of setting himself ends." In this ascent, it is not feeling but "the law" that is "the incentive" to our actions; and while moral feeling can follow from doing our duty, it is a fundamental mistake to suppose that such feeling "could precede reason."[114] Again, "the determining ground of the will" in matters of morality is no feeling, not even practical love, but rather "the representation of the law in itself."[115]

And even though moral feeling presupposes this representation of the law, Kant asserts that such feeling is in us "originally" and that without it "humanity would devolve (by chemical laws, as it were) into mere animality." This original feeling "(like pleasure and displeasure in general) is something merely subjective, which yields no cognition."[116] What Kant does not explain is how a being that is essentially the same as a nonhuman animal when viewed as *homo phaenomenon* can possess a natural capacity (after all, moral feeling according to Kant is part of our natural rather than our noumenal constitution) that so radically distinguishes us from nonhuman animals. To attribute moral feeling to humans and deny it to nonhuman animals is to do something with much more profound implications than, say, attributing a particular physical attribute exclusively to human beings; it is to take the position that nature, and not merely our "noumenal" nature, marked human beings out for a special, higher form of existence. In this respect, Kant's treatment of moral feeling recalls his suggestion that nature contains the hidden *telos* of leading humanity to the goal of perpetual peace.[117] Thus it is not only in terms of our moral being (*homo noumenon*) that we differ in kind from nonhuman animals, but in terms of our natural being (*homo phaenomenon*) as well. From there it is a short step to Kant's proclamation that the human being is "the titular lord of nature."

In one place, Kant describes the point of crossover between the two realms in the following terms. Addressing the question how we attain orientation in speculative thought, Kant writes that "reason does not feel. It perceives its own deficiency and produces a feeling of need through the *cognitive impulse*. The same applies in the case of moral feeling."[118] This is a somewhat different account of moral feeling or practical love than Kant offers in *The Metaphysics of Morals*; there he states that moral feeling is "in us originally," whereas here he states that reason itself produces the feeling on the basis of a perceived lack in us. I believe that this tension or inconsistency in Kant's account of moral feeling is a product of his effort to acknowledge the embodied aspects of our nature while maintaining the possibility of transcending the limitations of that nature: Kant recognizes that certain varieties of feeling may merit a different kind of treatment than the more pointedly selfish varieties, but in his haste to distinguish us from nonhuman animals he finds himself pinned in a corner where he cannot ultimately make sense of the idea of moral feeling. Kant wants to treat it as a kind of feeling, indeed one that is vital to the cultivation of our humanity, but he has to jump through some hoops to avoid the implication that moral feeling forms any part of the basis of moral commitment.

Moral feeling, Kant states, is "unanalysable, basic, the ground of conscience," and "my final yardstick" in moral matters, just as "true and false is the final yardstick of the understanding."[119] Here Kant offers a view of moral feeling that is closer to my own view, one that recognizes the essential role played by at least some emotions in the very process of moral commitment. In calling moral feeling "my final yardstick," he seems to mean not that moral feeling is the determining basis of right and wrong, but rather that the moral feeling that accompanies actions and judgments with moral significance functions as a kind of litmus test for the legitimacy of our moral judgment. In this respect, Kant is committed to a coordination between the phenomenal (physical, empirical, embodied) and the noumenal (a priori, disembodied, wholly rational) aspects of our being: We just happen to be fortunate in having been endowed by nature with a capacity for feeling that lends support to our efforts to act in accordance with what the moral law requires. Kant is consistent in maintaining that "moral feeling" is not "qualified to give universal laws of morality."[120]

But Kant equivocates on the question whether it is moral feeling itself, or instead detached rational insight, that *motivates* us in such matters.

> The moral feeling is a capacity for being affected by a moral judgment. When I judge by understanding that the action is morally good, I am still very far from

doing this action of which I have so judged. But if this judgment moves me to do the action, that is the moral feeling. Nobody can or ever will comprehend how the understanding should have a motivating power; it can admittedly judge, but to give this judgment power so that it becomes a motive able to impel the will to performance of an action—to understand this is the philosopher's stone.[121]

A great many philosophers who have developed their views about morality under Kant's influence have proceeded on the assumption that judgment has a fundamental priority over feeling in matters of moral motivation: It is "intelligence," not emotion, that truly moves us, according to Dewey.[122] Desires that motivate us, Rainer Forst suggests, are based on reasons, not on other desires.[123] According to Bernard Williams, those motivating reasons must be internal rather than external.[124] And on Thomas Nagel's view, even though a desire is always present when reasons motivate, that desire does not by itself account for the motivation; "ultimately something besides desire explains how reasons function" and why they motivate.[125] Just how these thinkers go about making a case for the motivating power of reasons is revealing, as is the rationale they employ in making their case.

Implicitly endorsing what Scanlon aptly calls "reasons fundamentalism," Nagel seizes upon a central commitment of the anthropocentric tradition in ascribing centrality to the notion of *timelessness* as a crucial feature of the kinds of reasons that properly motivate action.[126] Nagel is exemplary of views indebted to Kant in basing his entire account of moral motivation on a specific ideal of humanity. Nagel argues that "a central role in the operation of ethical motives" is "a certain feature of the agent's metaphysical conception of himself," a conception "which each person has of himself and of his relation to the world."[127] This self-conception is that of a self that persists through time and is able to gain a perspective on its life as a whole. Beings who count as selves in this sense are able to attain "a standpoint of temporal neutrality towards the events of our lives," and this capacity confers on us an obligation to "regard the facts of one's past, present, and future life as tenselessly specifiable truths about different times in the history of a being with the appropriate kind of temporal continuity."[128] This in turn suggests that "the reasons for a practical judgment," which are "reasons for *doing* or *wanting* something," transcend discrete moments of time and possess an authority that endures. A simple example might be this: If it was wrong for me to lie to you yesterday, then it is equally wrong today for me to have lied to you yesterday; it isn't as if a reason that was authoritative yesterday had a shelf life that expired last night at midnight, nor does the fact that I happen to feel differently today than I did yesterday about lying to you have (on the Kantian approach, at any rate) any implications for the legitimacy of the reasons that determine the rightness or wrongness of my action. What transcends the vicissitudes of my personal feelings is an edifice of reason whose authority transcends time and perspective in virtue of finding its foundation in an enduring ideal of humanity.

Nagel is far from alone in basing his view of moral judgment on an appeal to an ideal of humanity; such an appeal is present implicitly if not explicitly in virtually all the thinkers I have been discussing, both anthropocentric and non-anthropocentric. Nagel's appeal has a great deal in common with those made by Aristotle, Seneca, and Kant: All argue for the primacy of reason in morality, all deny rational capacity in nonhuman animals, and all argue for an ideal of reason as timeless. Where Aristotle lauds the liberating power of pure, godlike contemplation; where Seneca maintains that we are most godlike precisely where we most effectively deny the influence of bodily impulses on our conduct; and where Kant asserts that it is in virtue of our rational capacity that we can escape the pernicious influence of inclination and ascend to the status of "titular lords of nature," Nagel invokes "the impersonal standpoint" as essential to the process of moral judgment.[129]

I noted at the beginning of this chapter that taken at face value, the defenses of reason offered by the thinkers I examine here seem straightforward and unobjectionable. Nagel's invocation of the impersonal standpoint seems to fit this description: After all, we all know that basing moral judgments on personal affiliations and preferences is very dangerous and often leads to infringements on the inherent worth of beings possessing moral worth. Does this mean that we need to move to the opposite extreme, and lose all sense of personal identity in matters with moral significance? Or might there be a way to reintroduce personal affective dispositions into the process of moral judgment, and to do so in a way that not only does not threaten moral integrity but proves to be essential to its realization? That is a prospect that I will explore in the remainder of this book. But first, let's consider a little more closely the ideal of timelessness that informs the views of thinkers from Aristotle to Kant and Nagel.

We saw in examining Seneca's views that a central Stoic commitment is the ideal of constancy or equanimity, a state of freedom from the disturbing influence of bodily drives that permits a sense of detached satisfaction. That detachment is possible only for us godlike human beings, as nonhuman animals are all moved by impulse. Nonhuman animals are imprisoned in an eternal present and, on Seneca's view, are not even capable of emotional states. Thus one is left to wonder what the subjective life of a nonhuman animal is like, if indeed one is willing to acknowledge anything like an inner life in a being that lacks the specifically human form of linguistic rationality. While they do not commit themselves explicitly to this view, it seems implicit in the characterizations of nonhuman animals offered by Aristotle, Seneca, and Kant that nonhuman animals are not subject to the pain and tribulations caused by the vicissitudes in emotional states that human beings experience, those difficulties being absent from the life of a being that (supposedly) lacks *logos* and thus (again, supposedly) has no ability to grasp and evaluate its immediate experiential states. To this extent, we have no cause to feel compassion for nonhuman animals—unless, that is, we follow Kant's indirect duties view and suppose that harming nonhuman animals is bad simply because it leads to what Aquinas called "the temporal hurt of man," or else (against the chief exponents of the tradition) we simply acknowledge that beings lacking human linguistic rationality have all sorts of rich subjective experiences and a freedom that we human beings may be unable to recognize and appreciate simply because it is unlike our own.[130]

Seneca is not alone in presenting an ideal of constancy based on the cultivation of reason. Kant, even though he is critical of the Stoics for excluding happiness as a component of the highest good, embraces the Stoic premium on constancy as a way of calming the empirical will's tendency to pursue ends that are inconsistent with moral integrity.[131] In seeking to live morally, we need "a contented outlook, peace of mind, and freedom from all reproach, true honor, respect for oneself and for others, indifference, or rather equanimity and steadfastness, in the face of all evil for which we are not to blame."[132] The notions of indifference [*Gleichgültigkeit*], equanimity [*Gleichmütigkeit*], and steadfastness [*Standhaftigkeit*] are all core aspects of the Stoic ideal of living, the term 'Standhaftigkeit' linking Kant's discussion to Seneca's by way of the Renaissance writer Lipsius's influential text *De constantia*, whose German translation is *Von der Standhaftigkeit*.[133] When Kant counsels "steadfastness," he is doing something of a piece with Aristotle and Seneca: He is arguing that the self needs protection from its reality as *homo phaenomenon*, specifically in the mode of rising above the one world in which we find ourselves and in which we must inevitably make our way.

Lipsius defines constancy as "a right and immovable strength of the mind [*animo*], neither lifted up nor pressed down with external or causal accidents. By 'strength' I understand a steadfastness not from Opinion, but from judgment and sound Reason."[134] On this view, the function of reason—and here Lipsius has in mind what Kant would later call practical (moral) reason—is to insulate us from the very vicissitudes of fortune that Seneca characterized as posing a threat to our well-being. To take such a stand on the life of embodiment is to embrace the spirit of *contemptus mundi* (contempt for the world) for which St. Augustine is famous. Even while we acknowledge that embodiment is essential to our being, we denigrate it by posing a sort of "higher world" of spiritual integrity. The promise of that world is twofold: to shield us from the destructive power of time (as we have seen, the anthropocentric ideal is an ideal of *timeless* reason) and thereby to substantiate a historical anthropocentric commitment to human exceptionalism or uniqueness (for after all, on the traditional view I have been examining, only human beings possess *logos* and thus have access to the "higher" world).

It will take the remainder of this book to explore fully the nature of this guiding anthropocentric commitment and diagnose its fatal shortcomings. Here I have done less to show its shortcomings than to shed light on the ideological motivations that have given rise to it. In subsequent chapters I will explore the question whether we really do best to think of our moral worth as being exclusively a function of detached rationality; I will offer the tentative conclusion that while the tradition has been entirely correct to characterize the human capacity for moral growth and integrity as involving an ideal of reason as timeless, rationality is a necessary but not a sufficient condition for the kind of moral agency that the thinkers I have been examining attribute to human beings and categorically deny to nonhuman animals.

In order to argue for this conclusion, I will need to share some remarks about the ways in which human "inclination" and nonhuman animal "impulses" or "drives" are like and unlike one another. Among other things, this will involve a consideration of the view, held by quite a few people in addition to the Stoics, that nonhuman animals do not "really" experience emotional states, as well as the view that nonhuman animals experience some emotional states but lack the cognitive apparatus necessary for "higher" sorts of emotions (such as remorse, regret, etc., which require the sense of time that traditional thinkers deny to nonhuman animals). It will also prove fruitful to reflect further on the nature of the term 'logos' and confront very squarely the question whether a being must have precisely those cognitive abilities that we attribute to human beings in order to count as a "rational" being. In the next chapter, I will examine some contemporary attempts to "historicize" reason by showing how one's location in time and place influences one's sense of what counts as objective or detached judgment; my aim there will be to determine whether a conception of reason that respects rather than seeks to transcend time is more suitable to a robust conception of moral judgment.

As a preliminary to that examination, I will conclude this chapter with three considerations that I believe shed important light on the anthropocentric ideal of reason that I have just sketched. The first is a famous passage in Nietzsche's *Twilight of the Idols* entitled "How the True World Finally Became a Fable," in which Nietzsche characterizes the higher world of spiritual purity as "unattainable, indemonstrable, unpromisable; but the very thought of it—a consolation, an obligation, an imperative."[135] Here Nietzsche brings into fine focus the central contradiction of the tradition's appeal to a higher world of spiritual integrity: It is essentially a fiction, a goal so utterly unattainable by mortals that it ought to be obvious that it is an empty promise, and yet we cling to it as if it were both a consolation for our sufferings and an imperative to pursue. Why can we not simply dispense with the pretense of escaping this world, the object of so much contempt, and resign ourselves to (perhaps even embrace) our embodiment, our essential animality, our frailty, our ordinariness? The bookend to Nietzsche's dismissal of "the true world" is his early suggestion that human beings were clever enough to have "invented knowing" but that sooner or later we, too, "[have] to die."[136] Death, as Gertrude observes in *Hamlet*, is the great cosmic equalizer.

A second consideration underwriting the endeavor to rethink the terms of the tradition's ideal of reason is Schopenhauer's revision of Kant's views about the faculties of understanding and reason. On Kant's view, as I have noted, understanding and reason both involve concepts, and to the extent that nonhuman animals lack conceptual ability they are incapable of both reason and understanding. Schopenhauer considerably broadens the notion of understanding so as to signify various sorts of reckoning with causes and contingencies in the external world, where these forms of reckoning have no necessary dependence on concepts. On

Schopenhauer's view, any being with eyes possesses knowledge.[137] But where non-human animals possess both sensibility and understanding (the faculty of knowledge), only human beings possess the capacity for conceptual abstraction that is distinctive of reason.[138] Thus whereas nonhuman animals are confined within an eternal present, only human beings, "by virtue of knowledge in the abstract, comprehend not only the narrow and actual present, but also the whole past and future together with the wide realm of possibility."[139] This means that only human beings can attain the status of individuality, form character states, take on responsibilities, and perform all the functions that require conceptual abstraction.

I have shown in previous work that Schopenhauer, in spite of a number of ideas and remarks that take him very close to a non-anthropocentric ideal of living, ultimately remains firmly situated in an anthropocentric standpoint according to which practices such as meat eating and vivisection are morally justifiable.[140] One cannot rely on a thinker such as Schopenhauer (nor one such as Porphyry, as I will note in a moment) to get beyond certain non-anthropocentric intimations and embrace a fully non-anthropocentric worldview. But at the same time, his notion that vision (or touch) itself confers knowledge is highly thought-provoking, and is not all that far removed from the views of traditional thinkers who express no apparent concern for the divide between human and nonhuman animals. I am thinking here of Bishop Berkeley, who counters Descartes's account of depth perception by appealing to experiential resources that are not unique to human beings. Where Descartes argues that we perceive depth in space by performing geometric calculations, Berkeley maintains that the perception of depth absolutely requires movement in space and the use of our sense of touch: we have to move around in the world and come into physical contact with things in order to grasp depth.[141] Although it appears to have been the last thing on his mind in proposing this, I believe that in offering this characterization of depth perception Berkeley presented a view of human reckoning with the environment that is considerably closer to the reality of embodied experience than the account offered by Descartes.

Schopenhauer, by the same token, demonstrates a willingness to see a fundamental continuity between the embodied experience of human beings and nonhuman animals, and I believe that arriving at a more adequate account of the human condition as well as of the nature of moral commitment will depend on this kind of willingness to step back (to use Heidegger's phrase) from the tradition's insistence on characterizing human experience as so fundamentally at odds with the experience of nonhuman animals.

And yet, as I have noted from the outset, there are some key differences between human beings and nonhuman animals that are traceable to the human possession of linguistic rationality. The ability to formulate, contemplate, debate, and revise general rules for social cooperation does appear to be reliant on the specific kind of rationality exhibited by human beings. What I have been arguing is that this difference has no implications whatsoever for moral worth: There is no compelling logic—no logic that would be compelling to anyone behind a Rawlsian veil of ignorance who did not know what species they would be assigned upon entry into the world—according to which beings possessing this capacity for rule-making "count" more in the moral scheme than beings who lack it. This is what Regan was getting at with the concept of the moral patient: Some beings, even though they lack (full) linguistic rationality, nonetheless possess the same inherent moral worth as beings endowed with linguistic rationality, as the case of infants, young children, and adult humans with significant cognitive impairments make amply clear. The signal distinction between moral agents and moral patients is not that one matters more than the other, but simply that the former have obligations that cannot be attributed to the latter—precisely because the ability to take on a moral obligation requires, just as Rawls maintains, a sense of what a rule or obligation is. Where thinkers such as Rawls go wrong is in supposing that only those beings who can conceptualize rights and duties can be considered beneficiaries of rights.

A final consideration in this connection is the medieval philosopher Porphyry's challenge to the tradition's narrow characterization of the notion of *logos*. Where the tradition characterizes it as a rarefied, linguistically-informed capacity that enables us to survey existence as if we stood outside of time (thereby asserting our affinity with the divine), Porphyry anticipates Schopenhauer's appeal to a considerably more quotidian conception of vision and knowledge as a kind of great equalizer between human beings and nonhuman animals. Porphyry states that "almost everyone agrees that animals are like us in perception and in organisation generally with regard to both sense-organs and to the flesh."[142] Nonhuman animals "know everything that is to their advantage" even if we are incapable of understanding the ways in which nonhuman animals express themselves.[143] Porphyry even goes so far as to argue for a natural condition of reciprocal justice that obtains between humans and nonhuman animals, although in the end, like Schopenhauer, he stops short of a full recognition of the moral worth of nonhuman animals when he sanctions meat eating for at least some members of human society.[144]

It is this conception of justice, one that includes all sentient creatures as its direct beneficiaries, for which I argue in the second half of *Animals and the Moral Community*, and for which I continue to argue in this book. The "step back" from willing recommended by Heidegger is a step back from the entire background ideal of living through the lens of which we see all nonhuman beings in the world as resources for the satisfaction of our needs and desires. To step back from this global perspective demands exactly the sorts of virtues that exponents of the tradition laud as specifically human achievements: generosity, humility, and the kind of constancy that results from detaching ourselves from the sorts of drives that tend to distort the value of the goods that we pursue. There is more than a little irony in the fact that the very capacities that we have seized upon as signs of our own superiority have been used, so widely and for so long, as means for inflicting harm on beings who pose little if any real threat to us. That harm, as I am attempting to show, is as much conceptual (think of our prejudices about "dumb beasts") as it is physical (think of those ever-increasing United Nations figures on the number of nonhuman animals killed annually for human consumption). If we are the beings cut out by nature to pursue and attain "perpetual peace," we have done a remarkably poor job of living up to that promise.

Notes

1 See René Descartes, *The Passions of the Soul*, book 1, sec. 48, René Descartes, *The Philosophical Writings of Descartes*, vol. 1, p. 347: The will's "proper weapons" against the distorting influence of the passions are "firm and determinate judgments bearing upon the knowledge of good and evil."

2 On Kant's racism, see Immanuel Kant, "Of the Different Races of Human Beings," pp. 82–97 and Emmanuel Chukwudi Eze, "The Color of Reason." On Hegel's privileging of European reason, see G.W.F. Hegel, *Introduction to the Philosophy of History*, pp. 20, 93–5. Descartes's remarks about "Turks and other infidels" also come to mind in this connection; see René Descartes, "Author's Replies to the Second Set of Objections," p. 105.

3 *Critique of Pure Reason*, A805/B833.

4 On Nietzsche's total critique of reason, see my remarks in *Animals and the Limits of Postmodernism*, pp. 21ff.

5 On the ancient Greeks' views about *logos*, see my *Anthropocentrism and Its Discontents*, Chapter 3; on Pope Francis's views in the 2015 encyclical *Laudato si'*, see my essay "The Moral Schizophrenia of Catholicism."

6 *Critique of the Power of Judgment*, p. 298 (Ak. 431: betitelter Herr der Natur).

7 *The Metaphysics of Morals*, "The Doctrine of Virtue," sec. 16, p. 192 (Ak. 442).

8 *Practical Ethics*, p. 92.

9 Thomas Kuhn, *The Structure of Scientific Revolutions*, p. 206.

10 In doing so, I will be departing from Kant, who treats "practical" (i.e., moral) reason differently than he treats reflective judgment; on Kant's view, reflective judgment is at work in our determinations about aesthetic and teleological matters.

11 See in particular Martin Luther's *Open Letter to the Christian Nobility of the German Nation Concerning the Reform of the Christian Estate* and the opening lines of René Descartes's *Discourse on the Method of Rightly Conducting One's Reason and Seeking the Truth in the Sciences*.

12 René Descartes, *Discourse on Method*, Part 6, *The Philosophical Writings of Descartes*, vol. 1, p. 142f. (translation altered). One sign of Descartes's belief that reason is timeless is his acknowledgment that the ancients, even though their morals were "very proud and magnificent palaces built only on sand and mud," nonetheless had a full and adequate appreciation of the truths of mathematics. *Discourse on Method*, Part One, p. 114.

13 *Nicomachean Ethics*, book 6, Ch. 7, Aristotle, *The Complete Writings of Aristotle*, vol. 2, p. 1802 (1141b2–4). Here Aristotle does not explicitly commit himself to the view that man is higher than nonhuman animals, but simply entertains the possibility; I argue in Chapter 3 of *Anthropocentrism and Its Discontents* that Aristotle ultimately does treat nonhuman animals as inferior to humans, one indication of which is his willingness to classify them as resources (living instruments).

14 *On the Soul*, book 1, Ch. 4, *The Complete Writings of Aristotle*, vol. 1, p. 651 (408b18–19); book 2, Ch. 2, p. 658 (413b25); book 3, Ch. 5, p. 684 (430a24).

15 See *Nicomachean Ethics*, book 10, Ch. 7 at 1177a28–1177b5.

16 *Nicomachean Ethics*, book 10, Ch. 8 at 1178b8, 1178b22–30, *The Complete Writings of Aristotle*, vol. 2, p. 1862f.

17 *Politics*, book 7, Ch. 13, *The Complete Writings of Aristotle*, vol. 2, p. 2114 (1332b4–5).

18 *Physics*, book 2, Ch. 8, *The Complete Writings of Aristotle*, vol. 1, p. 340 (199a20–30).

19 Urs Dierauer, *Tier und Mensch im Denken der Antike*, p. 143.

20 Seneca, Letter 121.6, 121.22, *Epistles 93–124*, pp. 399, 409.
21 *Nicomachean Ethics*, book 6, Ch. 2 at 1139a18–33, *The Complete Writings of Aristotle*, vol. 2, p. 1798.
22 *Nicomachean Ethics*, book 3, Ch. 3 at 1112b12–16, *The Complete Writings of Aristotle*, vol. 2, p. 1756.
23 *Nicomachean Ethics*, book, 3, Ch. 2 at 1111b7–18. See also book 3, Ch. 8 at 1116b24–1117a5 and *Eudemian Ethics*, book 2, Ch. 10 at 1225b26–7, where Aristotle states that nonhuman animals also behave on the basis of passion or anger (*thymos*).
24 *Nicomachean Ethics*, book 3, Ch. 8, *The Complete Writings of Aristotle*, vol. 2, p. 1763–4 (at 1116b34–1117a21).
25 *Nicomachean Ethics*, book 3, Ch. 12, *The Complete Writings of Aristotle*, vol. 2, p. 1767 (1119b15–17). See also book 7, Ch. 6, p. (1149b30–34), where Aristotle states that "we call the lower animals neither temperate not self-indulgent except by a metaphor" inasmuch as they "have no power of choice or calculation."
26 *Politics*, book 1, Ch. 5, *The Complete Writings of Aristotle*, vol. 2, p. 1990 (1254b7–8); book 10, Ch. 5, p. 1858 (1176a1–3).
27 *On the Soul*, book 3, Ch. 12, *The Complete Writings of Aristotle*, vol. 1, p. 690 (434a12).
28 See Michael MacKinnon, "Animals, Economics, and Culture in the Athenian Agora."
29 See *Anthropocentrism and Its Discontents*, pp. 77–92.
30 Epictetus, *The Discourses as Reported by Arrian Books I-II*, 1.6.18, 1.16.1–5, 2.8.6–8.
31 *The Discourses as Reported by Arrian Books I-II*, 2.10.2–3; on Diogenes of Synope's ideal, see Diogenes Laertius, *Lives of the Eminent Philosophers*, vol. 2, p. 65 (6.63).
32 *The Discourses as Reported by Arrian Books I-II*, vol. 2, p. 263 (2.9.5–6).
33 Letter 124.1–2, *Epistles 93–124*, p. 437.
34 Letter 124.15–18, *Epistles 93–124*, p. 445.
35 See Arthur Schopenhauer, *The World as Will and Representation*, vol. 1, book 1, sec. 8.
36 Seneca Seneca, *On the Happy Life*, 10.1, 14.3–15.2, Seneca Seneca, *Moral Essays*, vol. 2, pp. 123, 137.
37 Seneca, Letter 71.33, *Epistles 66–92*, p. 93.
38 Letter 76.21, 76.26, *Epistles 66–92*, pp. 159, 162f.
39 Letter 76.9, *Epistles 66–92*, p. 151.
40 Letter 92.32, *Epistles 66–92*, p. 467.
41 *On the Happy Life*, 8.2–5, *Moral Essays*, vol. 2, p. 119.
42 On Aristotle's zoological texts, see my remarks in *Anthropocentrism and Its Discontents*, pp. 69–76.
43 *On the Happy Life*, 4.4–5, *Moral Essays*, vol. 2, p. 111.
44 *On the Happy Life*, 14.2–6, *Moral Essays*, vol. 2, pp. 137–9. See also Seneca, *On Firmness*, 5.4–5, Seneca, *Moral Essays*, vol. 1, p. 61.
45 *On the Happy Life*, 16.1–3, *Moral Essays*, vol. 2, p. 141.
46 *On the Happy Life*, 10.3–11.1, *Moral Essays*, vol. 2, p. 125.
47 *On the Happy Life*, 4.2–3, *Moral Essays*, vol. 2, pp. 109–11.
48 Seneca, *On Anger*, book 2, 16.1–2, *Moral Essays*, vol. 1, p. 201.
49 *On the Happy Life*, 2.2, *Moral Essays*, vol. 2, p. 103; *On Anger*, book 3, 27.3, *Moral Essays*, vol. 1, p. 323.
50 *On the Happy Life*, 4.4, *Moral Essays*, vol. 2, p. 111.
51 *On Anger*, book 2, 8.3, *Moral Essays*, vol. 1, pp. 181–3.
52 Seneca, Letter 59.14, *Epistles 1–65*, pp. 419–21.
53 Letter 76.20, *Epistles 66–92*, p. 159.
54 *On Anger*, book 1, 17.2, 18.1, *Moral Essays*, vol. 1, pp. 151–3.
55 See *Lives of the Eminent Philosophers*, 7.105, vol. 2, p. 211; Cicero, *On Ends*, 3.15.50–51, pp. 269–71.
56 *On the Happy Life*, 21.4, *Moral Essays*, vol. 2, p. 155.
57 *On the Happy Life*, 27.4, *Moral Essays*, vol. 2, p. 177.

58 *On Firmness*, 2.1, *Moral Essays*, vol. 1, p. 51. See also 8.3: "Death is not an evil and therefore not an injury either." *Moral Essays*, vol. 1, p. 73.
59 *On Firmness*, 19.2, *Moral Essays*, vol. 1, p. 103.
60 *On Anger*, book 3, 43.5, *Moral Essays*, vol. 1, p. 355.
61 Another difference between Aristotle and Seneca is that whereas Seneca offers a negative account of the emotions and pleasure, Aristotle gives an account of eudaimonia or the life well led that makes a place for emotion and pleasure. Moreover, where Seneca characterizes passions as involving choice, Aristotle states that our passions are not subject to praise or blame inasmuch as they (in contrast with the virtues) do not involve choice. *On the Soul* 2, 5 at 1105b29–1106a3.
62 *On Anger*, book 1, 3.6–7, *Moral Essays*, vol. 1, pp. 115–7; see also book 2, 16.1, p. 201. On the Stoic notion of *lekta*, see *Lives of the Eminent Philosophers*, 7.63 and my remarks in *Anthropocentrism and Its Discontents*, pp. 78–9.
63 See *On the Happy Life*, 5.1, *Moral Essays*, vol. 2, p. 111.
64 *On Anger*, book 2, 30.2, *Moral Essays*, vol. 1, p. 233; see also book 3, 3.6, p. 261.
65 Seneca characterizes anger as "an active impulse" accompanied by "consent of the will" or "the cognizance of [one's] mind." *On Anger*, book 2, 3.4, *Moral Essays*, vol. 1, p. 173.
66 *On Anger*, book 1, 3.8, *Moral Essays*, vol. 1, p. 117.
67 *On the Happy Life*, 14.2, *Moral Essays*, vol. 2, p. 135.
68 *On the Happy Life*, 10.3, *Moral Essays*, vol. 2, p. 125. See also 8.1–2, p. 121: Like Aristotle, Seneca allows for pleasure as a "by-product" of virtue.
69 *On Anger*, book 2, 3.5, *Moral Essays*, vol. 1, p. 173 (anger involves "the mind's assent"); book 1, 3.4–6, p. 115 (nonhuman animals cannot experience emotions because they lack reason and are moved by "impulses").
70 *On Anger*, book 2, 8.3, pp. 181–3 and book 3, 4.3, p. 263.
71 *On Anger*, book 1, 10.4, *Moral Essays*, vol. 1, p. 133.
72 *On Anger*, book 1, 6.5, *Moral Essays*, vol. 1, p. 123.
73 Or, more accurately, many people do treat nonhuman animals with substantial if not complete disregard for their welfare, but the official position of the indirect duties view is that such treatment is unacceptable—again, not because it constitutes an infringement on the moral integrity of nonhuman animals but because it constitutes an affront to *human* dignity.
74 René Descartes, "Dedicatory Letter to the Sorbonne," p. 8.
75 Löwith, "Der Weltbegriff der neuzeitlichen Philosophie."
76 *The Metaphysics of Morals*, "The Doctrine of Virtue," sec. 11, p. 186 (Ak. 6.434).
77 *Descartes as a Moral Thinker*, pp. 14, 153, 176, 200, 202.
78 *Lectures on Ethics*, p. 12 (Ak. 27:25).
79 *The Metaphysics of Morals*, "The Doctrine of Virtue," Concluding Remark, p. 232 (Ak. 6.491).
80 *Critique of Pure Reason*, pp. 88, 90 (B68, B72).
81 *The Metaphysics of Morals*, "The Doctrine of Right," sec. 46, p. 92 (Ak. 6:314). Here Kant gives the examples of apprentices, domestic servants, minors, and "all women."
82 *Grounding for the Metaphysics of Morals*, p. 35 (Ak. 4.428).
83 "When anatomists take living animals to experiment on, that is certainly cruelty, though there it is employed for a good purpose; because animals are regarded as man's instruments, it is acceptable." *Lectures on Ethics*, p. 213 (Ak. 27:460). However, "agonizing physical experiments for the sake of mere speculation, when the end could also be achieved without these, are to be abhorred." *The Metaphysics of Morals*, "The Doctrine of Virtue," sec. 17, p. 193 (Ak. 6:443).
84 *Anthropology from a Pragmatic Point of View*, p. 369 (Ak. 7:268); *Lectures on Ethics*, p. 381 (Ak. 27:641). Cf. *Nicomachean Ethics*, 8.11 at 1161b2–3; St. Augustine, *The Catholic and Manichaean Ways of Life*, book 2, Ch. 17, sec. 54 and 59.

85 *The Animals Issue*, p. 165.

86 *Critique of Pure Reason*, p. 635 (A805/B833).

87 *Anthropology from a Pragmatic Point of View*, p. 369 (Ak. 7:268).

88 Regan defines a moral patient as a being who merits direct moral concern in spite of lacking the capacity "to control their own behavior in ways that would make them morally accountable for what they do," such as "human infants, young children, and the mentally deranged or enfeebled of all ages," as well as nonhuman animals "who have desires and beliefs, who perceive, remember, and can act intentionally, who have a sense of the future . . . a psychophysical identity over time. . . . [and] a kind of autonomy (namely, preference autonomy." *The Case for Animal Rights*, pp. 152–3.

89 See *Grounding for the Metaphysics of Morals*, pp. 29–30, 36, 43 (Ak. 4:420–1, 429, 438).

90 *Grounding for the Metaphysics of Morals*, p. 10 (Ak. 4:397).

91 *Grounding for the Metaphysics of Morals*, p. 9 (Ak. 4:396).

92 *Grounding for the Metaphysics of Morals*, p. 13 (Ak. 4:400).

93 *Grounding for the Metaphysics of Morals*, p. 30 (Ak. 4:421).

94 *Grounding for the Metaphysics of Morals*, pp. 36, 43 (Ak. 4:429, 438).

95 *Lectures on Ethics*, pp. 234, 125 (Ak. 29:611, 27:344).

96 *Grounding for the Metaphysics of Morals*, p. 43 (Ak. 4:438).

97 Thomas Nagel, "What is It Like to be a Bat?" p. 166.

98 *The Case for Animal Rights*, pp. 152–3, 243.

99 Immanuel Kant, *Critique of Practical Reason*, pp. 128–9 (Ak. 122–3).

100 *Critique of Practical Reason*, pp. 132–3 (Ak. 126–7).

101 *Grounding for the Metaphysics of Morals*, p. 36 (Ak. 4:429).

102 *On Anger*, book 1, 10.1–2, *Moral Essays*, vol. 1, p. 131.

103 *On Anger*, book 1, 7.74, *Moral Essays*, vol. 1, p. 125.

104 *Grounding for the Metaphysics of Morals*, pp. 14, 37 (Ak. 4.402, 4.430). On hypothetical as opposed to categorical imperatives, see p. 25 (Ak. 4.414).

105 St. Augustine, *Confessions*, book 7, sec. 10 and 20; book 10, sec. 3 and 33.

106 *Grounding for the Metaphysics of Morals*, p. 12 (Ak. 4:399).

107 *Critique of Practical Reason*, pp. 77, 79 (Ak. 74, 76).

108 *Critique of Practical Reason*, p. 82 (Ak. 79).

109 *The Metaphysics of Morals*, "The Doctrine of Right," p. 15 (Ak. 6:221).

110 *The Metaphysics of Morals*, "The Doctrine of Virtue," p. 162 (Ak. 6:402).

111 *The Metaphysics of Morals*, "Preface" to "The Doctrine of Virtue," p. 141 (Ak. 6:376).

112 *Critique of Practical Reason*, p. 77 (Ak. 74).

113 *The Metaphysics of Morals*, "Preface" to "The Doctrine of Virtue," p. 142 (Ak. 6:377).

114 *The Metaphysics of Morals*, "The Doctrine of Virtue," p. 151 (Ak. 6:387).

115 *Grounding for the Metaphysics of Morals*, p. 13 (Ak. 401).

116 *The Metaphysics of Morals*, "The Doctrine of Virtue," p. 160 (Ak. 6:399–400).

117 Immanuel Kant, "Perpetual Peace," p. 109; Immanuel Kant, "Conjectural Beginning of Human History," p. 173.

118 Immanuel Kant, "What is Orientation in Thinking?," p. 243n (italics in original).

119 *Lectures on Ethics*, pp. 4–5 (Ak. 27:5, 27:8).

120 *Lectures on Ethics*, p. 267 (Ak. 27:500).

121 *Lectures on Ethics*, p. 71 (Ak. 27:1428); see also p. 88 (Ak. 27:296).

122 John Dewey, "Moral Theory and Practice," p. 197.

123 Rainer Forst, *Das Recht auf Rechtfertigung*, p. 123.

124 Bernard Williams, "Internal and External Reasons," p. 111.

125 Thomas Nagel, *The Possibility of Altruism*, p. 31.

126 See T.M. Scanlon, *Being Realistic About Reasons*, p. 2.

127 *The Possibility of Altruism*, pp. 14, 18; see also p. 97.

128 *The Possibility of Altruism*, pp. 47, 60–1.

129 *The Possibility of Altruism*, p. 101.
130 Saint Thomas Aquinas, *Summa Contra Gentiles*, vol. 3, Ch. 92, Saint Thomas Aquinas, *Basic Writings of Saint Thomas Aquinas*, vol. 2:904.
131 Kant criticizes the Stoics for having "refused to accept the second component of the highest good, i.e., happiness." *Critique of Practical Reason*, p. 133 (Ak. 127). Here Kant also criticizes the Epicureans for having "raised a wholly false principle of morality, i.e., that of happiness, into the supreme one." As I noted above, Kant considers the highest good to consist in virtue together with happiness in proportion to one's worthiness to be happy, where happiness is never the motivation for moral conduct but rather is a reward for having acted morally.
132 *Lectures on Ethics*, p. 217 (Ak. 27:465).
133 See for example Justus Lipsius, *De constantia*.
134 Justus Lipsius, *On Constancy*, book 1, Ch. 4, p. 37.
135 Friedrich Nietzsche, "Twilight of the Idols," p. 485. The problem I address in the remainder of this book is how to defend what I have called an "immanent" critique of reason, in sharp contrast with the "total" critique of reason recommended by Nietzsche. The task is to understand how reason is engaged in a dialectical relationship with states of embodiment, rather than to ascribe complete primacy to one over the other.
136 Friedrich Nietzsche, "On Truth and Lies in a Nonmoral Sense," p. 79.
137 *The World as Will and Representation*, vol. 1, book 1, sec. 7, p. 30.
138 *The World as Will and Representation*, vol. 1, book 1, sec. 7, pp. 34–5.
139 *The World as Will and Representation*, vol. 1, book 1, sec. 16, p. 84. Here Schopenhauer notes that animals lacking eyes can attain knowledge through the sense of touch.
140 *Anthropocentrism and Its Discontents*, pp. 188–9.
141 René Descartes, *Optics*, Discourse 6, *The Philosophical Writings of Descartes*, vol. 1, pp. 170–3; Berkeley, "The Principles of Human Knowledge," Part 1, sec. 43–44, pp. 98–9.
142 *On Abstinence from Killing Animals*, 3.7.2, p. 84.
143 *On Abstinence from Killing Animals*, 3.9.4, 3.3.4, pp. 86, 81.
144 *On Abstinence from Killing Animals*, 3.12.3, 2.3–4, pp. 87, 55.

3

HISTORICAL IDEALISM AND THE PROCESS OF CRITICAL REFLECTION

> Man needs a new revelation, and this can only come from historical reason.
>
> Ortega y Gasset, *Historical Reason*

1. Rationality: Rethink or Reject?

The traditional ideal of reason that I have examined in the first two chapters is one according to which the ability for abstract reflection counts as "highest" among the various experiential faculties exhibited by sentient creatures (reason, understanding, will, emotion, sensation, etc.), and according to which this capacity is the exclusive possession of human beings. Numerous thinkers in the tradition have presented a view of reason as the central unifying faculty of human experience—unifying in the sense that it coordinates and exercises control over all our other experiential faculties. Aristotle's elevation of *proairesis* (rational choice), Seneca's characterization of timeless things as higher than temporal ones, Descartes's call to invoke reason as the will's "proper weapons" to combat our passions, Kant's categorization of nonhuman animals as "things," and the Epicurean-Rawlsian claim that nothing we do to nonhuman animals can possibly be considered an injustice—each one of these commitments relies explicitly on the notion of a cosmic hierarchy that places human beings above all nonhuman earthly beings. And it is the very faculty of reason that, on the view of each of these thinkers, *discloses to us* that we are the highest sublunary beings and that reason is the "highest" experiential faculty.

So far I have lodged a challenge to this ideal of reason on the grounds that we human beings, in virtue of our finitude, fundamentally lack access to the sort of cosmic standpoint from which one might reasonably conclude that reason is the "highest" experiential faculty. In challenging the traditional anthropocentric ideal, I am challenging not only the claim of human superiority but also the claim that

DOI: 10.4324/9781003425595-4

reason, at least when it is operating optimally, gives us access to the undistorted truth about the world around us. We did not need to wait for Darwin to inform us that that world is in important respects mysterious and threatening to us. What we do seem to have needed to wait for was an effort on the part of a number of twentieth century thinkers to analyze the nature of human experience so as to call seriously into question the notion that the world is simply "out there," waiting to be discovered. The result of this effort has been to divert the focus of reflections on reason away from questions such as "what *is* the world?" and toward the question "how does the world *get disclosed* to us?" The shift from "what is out there, beyond us?" to "how do things show up in our experience" is a decisive one, a shift brought about through a series of steps in the history of Western philosophy culminating in Kant's transcendental turn.

What I would like to do in this chapter is trace out the motivations, central commitments, and conclusions of some influential twentieth-century philosophers, with an eye toward revisiting the questions of what reason actually is, how it functions, what it can tell us, and whether we might ultimately do best to dethrone it from its traditional place of pride in our account of experiential faculties. Some of these thinkers seek to defend reason, while others undertake a total critique of reason and appeal to human discursive practices (what Wittgenstein and some other thinkers call "forms of life") as the locus of truth. Where thinkers such as Ortega, Santayana, and the important but obscure philosopher John William Miller seek to defend reason by situating its operation within the course of time, comparatively pragmatist-minded thinkers such as William James and Richard Rorty shift the emphasis away from reason altogether and focus instead on a conception of truth as, in James's words, whatever is "profitable to our lives."[1] The central conclusion for which I will argue in this chapter is that thinkers of the latter variety, who undertake a total critique of reason, fail to recognize the at least limited autonomy and vital role played by reason in moral judgment; these thinkers ultimately present us with a picture of human choice according to which contingency plays too much of a role and reason too little in the process of human self-understanding and critical judgment.

The real puzzle of human experience in this connection is how to do justice to the ways in which reality "resists" our efforts to know it while not sacrificing altogether the capacity of reason to unify our experience and provide us with a sense of orientation in the midst of the endless contingency that surrounds us. The tradition sought to resolve this difficulty by suggesting that reason can identify a realm of necessity lying behind the flux of contingency, as when the thinkers in the scientific revolution (Copernicus, Tycho, Kepler, Galileo, Descartes, and Newton) developed a view according to which contingency in nature is merely apparent and in fact is governed by universal, necessary, rationally knowable laws of nature. Descartes, for example, maintained that the laws of nature are innate, that they are essentially identical with the laws of mechanics, and that they hold for all possible worlds.[2] It was in this manner that Descartes sought to overcome the threat posed by the "obscurity and confusion" of unreflective experience through an appeal to "clear and distinct" insight.

Quite a bit of ink has been spilled over the question of just how much Cartesian certainty we can actually attain. Cut as it is to the measure of truth conceived *sub specie aeternitatis* (from an absolute standpoint outside of time), Cartesian certainty appears to be possible only with regard to things untouched by the destructive power of time. This would seem to leave only fields such as formal logic and mathematics as areas in which we might arrive at stable, enduring, necessary truth. Does this mean that everything else that passes for truth is ultimately an invention that might just as well have been invented some other way? I.e., is "truth" beyond the realm of mathematics and logic simply a contingent matter? And if it is simply contingent, does that mean that it exercises no real authority over us? Or does it mean instead that we need to rethink at a fundamental level what notions such as truth, reason, and authority can mean in our lives?

Near the end of Chapter 2 I offered some remarks about Nietzsche's parable "How the True World Finally Became a Fable." There Nietzsche succinctly diagnoses the fateful crossroads at which humanity finds itself once it has dispossessed itself of the historical fantasy of human access to a "higher" world of eternal, necessary truth and goodness. This parable complements Nietzsche's declaration of the death of God, which signifies, as Heidegger points out, "the impotence not only of the Christian God but of every transcendent [*übersinnlichen*] element under which men might want to shelter themselves."[3] Like it or not, we find ourselves in a flux of historical contingency, with seemingly no means of escape to a higher, transworldly, "übersinnlichen" place of refuge from the obscurity, confusion, strife, and uncertainty that characterize the world that is in fact our home.

The central thrust of Nietzsche's parable is to debunk the pretension of traditional metaphysical thinkers to arrive at a detached, neutral account of reality. Thinkers who saw reality to coincide with permanence rather than with change, with necessity rather than sheer contingency, sought to provide an objective, timeless account of the way things are, quite independent of the role played by human experience. These thinkers offer a view according to which, of all our experiential faculties, it is reason and reason alone that can elevate us above the realm of contingency and disclose for us an insight into the eternal. Recall that ancient thinkers such as Aristotle and Seneca take such a position, asserting an affinity between human rational capacity and the divine standpoint. A key feature of that supposed affinity is our ability, at least in abstract thought, to detach ourselves from practical purposes and commonsense background assumptions about the concrete material reality into which we have been cast.

The crux of Nietzsche's critique of the metaphysical impulse is that such a "higher," "pure" world of permanence—a world that would promise escape from our finitude and the incessant threats posed by the world of contingency—is an utter fiction, one whose invention reflects human frailty and our inability to accept the ineluctable terms of existence: uncertainty, suffering, and death. Nietzsche, for his own part, saw the tradition's elevation of reason to a place of pride among human experiential capacities as a pyrrhic move of self-assertion in the face of the regrettable facts with which reality confronts us. It was on this basis that Nietzsche "demoted" reason by characterizing it as just one more bodily drive rather than as an autonomous faculty that might regulate our drives and instincts. On Nietzsche's view, our rational impulses (recall here Seneca's use of the term 'impulse'), like all drives, serve the will to power. Thus Nietzsche concludes that the entire account of freedom presented by Kant amounts to "moral Tartuffery," and that freedom correctly understood (assuming we can still speak meaningfully of "correctness") is really nothing other than "the affect of superiority in relation to him who must obey."[4]

Nietzsche's critique of traditional philosophical commitments has been enormously influential, and we are still working through both the terms and the implications of his critique. There are profound implications for the notion of community that has, at least until Nietzsche, been essential to our understanding of morality: Nietzsche drew from his critique the conclusion that notions such as democracy and equality are pernicious to the will to power—that, for example, the classical liberal (Lockean, Rawlsean) ideals of equal political empowerment and mutual recognition are predicated on a false picture of human beings as essentially equal, when in fact there prevail all sorts of "natural" *in*equalities to which we ought to do justice through our valuing and willing.[5] In arguing for such a conception of the human—indeed, in arguing for an entire conception of "value" as tied ineluctably to the "natural" aim of enhancing power—Nietzsche in effect proceeds on the basis of a background ideal of living that I did not examine in Chapter 1: an ideal according to which human beings are not divine but instead are "mundane" like other beings, and according to which the only appropriate measure of whether the human species has advanced is the extent to which certain individual exemplars of humanity exhibit advancement.[6]

To this extent, Nietzsche embraces a form of naturalism according to which human beings occur in nature along with other beings, and according to which we have no access to any "higher" truths that might distinguish us from other beings and elevate us above the flux of contingency.[7] It is on this basis that Nietzsche advocates a "gay" [*fröhliche*: glad, cheerful] science, one that acknowledges and celebrates contingency rather than purporting to unlock the timeless secrets that traditional metaphysics and modern natural science had sought to ascertain behind the veil of appearances. It is on this same basis that Richard Rorty concludes that Nietzsche did not so much overcome the tradition as simply invert the old relationship between reason and passion. Rorty shares with Heidegger the view that what Nietzsche offers us is an "inverted Platonism—the romantic attempt to exalt the flesh over the spirit, the heart over the head, a mythical faculty called the 'will' over an equally mythical one called 'reason'."[8] What does Rorty offer us in place of Nietzsche's "mythic" naturalism? That is an important question to which I will return later in this chapter. Here I will simply offer a preview by noting that

Rorty takes an approach that eschews appeals to "faculties" altogether and seeks to ground all human choice in language (discursive practices). In doing so, I will suggest, Rorty undertakes a simplistic inversion of his own: one that reverses the necessity-contingency relationship so as to ascribe absolute primacy to contingency. The result, I will argue, is that Rorty finds himself in the untenable position of appealing to principles of liberal democratic empowerment as if they were clearly the "best" or "true" principles for organizing human life, at the same time that he argues for the sheer contingency of all truth claims.[9]

I argued in *Animals and the Limits of Postmodernism* that this sort of contradiction is inevitable for any philosophical position that expresses deep concern for matters of injustice while at the same time ascribing absolute primacy to contingency or "the irreducible multiplicity" of *différance*. How are we to get from the proposition that what we call "reality" is essentially an infinite, irreducible play of X (where 'X' is typically characterized in terms of language, *différance*, metaphor, the trace, etc.) that in principle cannot be grasped rationally, to *authoritative* assertions about what constitutes justice and injustice? In *Animals and the Limits of Postmodernism*, I focused on Derrida and some thinkers writing under his inspiration and argued that contemporary animal studies exhibits many instances of this tragic limitation. In the absence of criteria, in the absence of any sort of authoritative rational abstractions (which among other things permit us to compare and assess different situations), *on what basis* are we to conclude that a given practice or set of practices is unjust, for example, that our treatment of nonhuman animals amounts to what Charles Patterson (echoing Isaac Bashevis Singer) called an "eternal Treblinka"?[10]

This notion of a basis for moral commitment has preoccupied me for many years. I concluded my doctoral dissertation (in which I examined Descartes's and Heidegger's respective views about the basis of morality) with a concern about what political philosophers call "decisionism," which is essentially the problem of arbitrariness in judgment. Any thinker who undertakes a total (as opposed to merely immanent) critique of reason is beset at the most fundamental level with this problem. Remember: undertaking a total critique of reason does not mean that reason has no place whatsoever in a philosopher's account of human action; what it does mean is that that thinker denies any special authority to reason in arriving at questions of truth and value. On what basis, then, are we to make determinations and choices that are more than merely arbitrary?

Any thinker concerned with matters of truth and justice has to provide an answer to this question. Thinkers such as Rorty and William James seek to answer it without having recourse to reason as it was conceived by the tradition; others such as Ortega and Miller seek to answer it by developing a new conception of reason as inscribed firmly within time. Both approaches are beset with epic difficulties, although in the end I will argue for the latter approach, which Ortega aptly refers to as "historical reason." In this chapter I will provide an overview of each of these approaches, explain why I favor the appeal to historical reason, and provide a preview of the way in which I think Ortega's approach needs to be supplemented in order to address what I take to be a signal limitation in it. In Chapter 4 I will argue that our embodied, affective constitution is essential for providing reason with the *orientation* needed to ground and guide moral judgment. In doing so, I will examine historical suspicions about the influence of emotion on moral judgment and argue that only a dialectical conception of the relationship between reason and emotion can do justice to the nature of moral judgment.

First, however, an additional remark about the charge of decisionism. This term is commonly associated with the German legal theorist Carl Schmitt, who advocated an absolutist model of political authority and argued that the sovereign must be free to make on-the-spot decisions at every instant, without having to respect any rules or principles that stand above the sovereign. Schmitt's reasoning was Hobbesian at its core: The world is full of extreme danger, each nation must view all other nations as enemies, and the only way for the sovereign to act so as to afford maximum protection to the citizenry is to be the absolute authority, answerable to no one. Schmitt sees each and every moment of time, viewed from the standpoint of the sovereign, as a "state of exception," which is to say that each individual moment is unique and incomparable to other moments. This in turn means that general rules for conduct are of no use to the sovereign, that "the exception . . . [is] more important . . . than the rule."[11] With this rejection of general rules as a basis for both the sovereign's actions and the citizenry's assessment of those actions comes an appeal to sheer contingency: "Looked at normatively, the decision emanates from nothingness [*aus einem Nichts*]."[12]

One might well ask here, what has this rarefied conception of political authority got to do with the nature of individual moral judgment? There is in fact a very close parallel between Schmitt's "state of exception" and the situation faced by anyone who embraces a total critique of reason: In both cases, there seem to be no general criteria or considerations on the basis of which to choose; in effect, one simply chooses, and once one has done so there can be no *rational assessment* of the whys and wherefores motivating the choice that might call the individual's decision into question. To this extent, the individual is not accountable to others, at least not in terms that can be shared, debated, discussed, revised, owned, or disowned in the space of public reason. In *Animals and the Limits of Postmodernism*, I argued that this picture of moral choice is virtually indistinguishable from Kierkegaard's Religiousness B, a state of being in which the individual has reached the highest form of human existence by having established a private relationship with God the terms of which are in principle uncommunicable to others in the community.[13] Kierkegaard's most vivid example of an individual who has reached faith in this sense is Abraham, whose "singular" (unique) relation to God accounts for his readiness to sacrifice his son Isaac on Moriah. Kierkegaard characterizes Abraham as being unable to communicate the meaning or motivation of his actions to others in his community, on the grounds that religiousness lies above language and the "universal" sphere within which human individuals can account for their actions and be held accountable to others.

It is not only Schmitt who has been (rightfully) charged with embracing a position with decisionistic implications. Not long after Schmitt advocated political decisionism, Heidegger conflated the personal and the political in *Being and Time* and argued for a conception of individual choice modeled explicitly on the Kierkegaardian "Augenblick" (moment of choice or decision), a conception whose pointedly decisionistic cast has been detailed by Ernst Tugendhat.[14] Division I of Heidegger's *Being and Time* has been enormously influential in shifting philosophical thinking away from a naïve conception of the human being as possessing some sort of "essence" on the basis of which it can enter the field of action, and toward a conception of the human as a process of interpretation. That process occurs in and through time, and in each case it is guided by an already-articulated understanding of being. That understanding is rooted in concrete historical situations and events that provide the individual interpreter with a sense of what has happened, what is important, what the world is like, how "one" behaves in given circumstances, and, most importantly, *what needs to be done* under the current circumstances. In terms that have become widely used in recent decades, Heidegger presents us with an account of the human individual as "socially constructed," i.e., as having become this particular individual through the sheer accident of having been born here and now. The contingency of this "here and now" affords me with a finite range of possibilities, and as a being oriented primarily on the future I have to make a choice as to how to live my life. Like Sartre after him, who argues that there is no context-free basis for making choices (should the young man stay home and care for his mother, or should he leave for England to fight with the Free French Forces? Sartre says only the young man can decide), Heidegger argues that the moment of choice, the "Augenblick" or "moment of vision" in which the individual chooses "authentically," is simply a pure, unmediated seeing accessible only to the individual in question and inexplicable to others—just like Abraham's choice.[15]

The troubling conclusion for which I argued at the end of my dissertation was that an account such as Heidegger's could be marshaled in the service of defending an individual as an "authentic Nazi." In the absence of *criteria* or *standards* for the assessment of conduct as just or unjust, justified or unjustified, etc., it would seem that any course of action could be seized upon as the "right" one. In particular, there seems to be no obstacle to elevating instrumental reason over practical reason (moral judgment), or even to collapsing the distinction between the two altogether. It is one thing to say that there can be no knock-down proof that a given choice is called for in a given set of circumstances; it is another thing altogether to move from that insight to the position that practical reason can play no authoritative role in the determination of conduct. Sartre makes this move when he suggests that there can be no cognitive certainty regarding choices with import for one's community: "What proof is there that that I have been appointed to impose my choice and my conception of man on humanity? I'll never find any proof or sign to convince me of that." On Sartre's view, in spite of the fact that there is no human "essence" and hence no timeless rules or criteria for human choice, when we do choose, our choice affects our community; this fact confers a burden of responsibility on each of us to choose in such a way that we serve as an exemplar of humanity. For Sartre, our choice is not arbitrary but has its basis in "a universal human condition," namely, "the necessity for [each human individual] to exist in the world, to be at work there, to be there in the midst of other people, and to be mortal there."[16]

If there is a thinker in the mid-twentieth century who could be expected to challenge the idea of an authentic Nazi, it would be Sartre. He did so not through an appeal to general criteria or moral rules, but through an appeal to the "universal human condition" or "situation." But simply noting that to be human is to be involved in a life-and-death struggle for survival, prosperity, recognition, dignity, and the like is not nearly enough to counter the charge of arbitrariness that Sartre explicitly rejects.[17] There are a great many ways to construe "the good life" if one is subject only to generalizations of this sort.[18] If one asks exactly how Sartre gets from the insight that self and world have no inherent meaning to the suggestion that when I choose, I take responsibility for all of humanity, the answer can only be that he, like every thinker who sketches a realm of value, is implicitly appealing to an entire background ideal of living that underwrites both the notion of responsibility generally and that of individual choice. Sartre's ideal is that of a universal human community, not a subset of humanity that arrogates to itself a superior status and the prerogative to use and demean other segments of humanity as if they were less-than-human. Sartre starts with Heidegger's notion of the socially constructed individual as "the null basis of a nullity" (i.e., as having no essential content but deriving its content from the accidents of its birth) and manages to arrive at a position that is considerably more value-laden than it might appear at first blush.

But are the values with which Sartre's—or anyone's—viewpoint is laden really only a matter of the free choice of the individual, or do they constitute a crucial historical *inheritance* that affords us at least part of the basis for making choices? Sartre acknowledges that one's situation—what Heidegger called one's "thrown-ness"—affords one with all the possibilities from which one can choose in ordering and directing one's life project. What accounts for the very different directions in which Heidegger and Sartre proceed is the difference in background ideals of living embraced by the two, Heidegger embracing one that ascribed pride of place to the Germans as leaders in world history and Sartre embracing another that stressed a considerably more encompassing ideal of human dignity.[19]

It would not be an exaggeration to say that for some time there has been a standoff in the philosophical community on the question whether a total critique of reason is called for. Those who believe that it *is* called for tend to appeal to the irreducible multiplicity of individual phenomena in the field of experience and argue that neither reason nor any other human faculty can get "outside" that field of experience and attain a neutral standpoint that could afford us any enduring truth about ourselves or the world. Some, like Heidegger and Sartre, retain a conception of vision according to which we can establish some sort of distance from our experience—not enough to arrive at anything like Kantian "intellectual intuition," but enough to provide the individual with an authoritative basis for choice. Yet others move in a pragmatist or neo-pragmatist direction and argue for an ideal of human community. In the next section of this chapter, I will examine Richard Rorty's effort to redirect philosophical attention away from confident proclamations of the power of reason and toward an ideal of solidarity that is based on the sheer contingency of the human condition.

Thinkers who believe that a total critique of reason is not called for—either because it fails to respect the power of reason, because it terminates in an insurmountable nihilism, or because it entails both—tend to fall into one of two categories: those who seek to defend a traditional, very strong conception of reason as affording us enduring truth (Kant and many of his followers immediately come to mind) and those who recognize some key insights into the limits of the traditional conception of reason and seek to revise our very idea of reason so as to preserve its significance for human life while avoiding the mistakes of the tradition. In the concluding sections of this chapter, I will examine several interrelated such approaches to the rehabilitation of reason; there I will focus on Ortega's ideal of "historical" reason and John William Miller's "actualism," both attempts to inscribe reason within the larger context of human action. My aim by the end of this chapter is to offer a qualified defense of this turn to actualism and historical reason.

2. Rorty's Challenge to Reason and Criteria

For contemporary thinkers on both sides of the debate over the terms of our critique of reason (total versus immanent), the central concern is *action*. The thinkers I have named previously seek to move away from an "Eleatic" conception of reason, oriented as it is on permanence rather than change, and toward a conception of reason as always emerging from and being directed at the field of phenomena that we encounter in our experience. Ortega characterizes this form of reason as "Eleatic" inasmuch as it follows the lead of the pre-Socratic philosopher Parmenides of Elea, who considered reason to be a "pure" faculty in the sense that it could identify entities and relationships "governed by the rules of thought," thereby establishing the reality of those entities and relationships.[20] The Eleatic view presupposes a clear correspondence between the rules of thought and the structure of reality, as when Aristotle assumes that *nous* is eternal and that the teleological cast of the human mind corresponds to the structure of nature, and Descartes suggests that material reality is structured in accordance with clear and distinct ideas and that by viewing nature in these terms we can render ourselves "the masters and possessors of nature."

The attraction of this Eleatic conception of reason is clear: it promises to neutralize the resistance and opacity offered to unreflective consciousness by affording us something approaching the "intellectual intuition" that Kant associates with the

divine mind. Thus it is not surprising when Ortega characterizes the world putatively disclosed by Eleatic reason as "a divine-seeming world."[21] For even if it does
not purport to give us the kind of access to reality that a divine being would have, it
is "divine-seeming" in purporting to take the human mind beyond its own purview
and disclose something *essential* about the reality with which we find ourselves in
contention. The problem with this conception of reason, Ortega notes, is that reality exhibits a conspicuous resistance to our attempts to reduce it to anything like
Cartesian clear and distinct insight. Thus, Ortega proposes, "the true task begins
when thinking attempts to adapt logic, which is *intelligence*, to the illogical thing
that *reality* is."[22] In the face of attempts to cling to Eleatic reason as well as efforts
to dispense with reason altogether (or simply dethrone it from its prior place of
pride among the human faculties), Ortega recommends a middle course: one that
retains faith in reason while dispensing with the pretension to have arrived at an
account of reason as eternal. The key to Ortega's insight into the power of reason is his recognition that *reality does not share the logical structure of thought*
(a notion that would have been inconceivable to Aristotle) but instead is inherently
"illogical"—which does not mean that it cannot be understood, but simply that we
cannot understand it definitively or in its entirety.

Ortega's is a path that Rorty explicitly refuses to take. Consider statements of
the following kind:

> Philosophy, as a discipline, makes itself ridiculous when it steps forward at
> such junctures [e.g., when someone asks, "how do you know that?"] and says
> that it will find neutral ground on which to adjudicate the issue. It is not as if
> the philosophers had succeeded in finding some neutral ground on which to
> stand. . . . There are as many ways of breaking the standoff as there are topics
> of conversation.

Rather than a "fixed order of discussion," what we are faced with is an "an indefi-
nite plurality of standpoints" that admit of no resolution through appeals to reason, transcendence, or anything else.[23] Rorty proposes that we "drop the idea of
[philosophical] foundations" altogether, acknowledge Wittgenstein's insight that
"vocabularies . . . are human creations," and open ourselves to "a poeticized culture" that "would not insist we find the real wall behind the painted ones, the
real touchstones of truth as opposed to touchstones which are merely cultural arti-
facts."[24] For thinkers of this persuasion there is no "outside." In this respect, Rorty
shares something crucial in common with Derrida's conviction that "there is nothing outside the text."[25] Nothing, in other words, beyond the shadow-play on the
walls of Plato's cave.

A viewpoint such as this bears the imprint of the young Nietzsche's conception
of language: that what we call "truths" are nothing more than metaphors that have
come to become "fixed" in the sense of having gained wide acknowledgment *as*
true.[26] For the young as well as the mature Nietzsche, life is what Nietzsche would

later call *will to power*: all actions and relations are directed toward the enhance-
ment of a being's power, not simply toward its preservation. Here the meaning
of truth becomes completely untethered from its Eleatic moorings in timeless
essences, and becomes reduced in effect to "a moveable host of metaphors, meton-
ymies, and anthropomorphisms" subject to manipulation in ongoing struggles for
empowerment.[27]

I noted a moment ago that on Rorty's view, what "philosophy" has to offer
humanity is ultimately nothing more than the empty promise of "neutral ground"
for the resolution of disputes. This certainly holds for the strict Eleatic conception
of reason. What is less clear is whether, and to what extent, this criticism holds for
relatively recent attempts such as Ortega's to rehabilitate reason instead of jettison-
ing it altogether. Husserl's own view on this matter is instructive: Even though we
now know that "the rationalism of the eighteenth century . . . was *naïve*," we ought
to stop and ask ourselves whether this should lead us to abandon reason altogether
in favor of "irrationalism," or whether instead such an impulse is a product of
"the rationality of lazy reason."[28] If Husserl and Ortega are right to defend the
potential of reason in spite of the Eleatic misconception that has prevailed for
millennia—and I believe they are—then we need to evaluate very carefully Rorty's
implicit assumption that philosophical reason is all one thing and that the only
viable alternative is to jettison reason altogether in favor of linguistic forms of life
such as poetry.[29]

There is a reductive sort of binary thinking at work here: Either reason as traditionally conceived is authoritative without qualification, or reason is *in principle*
just another product of language and its groundless vicissitudes. Given the various
creative ways violence and cruelty have been perpetrated in the name of "reason,"
it is not difficult to understand why so many thinkers have embraced a total rather
than a merely immanent critique.[30] This concern for the use of reason as a weapon
for inflicting violence and oppression is genuine and should not be swept aside in
the name of some "cloud-cuckoo land" ideal of neutrality and rational purity.[31] So
once again: does this mean that we should reject reason, or that we should seek to
revise it? In this connection, it is important to note that when I say that thinkers
such as Nietzsche or Rorty "reject" reason, I do not mean that they make no place
for it in their accounts of human agency; instead I mean that they deny it the authority traditionally ascribed to it and treat it as essentially comparable to other forms
of life or language games.

Rorty does not so much provide an argument for his view as sketch its general
parameters. A further reflection on the nature of language sheds some light on his
underlying motivation. Permit me to offer a brief just-so story about the evolution
of the philosophical tradition's conception of language. Aristotle, as I noted a short
while ago, presents a view according to which thought (mediated by language) and
external reality share a common structure; thus on Aristotle's view, achieving a
correspondence between thinking and being is in principle unproblematic, at least
for a being with a rational soul. The essence of language in Christian tradition can

be seen in the opening verse of the Book of John: "In the beginning was the word [*logos*]." God, in other words, thinks or utters the world into existence. This establishes the measure for truth and value, so that to speak truly is to repeat after God (or, to adhere to the divine dictates). Heidegger offers a succinct analysis of this conception of truth: Reality finds its measure in the creative intellect of God (*adequatio rei ad intellectum*: the correspondence of things to the divine mind), and subsequently human thought finds its measure in the reality of things created by God (*adequatio intellectus ad rem*).[32] By the eighteenth century, when doubts about the Eleatic conception of reason had begun to manifest themselves, some thinkers began to question the thesis of the divine origin of language and sought alternative accounts, Condillac and Herder proposing that language originated in the imitation of natural sounds.[33] Condillac and Herder were at pains to acknowledge a natural continuity between human and nonhuman animals, in effect de-divinizing human beings.

The decisive step in this evolution was to dispense once and for all with the notion that language is simply a tool for the communication of ideas that human beings first formulate without the use of language, and to invoke in its place a conception of language as an ultimately mysterious force or phenomenon that in principle exceeds our grasp. This view of language has steadily gained acceptance over the past century, inspired by the young Nietzsche and facilitated by the writings of thinkers such as Wittgenstein, Heidegger, and Derrida. I have examined Derrida's views about language and agency at length in *Animals and the Limits of Postmodernism*, so here I will simply refer the reader to my discussion there of Derrida's notions of the "trace" and "the play of *différance*," as well as his conclusion that meaning is ultimately "undecidable."[34] Derrida himself was influenced by Heidegger, who suggested that "language is the house of being" and that "the human being is the shepherd of being."[35] For both the early and the later Heidegger, being "is the *transcendens* pure and simple."[36] To say that language is the house of being is to say that it is the medium through which being comes to presence for humans; to characterize the human being as the shepherd of being is to assign a fundamental ethical obligation to human beings to steward this process of coming-to-presence. This conception of language and our relationship to reality is a far cry from the Eleatic notion that reason provides us with more or less unfettered access to reality; it acknowledges human finitude by dispensing with the old idea that language is simply a "tool" that enables us to know reality "in itself."

A key inspiration behind the move to this most recent conception of language was Saussure's characterization of linguistic signs as a closed system in which signs refer not to anything in the world but rather to each other. On this view, to speak, say, of a writing pen is not to establish a connection between a signifier and a physical object outside the realm of discourse. It is to refer to a signified concept that itself derives its meaning from its relationship to other units of meaning within the realm of discourse: for a pen to be a pen, there have to be other units of meaning such as paper, ink, the institution of writing, etc. Even to speak of an "external object" onto which we project the meaning 'pen' is to presuppose exactly what the critique of Eleatic reason denies: It is not the case that I first survey the entities in my environment and only subsequently attribute meaning to them; instead, as even the defenders of historical reason recognize, I first find myself in an active encounter with an entire field of meaningful interrelationships and only subsequently employ rational scrutiny to isolate individual entities, relationships, properties, etc. To propose, in line with the tradition's adherence to a correspondence theory of truth (i.e., correspondence between proposition or representation and external object), that my utterance 'this pen' refers to a physical object outside the realm of discourse is, on the view of those who (like Rorty) ascribe a strong priority to language, to express a basic misunderstanding about the way language functions. It is to fail to recognize that even terms such as 'physical object' are products of human linguistic activity rather than referring to objectively given entities that lie beyond human experience and linguistic activity.[37] Readers familiar with Heidegger's analysis of "equipmentality" in *Being and Time* will recognize the influence of this approach to language.

A central implication of this approach is that rather than being masters of language, we ourselves are mastered by it. Many exponents of this conception of language argue that it is incompatible with the traditional ideal of individual autonomy, and for good reason: The view of rational autonomy advanced by Kant seems impossible to square with the proposition that what we call "rational processes" occur within a medium that we neither fully understand nor have any hope of ever mastering. From here there is a range of options: one can capitulate to nihilistic hopelessness, one can push back on the primacy-of-language claim (I will note a number of thinkers later in this chapter who do exactly this), or one can look for forms of language use other than rationality that hold the promise of enhancing our insight into ourselves and the world.

Rorty pursues the third option and invokes the power of poetic discourse to move us. His commitment to the sheer contingency of language leads him to reject not only metaphysics but the entire idea of what Kant called "the public use of reason," i.e., the capacity to use reason to communicate one's thoughts to others in the process of striving to achieve mutual understanding and attain a truly "cosmopolitan" society.[38] Rorty states that he wants "not to update either universalism or rationalism but to dissolve both and replace them with something else."[39] This dissolution demands a rethinking of the notion of truth. Where the tradition tended to conceive of truth as correspondence with external reality, Rorty states that "I do not think there are any plain moral facts out there in the world, nor any truths independent of language, nor any neutral ground on which to stand and argue that either torture or kindness are preferable to the other."[40] Standards and criteria are neither absolute nor timeless but instead (at least if one is a Rortyean "liberal ironist") are "never more than the platitudes which contextually define the terms of a final vocabulary currently in use."[41] What, then, counts as "true" discourse? On Rorty's view, "true" is "whatever view wins in a free and open encounter."

Likewise, "good" is defined as "whatever is the outcome of free discussion."[42] And that itself is a contingent matter through and through.

But at the same time, Rorty categorically rejects the charge of relativism. The reason he gives for doing so is that the charge would make sense only on the assumption that at least some "statements were *absolutely* valid."[43] Such validity would be possible only if it were the case that (some) statements could reach beyond the circle of language and establish an authoritative connection with external reality. To the extent that language is in principle incapable of establishing such a connection, the charge of relativism is not what a thinker such as Carnap would have deemed "cognitively meaningful." This response to the charge of relativism seems to me to miss the mark, for reasons that become clear when we consider the *basis* of Rorty's appeal to notions such as individual autonomy, liberal irony, and solidarity.

When philosophers lodge the charge of relativism, particularly in discussions of morality, they are expressing a concern that a particular standpoint is a product not of rational considerations that could be discussed, debated, and revised publicly, but rather a product of contingent, culturally specific considerations that do not pass the test of public reason. Imagine a society socially structured in terms of two genders in which individuals of one gender are customarily given privileged status over the other gender (a phenomenon that occurs even in societies that purport to be committed to gender equality) and another society that is deeply committed to gender parity. Which society is "right"? To provide a definitive answer to that question would require standards, criteria, or forms of justification that transcend both societies (including the respective linguistic traditions and practices of both)—which is precisely what Rorty claims is impossible. How, then, can Rorty maintain that there is no "neutral ground on which to stand and argue that either torture or kindness are preferable to the other," that there are "no plain moral facts," *and also* strongly intimate that there is no "social goal more important than avoiding cruelty"?[44]

The answer to this question is that Rorty, like all thinkers who advance ethical claims—and his claims do indeed have deep ethical import—implicitly appeals to a global background ideal of living that sets the parameters for the substantiation of beliefs and commitments. Rorty acknowledges this in *Contingency, Irony, and Solidarity*, and it would be obvious even if he hadn't done so. It is particularly clear in the book's epigram, a quotation in which Milan Kundera suggests that "what is most precious" about "modern Europe" is "its respect for the individual, for his original thought, and for his right to an inviolable private life."[45] This stress on the individual is the focal point of Rorty's entire discussion in *Contingency, Irony, and Solidarity*, and it brings with it an entire set of commitments that have emerged historically from out of the classical liberal tradition of political philosophy and that in a number of respects are difficult to distinguish from many of the commitments to liberal individualism expressed by Kant and Rawls. Notwithstanding some clear differences, Rorty shares with these thinkers the same governing commitments about the worth and dignity of the autonomous individual—and these commitments bring with them an ideal of reason that Rorty is at pains to avoid acknowledging.

In advancing these commitments about the importance of the individual, Rorty makes it clear that he is not offering any argument or justification for the proposition that liberal individualism is the "best" form of social and political life; even if, as I hope to show, Rorty's liberal ideal has its own internal logic and criteria for justification, Rorty is right to suggest that one's commitment to the liberal ideal is not itself a product of reason—or at least, not of reason alone. If we accept "Wittgenstein's insistence that vocabularies . . . are human creations," then we must dispense once and for all with "the idea that liberalism could be justified, and Nazi or Marxist enemies of liberalism refuted, by driving the latter up against an argumentative wall."[46] Here Rorty poses an important challenge to the sort of position that I am arguing for in this book, a position that expresses a strong sense of moral obligation (toward nonhuman sentient life) and calls on the entire human community to acknowledge that obligation. Although it might outwardly look as if Rorty himself is advocating such a universalistic position, in fact he is not. Not only does he explicitly reject the sort of universalism sought by Habermas, but he makes it very clear that there is no reason you can give to a person that will by itself move them to embrace any given moral commitment.[47] Indeed, Rorty states that "the distinction between morality and prudence, and the term 'moral' itself, are no longer very useful" inasmuch as 'moral' essentially means no more than a value that has arisen contingently and become ossified.[48] Rorty follows Nietzsche in dispensing with the idea of a universal human community on the grounds that the real starting point of intersubjective interaction, at least in a liberal society, is "an increasing sense of the radical diversity of private purposes."[49] On this view, each individual's moral struggle is private and "idiosyncratic."[50] If there is any accord among members of a society, it will have to rest not on reasons or arguments but on prevailing shared ways of doing things: "The idea that liberal societies are bound together by philosophical beliefs seems to me ludicrous. What binds societies together are common vocabularies and common hopes."[51]

Although Rorty is not entirely clear about it, given his appeal to language it would appear that on his view, solidarity in *all* societies would have to be based on shared vocabularies and hopes. After all, it isn't as if language operates in a Nietzschean-Wittgensteinian manner only in liberal societies; what thinkers in this tradition assert is something about the way language operates in all places and at all times—naturally an interesting accomplishment for thinkers who argue so strongly against the Eleatic conception of reason and the notion of abstract universals. I noted a moment ago that, on Rorty's view, there is no rational argument that, say, liberalism is "better" than Nazism (or Marxism). I noted shortly before that that I have long been concerned with the problem of the "authentic" Nazi. If only against his own intention, Rorty helps to shed light on what I take to be the core problem with an approach that grounds ethical commitments in whatever vocabulary and set of hopes just happen to prevail in a given society at a given time: There are good reasons to prefer liberalism to forms of political organization such as Nazism (Marxism is a more thorny and contentious matter, so I will leave it aside here), reasons that rely on exactly the kind of universalistic conception of the human being that Rorty treats as a fiction even though it was proclaimed as an ideal at least as early as Diogenes of Synope's assertion, when asked of which country he was a citizen, that he was "a citizen of the world": *kosmopolites* [cosmopolitan].[52] What will become clear in the remainder of this chapter and the next chapter is that to say that there are "good reasons" for embracing the liberal ideal does not mean that pure practical reason in the Kantian sense is capable by itself of grounding a commitment of this kind, but rather that reason performs the crucial function of reflecting on and substantiating prior commitments that first arise not cognitively but affectively, i.e., through our embodied, felt connection with others and the world around us. Some of these prior commitments merit reaffirmation and some do not; practical reason, properly understood, is the process of distinguishing between the two.

I have acknowledged that Rorty poses an important challenge to the sort of position I am developing here. The challenge is one that I noted earlier, in my discussion of Descartes's views on morality: There I suggested that reasons do indeed run out in an attempt to justify a given moral proposition, and that we ultimately have to appeal to some "extra-rational" source of meaning in order to provide a basis and guidance for our moral strivings. For Rorty, in such an endeavor I am confined to the prevailing vocabulary and hopes that I have inherited from my culture as an entirely contingent matter, and I then (at least if I am a "liberal ironist," an advocate of individual autonomy and dignity) either work within the prevailing ethos or I seek to modify it through acts of "continual redescription."[53] 'Redescription' here takes the place of rational considerations and is open-ended in the sense that it is guided by no criteria or principles that transcend the time and the place in which these acts of redescription take place. For those of us lacking in sufficient imagination to undertake this process of redescription entirely on our own, we have the benefit of "writers with very special talents, writing at just the right moment in just the right way" to provide us with "redescriptions which change our minds."[54] Moreover, at least in a liberal society, the very foundations of "we-consciousness" are "merely poetic."[55]

What is important about Rorty's challenge is that it reminds us of the difficulty of locating a source or basis for meaning and orientation somewhere outside of the arid landscape of detached rationality. Rorty's appeal to poetry recalls Heidegger's own turn to poetry as a vital source of meaning and orientation, a turn that Heidegger made in the face of what he took to be a deeply nihilistic culture. Having undertaken a total critique of reason, Heidegger saw the need to seek the prospect of redemption in poets such as Trakl and Hölderlin on the grounds that whereas "the thinker says being . . . the poet names the holy."[56] Heidegger frames the tasks of the thinker and the poet in terms of the need to find a *measure* for thinking, and he proposes that "poetizing is truer than the exploration [*Erkunden*] of beings" undertaken by rational inquiry.[57] Thus Heidegger, like Rorty after him, ascribes a power to poetry to disclose important insights to us that are denied to pure (practical) reason.[58]

One such insight, I would like to argue, is that we are part of a much larger world that tends to get ignored or at least substantially demoted in significance by those who ascribe a strong priority to language in the way that Rorty does. Language is a mysterious and pervasive phenomenon, and it is the medium in which thought occurs. Does that by itself mean that thought is inherently subject to the vicissitudes of time and accident, or is it possible for reason to gain a foothold in the sea of contingency long enough to arrive at insights that merit being retained and honored—not only beyond this moment and this place, but beyond this particular culture and its contingent vagaries? The truth is that nobody can rightfully claim to have a definitive answer to this question, since to arrive at one would require the ability to render language an object of detached rational scrutiny. After looking a little more closely at some of Rorty's central commitments about liberalism, I will return to this question and explain why I believe that the wisest approach is to retain at least limited faith in reason to enhance our sense of orientation.

Rorty sketches an ideal of liberal society according to which the individual is central. Rorty's ideal of the individual shares much in common with the classical liberal ideal advanced by Locke, Kant, and Rawls. In particular, Rorty stresses the importance of individual autonomy and "the demands of self-creation."[59] The Rortyean individual is a "liberal ironist" who recognizes that these demands never end and that we must always assess and respond to them as current circumstances seem to require; the process of self-creation is endless, which means that it cannot be reduced to a simple formula, at least not one whose hold on us could be expected to last for very long. In this respect, the life of the individual is private—and, Rorty maintains, it is ultimately incommensurable with the social goal of "human solidarity."[60]

This strict bifurcation appears to be an unavoidable consequence of Rorty's total critique of reason: Along with that critique comes a flat rejection of the idea of humanity in general, and a turn toward the sheer contingency of time and circumstance as the only possible source of meaning. Rorty's affirmation of individual autonomy is so strong that it reduces morality to a strictly private, "idiosyncratic" affair that is every bit as "valid" as our social concern for solidarity even though, on Rorty's view, the public and the private are ultimately "incommensurable."[61] When Rorty touches on the ideal of solidarity, he focuses his remarks on aims of reducing suffering and avoiding cruelty.[62] Rorty says nothing about the relationship between the public and the private beyond noting that both are "valid"; he does not consider, for example, that the same sorts of considerations that make us favor individual autonomy might also underwrite our commitment to reducing suffering and avoiding cruelty—namely, a respect for the individual *in general*, whether in one's own person or in another.

Rorty's vision is that of a society of "liberal ironists," each of whom recognizes the wholly contingent basis of selfhood and choice, affirms their own individual worth and autonomy, and expresses concern for the suffering and cruelty inflicted on other individuals. What is distinctive about the values expressed in this vision is not their pointedly liberal cast, but rather Rorty's account of the *basis* of these values: Tersely stated, there is no basis for them. They are not derived from reason. And even if they are derived from past cultural practices and values, and even if it is the case that we are forced by dint of necessity to appeal to the past in order to gain a sense of orientation that will permit us to think our way into a better future, the fact that there are no criteria or standards—no concrete *measure*—for the choice of specifically liberal ideals means that Rorty's advocacy of liberalism is ultimately arbitrary. I have already noted Rorty's assertion that liberalism cannot be justified as preferable to Nazism or Marxism. The question whether liberalism can be justified is one that I will examine (and seek to answer in the affirmative) in the remainder of this chapter. Here I will simply note the implications of Rorty's statement. It amounts to what Kant would call a "hypothetical" as opposed to a "categorical" imperative: *If* you want X, *then* do Y. Here the 'X' is the realization of a liberal society, including room for individual self-invention and the solidarity that underwrites an interest in minimizing suffering and cruelty, and 'Y' is a set of measures including each individual taking on the role of liberal ironist, dispensing with the goal of uniting public and private, etc. Had he chosen to do so, Rorty could have put the point in these terms: "I can't say whether liberalism is a good idea or not. In fact, depending on your sensibilities, you might just prefer to be a Nazi; that's entirely up to you. If you think liberalism a good idea, then do Y (be a liberal ironist, etc.); if you don't think it's a good idea, then go do something else."

An account along these lines of why an individual embraces liberalism strikes me as a sincere but nonetheless underdetermined account of what makes someone embrace liberalism. Rorty is unremitting in his insistence that terms such as 'individual' and 'solidarity' have no essential or 'deeper' meaning, but are products of sheer linguistic and historical accident. His rejection of "plain moral facts" makes it impossible for him to assert categorically that cruelty and the gratuitous infliction of suffering are an affront to the individual; he is forced to argue that this conception of concern for others is an entirely contingent commitment with no essential validity outside the circle of people who happen to share it. In effect, there is nothing we can say to a person who is not already committed to liberal values that will convince them that they should embrace liberalism—or, if it does convince them, then the act of convincing was a sheerly contingent matter and could have played out a different way.

A key question in the evaluation of Rorty's viewpoint is whether, in spite of the contingency of language and history, human beings possess a capacity for rational scrutiny that gives us the kind of critical distance from the unfolding of time that could enhance our sense of orientation within this large sea of contingency. This would be a rational power proceeding from and directed at time rather than at any

Eleatic dream of absolute permanence. A related question is whether such a rational power might help us to reestablish a connection with the natural world around us, particularly with the nonhuman animals whom we have done so much to harm and who have traditionally been excluded from full membership in the moral community. (After all, like us, they are living, sentient presences, not simply manipulable units in acts of signification.) These are my core concerns in the balance of this chapter and the next.

3. The Ideal of Critical Detachment Revisited

Where a thinker such as Rorty characterizes an "appeal to reason" as nothing more than "an appeal to a widely shared common ground by reminding people of propositions which form part of this ground"—a view that bears the clear imprint of the young Nietzsche's account of metaphor—a number of other thinkers have sought to revise our conception of reason so as to divest it of its Eleatic excesses. The insight that these thinkers share is that reason needs to be inscribed within time rather than being oriented on permanence. Their goal is to vindicate reason rather than to reject or demote it, to argue that the very operation of reason through time has demonstrated its at least limited validity. Reason, these thinkers maintain, is a vital function of an historically existing being, one that permits the recognition and cultivation of a trajectory of thought toward an ideal of living that claims "universal" validity—but where the meaning of 'universal' will have to be revised in accordance with the move from Eleatic to historical reason.

In Chapters 1 and 2 I provided an overview of the Western philosophical tradition's commitment to the proposition that human beings are unique and superior to all nonhuman animals in virtue of possessing some special quality. This is what thinkers in animal studies call the doctrine of human "exceptionalism." For the leading exponents of the tradition, the special quality in question has been *logos*, a sort of cross between reason and predicative linguistic ability. In recent generations, a number of thinkers have attempted to identify some more specific quality that serves as an index of human uniqueness. Many thinkers have seized on the capacity for self-awareness, arguing that nonhuman animals cannot possess such a capacity inasmuch as it entails abstract notions such as 'self' (in general), 'other self', 'this self', and a whole set of other abstract notions detailing the respective entitlements and obligations of various selves that (on the accounts of these thinkers) nonhuman animals "obviously" lack. Others have appealed to the ability to have "a biographical sense of self," again arguing that nonhuman animals lack this capacity.[63] The shared aim of all these thinkers appears to be to attribute a capacity to human beings in virtue of which we are entitled to view ourselves as superior to beings who lack that capacity.

Rorty for his part seizes upon the capacity to experience humiliation as unique to human beings. When he does so, he frames the idea in terms of "what unites [each of us] with the rest of the species" and states that the capacity to be humiliated is a special case of the capacity all sentient creatures have to experience pain.[64] Several things should immediately jump out at the reader here: Rorty's appeal to a capacity that unites us with all of humanity; his endeavor to identify something about human beings that renders us unique; and the fact that when he asserts a clear dividing line between human beings and nonhuman animals, he employs some surprisingly insensitive terms to refer to the latter. A brief reflection on each of these features of Rorty's assertion will pave the way toward an examination of Ortega's notion of historical reason.

Rorty's appeal to species solidarity fits together with the classical liberal ideal considerably more seamlessly than it does with the proposition that every personal or political position is ultimately a matter of sheer contingency. Again: Is it really satisfactory to suppose that the notion of a universal human community, bound together in common cause, is just another idea? Thinkers such as Ortega and Miller beg to differ; the influence of Hegel brought both to recognize that the classical liberal notion of the individual—a being equal to others in the political and moral community, possessing both autonomy and inherent moral worth—is a distinctive historical achievement, notwithstanding the fact that it emerged from a specific historically-and culturally-situated tradition. To put the point bluntly: the fact that this notion arose contingently, and might not have arisen at all had the circumstances been different, has little if anything to do with the question whether this is a notion that the human community *ought* to embrace. After all, why care about suffering at all? Why not go in the direction advocated by Nietzsche, a thinker who saw tremendous creative potential in suffering and advocated the sort of hierarchically structured society I have already detailed? Ortega and Miller give us good reasons to be devoted to an ideal of universal human equality and dignity—even though, as I will argue in Chapter 4, those reasons presuppose access to what I have called "an extra-rational source of meaning."

Equally problematic is Rorty's investment in distinguishing human beings from nonhuman animals. In my own writings I have tentatively proposed that the tradition was right to seize upon certain capacities as (probably) unique to human beings; but in each case I have stressed that these differences have no actual significance for assessments of moral worth. On the view I have been developing for several decades, arrogating a higher moral status to ourselves on the grounds that we are "smarter" (more susceptible to humiliation, capable of self-awareness, etc.) makes about as much sense as saying that smarter human beings possess a moral worth superior to that of less intelligent human beings. The only way to embrace the former (human superiority) but not the latter (smarter humans count more than less intelligent ones) is by dint of speciesism—in this case, by using different criteria in the comparison of humans and nonhuman animals than we use when comparing different human beings with one another.

If only against his own intention, Rorty exacerbates this problem by referring to nonhuman animals on several occasions as "brutes" and "beasts," arguing that instead of seeking to establish an intimate connection with nonhuman animals we ought to "try to isolate something that distinguishes human from animal pain"—namely, the capacity to be humiliated.[65] Thus it comes as no surprise when Rorty offers the verdict that, instead of seeking to establish a connection with (or even acknowledge in any robust way the existence of) the world beyond the human, we should seek to realize a culture that "would have no room for the notion that there are nonhuman forces to which human beings should be responsible." For Rorty, this move is made incumbent upon us by the effort to "de-divinize" the notion of rationality, i.e., by coming to recognize that reason does not give us anything like Eleatic "intellectual intuition" after all.[66] Taken together, these demeaning references to nonhuman animals and the seeming abdication of responsibility to anything beyond the human add up to a pointedly anthropocentric background ideal of living—which shouldn't be surprising, given the fact that the turn to the primacy of language advocated by so many contemporary thinkers was inspired as much by Kant's transcendental turn as by anything else.

Even though the advocates of historical reason express no particular concern for the nonhuman, their approach holds the promise of helping us to re-establish the connection to the world around us that was lost through the turn to Eleatic clarity and determinacy. The Cartesian-Kantian approach to which we have become inured over the past four centuries is one according to which, as I have already noted, things "exist" only to the extent that they enter the purview of our experience, and according to which the "how" of their existence is measured and limited by *our ability to grasp* that "how." In the face of this way of conceiving our relationship to reality, what I believe we ought to do is acknowledge our rootedness in nature, in physical reality, even if we cannot arrive at a definitive description of that reality. *Reality is not simply a construction of the human mind*, even though our encounter with it is mediated through language and concepts. The predicament in which we find ourselves today is one in which we are uncertain how to re-establish the connection with reality that was lost through the turn to Eleatic reason.

We are not yet at the point where we see how to re-establish this connection; as I have argued from the start, I believe that *we will need to affirm something deep about our own animality in order to make a first step in the direction of this reconciliation with reality*. This effort has been complicated by the fact that many thinkers in animal studies, rather than developing a sufficiently fine-grained analysis of both the continuities and apparent differences between human and nonhuman animals, simply invert an old opposition: Where a great many defenders of anthropocentric ideology have maintained that human beings possess sophisticated forms of intelligence and creativity but that nonhuman animals lack these capacities, too many contemporary advocates of nonhuman animals insist that nonhuman animals really do possess these capacities.[67] My own view on this matter is that while we ought to be careful not to deny capacities to nonhuman animals that they may well possess, we should also be wary of the temptation to go from one extreme to the other and attribute capacities to nonhuman animals that do indeed appear to require linguistic abilities. Given our experiential limitations, it is a tremendous challenge to know when we are attributing too much to nonhuman animals, as well as when we are attributing too little to them. In the face of millennia of having attributed too little to nonhuman animals, I consider it incumbent on us to employ a carefully deployed principle of charity in assessing the experiential capacities of nonhuman animals. How much charity? Perhaps not enough to attribute the capacity to employ predication and concepts to nonhuman animals, but certainly enough to dispense once and for all with epithets such as 'brute' and 'beast'.[68]

In recent years, philosophers and ethologists have increasingly acknowledged the difficulty involved in attempting to plumb the depths of the nonhuman animal mind, the ethologist Douglas Candland observing that we are able to get a sense of the mental lives of nonhuman animals only by means of analogy to our own experience; to this extent, "we must deny the arrogance of thinking that we are objective and devote our attention to examining our own categories and thereby the power and weakness of our human natures."[69] Consider what a far cry this approach is from Morgan's Canon, the early twentieth century's guiding principle of ethology, according to which "we should not interpret animal behavior as the outcome of higher mental processes, if it can be fairly explained as due to the operation of those which stand lower in the psychological state of development."[70] Thinkers such as Candland reject Morgan's Canon as reflecting the prejudice of human exceptionalism, and they propose in its place an ethic of humility on the part of ethological researchers. Imagine how different a book such as *Contingency, Irony, and Solidarity* (this naturally can be said about a great many books) might have been if Rorty had taken such an approach.

Notwithstanding its anthropocentric cast, Rorty's approach has the great virtue of proceeding from the notion of *prior commitments*. Like the Pragmatists before him, Rorty recognizes that we do not first encounter the world in the mode of a detached Cartesian knower and only subsequently establish a sense of meaning; instead, in every instance we find ourselves in the midst of a global set of prior convictions and commitments that were not of our choosing. By the time any individual arrives on the scene, there is already a fully established prevailing set of ways of doing things (which followers of Wittgenstein tend to call "forms of life" or "language-games") that constitutes the range of possibilities open to each individual. The individual does not establish meaning but instead inherits it and proceeds to make choices on the basis of that previously disclosed field of meaning and possibilities. These choices are what move the process of thought, self-understanding, and action forward.

Rorty's point of departure, like Ortega's, is the current state of philosophical thought about the human condition and the need they both see to depart from the traditional Eleatic dream of clarity, determinacy, and total security. A central commitment coming out of that tradition of thought, as I have noted in the previous two chapters, is that human beings possess *logos* (linguistic rationality) and that nonhuman animals are at least comparatively *aloga* (bereft of *logos*). In a variety of ways, thinkers in the tradition have attributed a capacity for at least limited cognitive detachment to human beings and denied it to nonhuman animals. Where Aristotle characterizes that capacity in terms of *logos*, modern exponents of the tradition have variously employed terms such as 'Geist' (mind or spirit) and 'Besonnenheit' (reflection, deliberateness, contemplativeness) in an attempt to capture what is unique to human beings. Key questions in the assessment of the respective views of Rorty and Ortega include whether there is any capacity truly unique to human beings (not to mention, how we could know that it is unique, given the barrier I have noted to our insight into the mind of the nonhuman animal); whether, provided there is some fundamental difference between human beings and nonhuman animals, that difference has any moral significance; and whether the possession of such a capacity actually confers special responsibilities on human beings rather than special prerogatives to treat nonhuman animals as resources.

Some consideration of influential terms such as 'Geist' and 'Besonnenheit' is valuable because it will provide a basis for assessing Ortega's claim that reason can be vindicated in spite of the tradition's Eleatic excesses, as well as Rorty's claim that the process of decision-making is not guided by any (rational) criteria. Of the two terms, 'Geist' (mind or spirit) is considerably better known due to the centrality that Hegel placed on it. Like so many thinkers before him, Hegel appeals to thought as the capacity that distinguishes human beings from nonhuman animals; it is also the capacity (as it is for Kant) in virtue of which humans, but not nonhuman animals, possess freedom.[71] This freedom facilitates the apprehension of "thought as such" (which is to say, in its essence) and the overcoming of precisely those bodily impulses that Seneca identified as threats to our spiritual integrity.[72] That Hegel considered thought and freedom, the capacities central to *Geist*, to signify clear human superiority over nonhuman animals, is evident in his mockery of the Egyptians and Hindus for "rever[ing] animals as beings higher than themselves."[73] A related term employed by anthropocentric thinkers in the nineteenth century is 'Besonnenheit', which both Herder and Schopenhauer use to attribute rationality and freedom to human beings and deny it to nonhuman animals. Herder suggests that it is in virtue of our *Besonnenheit* (our capacity for detached reflection) that we can attain "*a progressive unity of all conditions of life*," a sense of unity with both scientific and moral applications.[74] In a similar vein, Schopenhauer attributes a certain "*sober reflectiveness*" [*Besonnenheit*] to human beings and categorically denies it to nonhuman animals, suggesting that

> the absence of reason restricts the animals to representations of perception immediately present to them in time, in other words, to real objects. We, on the other hand, by virtue of knowledge in the abstract, comprehend not only the narrow and actual present, but also the whole past and future together with the wide realm of possibility.[75]

Schopenhauer captures very succinctly what Derrida called the "logocentric" orientation of traditional philosophical thinking: Human beings are unique in being linguistic and rational, which opens us to a "wide realm of possibility." The nonhuman animal, in contrast, is imprisoned in an eternal present—a putative imprisonment in virtue of which thinkers such as Peter Carruthers maintain that the pain of nonhuman animals is "unconscious" and therefore not worthy of any direct moral concern.[76] Schopenhauer for his part fully acknowledges that nonhuman animals suffer, but he ultimately considers their suffering to count less in the cosmic scheme because he shares the tradition's conviction that nonhuman animals are essentially confined to an eternal present. It would later be in this spirit that Heidegger would characterize the nonhuman animal as "world-poor" in comparison with us "world-forming" human beings: *Besonnenheit* characterizes human *Dasein* (Heidegger's term for the kind of experience that he considers unique to human beings) whereas a comparative *Benommenheit* (numbness, drowsiness) besets nonhuman animals.[77]

What all these thinkers, historical and contemporary alike, have sought to locate in human beings is some capacity to *step back* from our involvements in the world with which we have been confronted and take a critical stand on those involvements. Notwithstanding Rorty's claim to reject criteria, I believe that this holds even for his view. The liberal ideals that underwrite his position provide criteria for the adjudication of a whole array of questions such as: is it permissible for one group within society to exercise state-sanctioned dominance over another? Rorty's position on all such questions is that the term 'moral' isn't very useful anymore, and that the moral struggle of each individual is "idiosyncratic." But are such statements truly compatible with the liberal notion of the individual, a notion that Rorty inherited and (it appears to me) sees *good reason* to retain? One has to ask whether it seems plausible to suppose that there really isn't any reason to embrace liberal individualism (or liberal ironism) and that in effect people are entitled to choose just any way of life that they happen to prefer, or whether instead the process of human reflection in the course of history can be seen as a progression toward the realization of the liberal notion of the autonomous individual. Ortega, as we shall see shortly, makes a case for the latter possibility.

The path that Ortega will sketch helps us to see that the notion of universality plays a crucial role in cultivating the sense of community that informs any truly moral outlook. Even for earlier thinkers in the tradition such as Aristotle, morality is about community. What has emerged in the course of time is an *increasingly inclusive* ideal of community, one founded on a universal notion of individual human dignity affirmed several generations ago by the United Nations Universal Declaration of Human Rights. That declaration refers to "the human family" and recognizes in each human individual a basic right not to be enslaved or tortured. Is such a declaration merely the expression of some "idiosyncratic" viewpoint that just happens to be shared by a large number of people, or does it reflect a long process of *critical reflection and judgment* about what it means to be human and what each of us is owed (and owes to others)?

On the view I am proposing, the U.N. Declaration is the historical culmination of a promise held out in antiquity by Diogenes's assertion that he was *kosmopolites*, a citizen of the world. My aim in arguing for a non-anthropocentric background ideal of living is to expand the scope of the moral community so that it encompasses not simply human beings but all sentient creatures. Even if Rorty is right that nonhuman animals are incapable of experiencing humiliation, that strikes me as a rather arbitrary criterion—yes, a criterion!—for circumscribing the scope of moral concern. Rorty's position on our relationship to the nonhuman is that we have no particular obligations to ameliorate the suffering of nonhuman beings; "faced with the nonhuman," we can do no more than "*recognize* contingency and pain."[78] I find a statement of this kind deeply troubling. In essence, it abdicates any responsibility to the nonhuman world around us on the grounds that "the meanings of [human beings'] lives" are derived exclusively from "other finite, mortal, contingently existing human beings."[79] In the face of contingency and pain, much of that pain being inflicted on nonhuman animals by human beings, all Rorty thinks we can do is "recognize" it, which is to say, not actually *do* anything about it—unless you happen to *feel* like doing something about it from within your idiosyncratic inner sanctum of morality.

The appeal to universality in the context of morality is an appeal to a shared sense of mission, purposes, and mortality. Rorty is highly effective in his appeal to literary figures such as Proust and Orwell in arguing for the power of poetic discourse; the price he pays for relying on these particular writers is a narrowing of the scope of moral concern to the strictly human and the relegation of the fortunes of nonhuman animals to what in effect is an afterthought. This is not a limitation unique to Rorty; as I have shown in previous chapters, the entire Eleatic ideal of reason bears the deep imprint of anthropocentric commitments. But there is this key difference between Rorty and his Eleatic forbears: having embraced Derrida's axiom that "there is nothing outside the text" (i.e., nothing outside the context of meaning in which human beings participate), Rorty effectively reduces the notion of "reality" to a linguistic construction. But is that really all that reality is? Are we really confronted with nothing in our experience other than what shows up within Nietzsche's "moveable host of metaphors"? And given that, as Kant recognized, we are incapable of arriving at a definitive grasp of reality due to the mediation of our sensibility and conceptual apparatus (later thinkers seize upon language itself), is there anything we can actually know about the reality that transcends us?

The inward turn taken by religious and philosophical views of consciousness has a long history that predates Descartes (with his turn to the interiority of clear and distinct insight) and Kant (with his appeal to transcendental conditions for experience and knowledge). There is evidence of it in the Gospels, as well as in an early formulation of the cogito ("I think") advanced by Saint Augustine.[80] It is also evident in the Platonic turn to dialectic as the method for making contact with the forms. Descartes and Kant do not produce a new way of thinking about our relationship to reality so much as they take a previously existing Eleatic one to its logical extreme—one according to which there is a fundamental gulf between the mind and external reality that needs to be bridged.

This commitment to the proposition that self and world are first separate and stand in need of unification is wholly at odds with the conception of "a precedent community of nature" advanced by Aristotle.[81] Aristotle sees nature as a teleological whole of which the human being, the *zoon logon echon* (the living being with *logos*), is but a part. We can separate ourselves in acts of abstract reflection, but the unity or community is prior. There is a sense in which even thinkers such as Rorty, who reject Eleatic notions such as that of a (teleologically structured) reality that exists completely independently of us, presuppose the truth of this principle: we do not first find ourselves to be lone entities in need of establishing a connection with the rest of (for thinkers such as Rorty, social) reality, but instead we first show up on the scene as members of an already-functioning community informed by a prevailing global sense of meaning. One thing the inward turn accomplished was to leave us (or certain influential philosophers, at any rate) with the impression that each of us is a metaphysically separate being; another thing it accomplished was to call radically into question whether and to what extent we can establish enduring knowledge about the world around us, not to mention about ourselves.

Notwithstanding our culture's historical move away from Aristotelian teleology and toward the modern ideal of *Geist,* the notion of a "precedent community of nature" has continued to exercise a significant influence on the way philosophers have conceptualized our relationship to the world around us. In the twentieth century, for example, the Gestalt psychologists stressed the priority of perceptual wholes over discrete bits of data abstracted from the field of experience; Heidegger developed his conception of "worldhood," according to which meaning comes not in discrete units but instead as an entire, historically disclosed context of possibilities; and Merleau-Ponty sought to undermine the Eleatic separation between self and world once and for all, appealing to a notion of "consciousness incarnate." In the next chapter, I will look more closely at the contributions made by Heidegger and Merleau-Ponty to this more holistic conception of human experience. First, however, it will prove worthwhile to push just a bit further on the question whether there is anything we can know about the reality that surrounds us.

Heidegger took up this question by contrasting the views of two philosophers, one ancient and one modern, on the relationship between self and world. In the context of a discussion of the dislocating power of modern technology, Heidegger argues for a fundamental shift in the meaning of the pre-Socratic philosopher Protagoras's statement that "man is the measure of all things, of those that are, that they are, of those that are not, that they are not" from antiquity to the modern age. Having declared the need to "overcome" Descartes's bifurcation of self and world as well as the model of subject-object representation that informs it, Heidegger sets up a contrast between Descartes and Protagoras intended to demonstrate the arch subjectivism at the core of the modern notion of self.[82] Protagoras's statement "man is the measure of all things," Heidegger suggests, may outwardly appear to proclaim a form of subjectivism, but in fact this is a basic misunderstanding due to our own contemporary Cartesian way of understanding ourselves and our relationship to reality; this becomes clear when we contrast the meaning of 'self' in the two thinkers. For Descartes, "*ego*" signifies "*ego cogito*": I think, where thinking establishes the measure for what does and does not exist. Whether things exist, and how they exist, is wholly relative to the Cartesian subject's capacity to characterize those things as objects structured in terms of clear and distinct insight. Descartes is a "subjectivist" in the sense that the world becomes reduced, in effect, to a set of correlates of consciousness. This is what Kant had in mind when he characterized Descartes's philosophical position as "problematic idealism."[83]

Protagoras, Heidegger suggests, conceives of self or ego in an entirely different, non-subjectivist manner. For Protagoras, "the ego tarries within the horizon of the unconcealment that is meted out to it always as this particular unconcealment."[84] Heidegger's notion of unconcealment complements the remarks I offered near the end of Chapter 1 about Heidegger's call to redress the contemporary problem of homelessness by taking a "step back" from traditional metaphysical thinking. One of Heidegger's central preoccupations is to debunk the correspondence notion of truth that I examined earlier in this chapter by arguing that correspondence presupposes a global disclosure of meaning that is not the work of human agency and that never manifests itself in Eleatic determinacy. Heidegger appealed to the notion of unconcealment (*aletheia*) in an attempt to restore a sense of modesty to human beings in our contemplation of what we can and cannot know, and of what we do and do not control in our experience. For Heidegger, truth at its most fundamental level is not correspondence but the always incomplete, gradual emergence of beings from out of an unknown origin that exceeds us. In associating Protagoras with this notion of unconcealment, Heidegger is intimating that we moderns have lost something essential in our connection to reality that prevailed in Protagoras's time: an ability to "let beings be" and thereby maintain a sense of *belonging within* a cosmos that vastly exceeds us. Heidegger argues that the shift away from this sense of belonging was initiated not by Descartes but by Plato (with the premium he placed on the idea as mediating between us and reality) and Aristotle (who posited *theoria* or detached contemplation as the mediating factor).[85] Heidegger's appeal to Protagoras is an appeal to an ideal of direct, undistorted, unmediated contact with what emerges from beyond the scope of our experiential capacities.

Heidegger's contrast between Protagoras and Descartes begs the question whether we are—or have ever been—the kind of beings who could enjoy this kind of unmediated contact with reality. One consequence of our culture's lengthy engagement with the Eleatic notion of reason is the insight that if such contact is possible, it does not occur (at least not primarily or exclusively) through acts of detached rational abstraction. A key lesson of Kant's transcendental turn was that, as finite beings, we necessarily employ sensory and conceptual apparatus proper to our own constitution, and this employment unavoidably mediates (which is to say, confers form on) our encounter with reality. This is what led Kant to classify even space and time as pure forms of sensible intuition rather than as entities or sets of relations that exist independently of us. More recent reflections have led many thinkers in the direction of accepting Kant's position regarding our scientific knowledge of reality while resisting the notion that we have no direct, unmediated contact with reality. These moments of resistance tend to focus on the nature of our embodiment rather than on our capacity for detached contemplation, and they tend to seize upon experiences such as physical pain.[86]

Here again we encounter a prejudice of the tradition that we are still in the process of overcoming: the prejudice that our embodiment constitutes the "lower" part of our nature, in contrast with our "higher" cognitive powers. In early modernity Descartes went so far as to aver that my embodiment is not really part of my essential nature, which he took to be thinking and thinking alone. It is not difficult to grasp the motivation that underwrites such a wildly counterintuitive notion: dread of the corruption and death of the body, and a desire to see ourselves as sharing in God's immortality. Even Descartes, who is widely considered to be a categorically secular thinker, writes to the learned fathers of the Sorbonne that in the *Meditations* he will demonstrate the immortality of the soul.[87] This points to an additional motivation, one that complements my thesis about the tradition's denial of *logos* in nonhuman animals, namely, a desire to arrive at a conception of ourselves as unique and superior to everything else in nature: by maintaining that the mind is separate from embodiment and does not reside in nature at all, Descartes is able to exempt the mind or soul from the reach of his physics, which reduces everything in nature to inert matter subject to external forces. This motivation is not unique to traditional defenders of the Eleatic ideal; I have demonstrated indications of it even in thinkers such as Rorty, who recommends that we (we liberal ironists, that is, not just anyone) take human humiliation deadly seriously but that all we need to do with regard to the pain of nonhuman animals is "recognize" it. To take human pain "deadly seriously" is not simply to recognize it but to *do something about it*. That willingness to take action presupposes a sense of community, one that on my view thinkers such as Rorty draw considerably too narrowly when they treat nonhuman animals as entities (part of those "nonhuman forces") to which we have no responsibilities. And I fear that this is a danger to which all thinkers proclaiming the primacy of language are susceptible.

Amid the great many thinkers in the tradition who ascribe categorical superiority to human beings over nonhuman animals, there are occasional voices of strong dissent. Two of the most conspicuous such voices are those of Porphyry in late antiquity and Schopenhauer in the nineteenth century. Each proceeds from an open recognition of the essential continuity between human and nonhuman animals, even though in the end (as I have already noted about Schopenhauer) each thinker views human beings as possessing certain prerogatives to use nonhuman animals as resources. Porphyry and Schopenhauer share the conviction that *any being capable of a perceptual encounter with the world possesses understanding*, even if that understanding does not take the form of Eleatic insight and is not linguistically structured. Each implicitly proceeds from the common sense proposition that perception provides vital information to a sentient organism about resources and threats in the environment, thereby enabling the organism to respond appropriately in its efforts to secure its existence.

Porphyry seizes directly upon the notion of *logos* that lies at the core of the tradition's anthropocentric commitments, and he argues that any being possessing voice (*phone*) possess *logos*.[88] In doing so, Porphyry rejects Aristotle's belief that many nonhuman animals possess *phone* but not *logos*, a belief that permitted Aristotle to characterize nonhuman animals as bereft of genuine understanding.[89] Where contemporary thinkers such as Gary Varner take the position that if nonhuman animals were truly intelligent we would recognize it (and that our failure to recognize it entitles us to assume that nonhuman animals are not actually intelligent), Porphyry emphasizes the conspicuous physiological and perceptual continuities between human beings and nonhuman animals, and he argues (as Marc Bekoff would later do) that many nonhuman animals possess not only intelligence but forms of wisdom and even a sense of justice.[90] Whereas the claim about justice in nonhuman animals is a vexed one, the claim that nonhuman animals know their way about in the world is considerably less controversial: As Porphyry observes, nonhuman animals exhibit a grasp of what is to their advantage or disadvantage, they know exactly how to use their bodies to their advantage (e.g., a bull knows to fight with its horns rather than with its tail), and they form and maintain communities.

Schopenhauer seizes upon many of the same considerations as Porphyry in arguing that understanding is by no means unique to human beings. From his earliest work, Schopenhauer maintains that "*motive* . . . controls animal life proper" and that "knowing . . . is the true characteristic of the animal. As such an animal moves toward an aim and end; accordingly this must have been *known* or *recognized* by the animal."[91] Schopenhauer cites the authority of Plutarch and Porphyry, the two most conspicuous advocates of nonhuman animals in antiquity, in support of the proposition that "sensation without understanding would be not merely a useless, but even a cruel, gift of nature" as well as the conclusion that when it comes to human beings and nonhuman animals, "the difference is only one of degree."[92]

In later work, Schopenhauer elaborates both on the notion of understanding and on his early suggestion that "on the path of representation, we can never reach the thing-in-itself."[93] Schopenhauer accepts Kant's proposition that we cannot know the thing-in-itself through cognition or understanding, which "always and invariably consists in the immediate apprehension of causal relations."[94] To experience a "representation" in this connection is to have a mental image or experience that has been shaped in important part by our *a priori* forms of sensibility, conceptualization, and unification. On Kant's view, the only route to knowledge is one that passes through these *a priori* forms of experience. Schopenhauer departs from Kant in seeing "quite a different path" that can lead us to contact with the thing-in-itself, a path that does not proceed through *a priori* forms of intuition, understanding, and apperceptive unification.[95] That different path will be taken not by our cognitive or intellectual apparatus but by *the will*.

Before examining this alternative path proposed by Schopenhauer, I want to say a bit more about the way in which Schopenhauer's conception of experience departs from Kant's. In *The World as Will and Representation*, Schopenhauer makes a fundamental distinction between two modes of access to reality: through the kinds of representation produced by perception, understanding, and reason; and those produced through a special operation of the will in which individuals suspend their engagement in practical activities and simply contemplate the eternal in nature. The former type of representation occurs in accordance with the Principle of Sufficient Reason, i.e., subject to the imposition of Kantian pure forms of experience (space, time, and causality). As such, it is a process of forming cognitive images derived from perception and directed at causal relations, all in the service of the motivations that Schopenhauer sees as driving all perceptual activity. In this respect, "the understanding is the same in all animals and in all men," although "the degree of acuteness and the extent of [different sentient beings'] sphere of knowledge vary enormously."[96] Schopenhauer's view is at one with Porphyry's: Any being possessing perceptual apparatus possesses it in order to understand and navigate its environment. This approach has the virtue of avoiding the tired cliché of relegating the behavior of nonhuman animals to the black box of blind "instinct." Indeed, on Schopenhauer's view, *any being that possesses eyes also possesses understanding*—which is not to deny the causal role played by instinct in behavior, but simply to assert that instinct is not the only such causal factor in the life of a sentient being.[97] All sentient beings start with embodied perception, and proceed to acts of discrimination, evaluation, and choice. This is what distinguishes a sentient creature from a machine—a distinction that Descartes, in virtue of his reduction of nature to inert matter subject to universal forces, fought hard to deny.[98]

Schopenhauer's departure from Kant is evident not only in his revision of the notion of understanding—now, instead of being concerned with formulating laws of nature, it focuses on individual causal relationships—but also in the way he characterizes reason. For Kant reason is an autonomous faculty that enables us to recognize the limits of the understanding (in particular, that the "nature" we construct with *a priori* forms of experience is not the same as the thing-in-itself) and arrive at authoritative guidelines for ethical conduct. For Schopenhauer, reason is a faculty of abstraction that operates on the data of experience provided by perception and the understanding; reason involves speech and comprehension, and permits us to arrive at "abstract knowledge in concepts," e.g., generalizations such as the laws of nature.[99] What is difficult to see at first blush, but turns out to be of vital significance, is Schopenhauer's contention that reason does not provide the basis of morality any more than understanding does.

On Schopenhauer's view, the route to morality passes not through understanding or reason but through *the will*, that dimension of our being that resides in embodiment and unites us with all sentient creatures. I noted in Chapter 1 that thinkers such as Aristotle deny choice to nonhuman animals altogether, on the grounds that genuine choice requires rational capacity. Schopenhauer recognizes that to be a sentient being is to be aware and actively engaged in the pursuit of ends. Such beings formulate cognitive representations in the course of negotiating their environments, which is to say that there is an intimate relationship in all sentient creatures between understanding and will. And not only does Schopenhauer believe that all sentient creatures possess will, but he provides a characterization of what he calls "the world will," which he identifies with all action in the world, not just that of sentient creatures. Even the roaring of the sea, or the tension between load and support exhibited in architecture, counts as a manifestation of the world will. From out of this primordial soup of anonymous willing, there emerge various sorts of beings, some of which possess sensory apparatus and understanding. Then there are some possessing reason (conceptual apparatus) that achieve the status of individuality, which on Schopenhauer's view constitutes one of the "higher grades of the will's objectivity"; he states that individuality manifests itself "especially in man" and that "in the animals this individual character as a whole is lacking."[100]

Schopenhauer's notion of individuality shares much in common with the traditional views about individuality that I have examined in previous chapters. In particular, Schopenhauer considers individuals, and individuals alone, to be capable of moral responsibility. Because nonhuman animals lack rational knowledge, they lack principles and any conscious sense of morality, although they can exhibit good and bad characters, etc.[101] What one needs in order to participate in moral agency is a sense of compassion, which on Schopenhauer's view is the true basis of morality.[102] Nonetheless, Schopenhauer considers nonhuman animals to be members of the moral community and to *merit rights*.[103] Thus he is harshly critical of Kant's classification of nonhuman animals as "things" as well as his indirect duties view.[104]

In appealing to compassion rather than reason as the basis of morality, Schopenhauer is appealing to something we feel through our embodiment that ties us directly (i.e., without the mediation of cognitive representations) to the world will. Schopenhauer goes so far as to suggest, contra Kant, that this world will is identical with the thing-in-itself.[105] Thus it is possible to gain knowledge of the thing-in-itself, although this is a very different kind of knowledge than the kind that arises through representation that occurs in accordance with the Principle of Sufficient Reason. Schopenhauer posits the ideal of a "pure, will-less, knowing subject" that removes itself from the context of causality and striving and gains insight into various "grades of the objectification of the will," the highest such grade being individuality. Taking his cue from the *Upanishads*, Schopenhauer presents a view of the individual self as being united with the world will and as appearing to be separate only by dint of Eleatic abstraction.[106] Proceeding from an implicit commitment to Aristotle's notion of "a precedent community of nature," Schopenhauer stresses our commonalities and shared mortality with nonhuman animals rather than our supposed differences—even though in the end he affirms some pointedly anthropocentric values such as the legitimacy of meat eating.

Schopenhauer makes a crucial contribution to the development of philosophical thinking about the relationship between self and world, one that has important implications for ethics. Schopenhauer, like Aristotle and Porphyry before him, proceeds from an acknowledgment of our fundamentally *embodied* condition, an acknowledgment that many Eleatically-minded thinkers would like to forget. But unlike Aristotle, who proceeded from this acknowledgment to the proposition that *nous* (reason) is eternal, Schopenhauer places reason on the side of representation and attributes to embodied contemplation (on the part of the individual, at any rate) the capacity for a fundamentally different type of contact with reality, one mediated not by the principle of sufficient reason (space, time, and causality, the building blocks of scientific knowledge) but by "the Platonic Ideas," which are representations we can experience in acts of detached contemplation. The sense of detachment here parallels but is not identical with the Eleatic notion of detachment: The Eleatic notion, as should be clear by now, is driven by an ideal of *control*. The detachment we undertake when we seek to contemplate the Platonic Ideas entails a suspension of our own individuality, which brings with it a complete suspension of the will.[107] It is in this sense that an individual can take that "step back" from willing that Heidegger would later counsel, thereby opening up an entirely differ-ent way of contemplating reality than the kind that prevails in Eleatic efforts to exercise control. That different mode of approach, Schopenhauer believes, gives us insight into certain eternal truths about reality—not truths such as $2+2 = 4$, but truths such as the insuperable connection between willing and suffering and the essential interconnectedness of natural phenomena (of which the emergence of the individual is a special case). Schopenhauer's (and Heidegger's) contention is that we have to lose our individuality, our involvement with practical endeavors, in order to achieve these sorts of insights. A sober insight into the nature of things demands the *quietism* of the will.

Of course, one can respond to this contention by observing that Eleatic con-templation, too, involves a suspension of practical tasks and a stepping back into contemplative repose. And in a sense that is true: Eleatic reason seeks to isolate constancies or regularities in the flux of phenomena, and this requires us to put our practical endeavors on hold so that we can simply study nature. I hope to have shown in the previous two chapters that exponents of Eleatic reason consistently maintain (and believe) that their conception of rational contemplation is entirely impartial in the sense that it is not driven by previously existing motives, but that this supposition proves highly dubious once we have taken stock of the motive of control that informs Eleatic endeavors—a motive directed in substantial part at asserting human superiority over nonhuman animals and the rest of nature. If you proceed from the (naïve) assumption that *all* your acts of detached reflection are "neutral" in the sense that they serve no pre-existing motives, it will be fairly easy to arrive at a conception of human and nonhuman animal experience according to which to be human is to be "higher" or "superior." (The same naturally holds, *mutatis mutandis*, for conceptions about gender and racial status.) What Schopen-hauer is drawing to our attention is the fact that many of our acts of representa-tion (namely, those in accordance with the Principle of Sufficient Reason) conceal ulterior motives.[108] The alternative he proposes is to adopt a standpoint of "pure will-less knowing" that recalls Meister Eckhart's ideal of "releasement" toward God, although for Schopenhauer it is not God but the world will itself toward which we ought to exhibit this gesture of modesty. The contemplation of the Platonic Ideas is to remind us of the mortality we share with all sentient life. In this respect it informs and reinforces the sense of compassion toward others (human and nonhuman animal alike) that Schopenhauer deems the basis of morality.

I will return to Schopenhauer's call for compassion in Chapter 5, when I look more closely at the resources our affective life can and must provide if moral judgment is to exhibit genuine authority. The fact that Schopenhauer himself succumbs to some pointedly anthropocentric prejudices (e.g., that meat eating and invasive experimentation, at least when done for more than trivial reasons, are perfectly acceptable) in spite of this effort to affirm our essential connectedness with the rest of nature and the need for an ethic of modesty, demonstrates a tragic bottom line about moral judgment: that it can never be secured once and for all, through even the most Olympian Eleatic efforts. That does not mean, however, that moral judgment is simply arbitrary, but rather that the Eleatic "correct versus incorrect" model that works for many scientific endeavors is ill-suited to the nature of moral judgment. In Chapter 5, I will propose a model of moral judgment that claims universality (i.e., binding authority over all moral agents) but that is subjective in the sense that it is not reducible to conceptual certainty. A core question in the evaluation of this model of moral judgment will be whether there is a form of contemplation (such as the form proposed by Schopenhauer) that affords us the kind of insight into the inner nature of things that would be necessary in order to provide a non-arbitrary *measure* for moral judgment.

In addressing this question, some thinkers have taken Rorty's and Derrida's general approach, which is to assert the primacy of language and to conclude that all such measures for judgment are ineluctably contingent. Others, as I noted earlier, have sought to invert the traditional picture by proclaiming emotion rather than reason to be the basis of morality (Fontenay's "pathocentrism," which I noted near the end of Chapter 1). One of my central aims in this book, as I stated in the Introduction, is to argue for *the necessarily dialectical relationship between reason and emotion in the process of forming and revising moral commitments*, rather than for the primacy or exclusive authority of one of them. Both are essential to moral judgment. Either employed in isolation from the other will unavoidably lead to judgments that are deficient in the sense that they privilege a particular, local set of interests or tendencies rather than viewing them in the larger life context in which we find ourselves. In the concluding chapter of this book I will return to this deep and persistent problem of local versus global (e.g., individual versus societal) interests and consider the prospects for achieving a more truly *inclusive* vision of the moral life.

Up to this point I have focused on a traditional anthropocentric ideal of reason and have attempted to demonstrate its dangers and limitations; in Chapter 5 I will subject our affective or emotional constitution to a comparable critique and argue that reason has the capacity to function as a vital corrective to the sorts of affective tendencies against which thinkers such as Seneca warn us. Thinkers who undertake a total critique of reason (namely, pathocentrists and critics of "logocentrism" alike) fail to appreciate the power of reason to inform the development of our emotional lives. But by now it should be clear that reason in its Eleatic incarnation is inadequate to this task. Thus it is necessary to rethink the ideal of reason. Where the tradition embraced an ideal of reason as eternal, some recent thinkers have taken a decidedly more Pragmatist approach and have sought to conceive of reason as inscribed *within time*. This turn, which Ortega y Gasset calls the turn to "historical reason," holds great promise for the rehabilitation of reason rather than its demotion or outright rejection.

4. Ortega's Turn to Historical Reason

In the course of the twentieth century, a growing number of thinkers have embraced the need to inscribe reason, along with the rest of human agency, within time rather than clinging to the Eleatic ideal of timelessness. To many, this move has been as counterintuitive as it would have been to Descartes, who repeatedly cites mathematics as the paradigm case of cognitive certainty and who chides the ancients for having failed to see how the power of (Eleatic) reason could be extended beyond mathematics not only to the natural sciences but even to morality itself.[109] That Descartes quite conspicuously failed to deliver his scientifically grounded morality is an indication that the scope of Eleatic insight is ultimately too limited to afford us authoritative insight into matters that arise in the course of human action rather than in the abstract.

I have already noted that this recognition of the limits of Eleatic reason has led thinkers such as Nietzsche, Derrida, and Rorty to despair of reason altogether and turn to other sources of meaning and orientation. Derrida's instantiation of a total critique of reason leads him to challenge the very idea of proceeding from an orientation, which begs the question: What constitutes the difference between an *authoritative* judgment (i.e., one with binding authority over the community) and one that simply reflects the idiosyncratic tendencies of a lone individual or subset of society?[110] Insight into the essential limitations of Eleatic reason brings with it the disquieting recognition that there can be no final certainty in moral (and a variety of other types of) judgment, that the process of judgment is in principle endless. Where thinkers such as Derrida and Rorty purport to abandon reason in favor of other grounds for action—assuming they would acknowledge anything like an enduring ground—others have taken a less extreme approach and have sought to rehabilitate or rethink rationality in a manner that befits our nature as temporal beings. The most promising such approach is taken by Ortega y Gasset, with his turn to historical reason.

Ortega develops his historical idealism by proceeding from the lapidary insight that human existence, unlike that of other beings in the universe, is "essentially problematic."[111] This statement shares much in common with the existentialist emphasis on the primacy of the will and the imperative to choose: Like the existentialists (and, for that matter, like the vast majority of philosophers writing about human agency), Ortega takes the orthodox approach of assuming that only human beings are confronted by the imperative to choose, nonhuman animals being tied to immediate biological imperatives and incapable of stepping back from them so as to be "free for activities which in themselves are not satisfaction of needs."[112] Let us leave aside for the moment this anthropocentric aspect of Ortega's thought and focus on the premium he places on human freedom. Like many thinkers before and after him (including even Rorty), Ortega emphasizes the primacy of the *individual* understood as a center of *agency*. *Ago, agere*: to act is central to human existence, indeed to such an extent that it is flat-footedly wrong on this view to conceive of the human being as an entity possessing certain properties that can enter into the field of action: On Ortega's view, which he shares with the Pragmatists, to be human is *already* to be engaged in action, where that engagement proceeds from out of a history of human action.

Ortega's emphasis on the priority of action reminds us of the fact that the world is not simply "out there," waiting to be discovered; instead, everything about reality that enters the purview of our consciousness is, as Kant recognized, *mediated*. Where Kant appeals to *a priori* conditions for the possibility of both scientific knowledge and authoritative ethical judgment as these mediating factors, by "de-divinizing" reason Ortega seeks to bring the basis and criteria for judgment from the heavens to the earth. Ortega proceeds from the axiom that our lived experience is prior to theorizing.[113] His point is not that theorizing has no power to contemplate and shed light on the human predicament, but rather that Eleatic reason arrives at abstractions that are false. The Cartesian notion of *res cogitans* (thinking thing), for example, is not a fruitful starting point for gaining an appreciation of what Ortega means by the "fundamental uncertainty" of human existence, nor for what he means when he suggests that human existence is a matter of "happenstance."[114] The kind of theorizing or contemplating that proceeds from an acknowledgment of our accidental, temporal nature rather than from a fanciful ideal of Cartesian-style certainty holds the promise of redirecting our efforts away from ideals of permanence and toward a genuine confrontation with history—with our historical legacy, with our current historical exigencies, and the question of which of our historical commitments merit retention and which stand in need of abandonment or revision.

William James provided inspiration for this approach to understanding the human condition and our relationship to the rest of reality when he argued for a pragmatic conception of truth as that which is "profitable to our lives" and attributed absolute centrality to "our acts."[115] Action, James argued, is *prior to* Eleatic generalizations about "a past eternity" and inherently subject to the vicissitudes of time—which is to say that our judgments and choices are "everlastingly in process of mutation." In the face of this acknowledgment of constant change, "the 'absolutely' true, meaning what no farther experience will ever alter, is that ideal vanishing-point towards which we imagine that all our temporary truths will some day converge."[116] Rather than rejecting the role of reason altogether, James posits the notion of the absolute as what Kant would have called a regulative ideal: It constitutes the end or aim toward which we strive in our efforts to attain clarity and make authoritative choices about the good. It is an "ideal vanishing point" in the sense that we continually seek to approach it even though we can never fully realize it. The very pursuit of this ideal, unfolding as it does in time, alters our sense of the ideal as well as our sense of how, under the current circumstances, we can best promote its realization. This is what James means when he states that "truth *happens* to an idea. It *becomes* true, is *made* true by events. Its verity is in fact an event, a process: the process namely of verifying itself, its veri-*fication*. Its validity is the process of its valid-*ation*."[117] No longer a static property or relation, truth is an *event*.

Ortega takes this event-character of truth as the starting point for his reflections on a conception of reason that acknowledges time rather than seeking to overcome it. Like his slightly younger contemporary Sartre, Ortega proceeds from the axiom that the human being is unique in having no fixed essence, that it is freedom that most fundamentally characterizes our being. This is the same freedom that the Renaissance thinker Pico Della Mirandola esteemed so highly that he suggested that even God and the angels envy us for our possession of it.[118] But unlike Pico and Descartes, who conducted their musings about the human condition against the background of a worldview that saw humanity as made in the divine image, Ortega appeals to the simplicity of "our unadorned life," unfolding through history, as the only available basis for judgment.[119] This shift from the Eleatic ideal to historical reason necessitates a lowering of our expectations regarding the prospects for establishing certainty: As fundamentally contingent beings who must seek meaning and purpose from out of the resources afforded us by our temporal nature, we cannot expect to attain absolute Cartesian certainty in social and moral matters but must rest satisfied with a different kind of certainty—a kind whose "absoluteness" is subject to endless revision. Indeed, not only can we not attain Cartesian certainty about such matters, but "our present doubt is even more radical than Descartes's" inasmuch as Descartes "unwittingly" retained faith in Scholastic philosophy as well as in (traditional ideals of) truth and logic.[120]

But this doubt, on Ortega's view, is not so radical that it is insuperable. Ortega places great confidence in our capacity to make meaningful choices based on the insights disclosed by our own historical consciousness, a consciousness that gains its inspiration and sense of orientation not from mute facts but from historical events, i.e., from *significant actions* in our past. "Man, in a word, has no nature; what he has is . . . history. Expressed differently: what nature is to things, history, *res gestae*, is to man. . . . Man . . . has no nature other than what he has himself done."[121] *Res gestae*—human achievements—do not occur in an unforeseen, arbitrary manner but instead trace out a progression based on reflection on the past and anticipation of the future. Ortega expresses suspicion of the Enlightenment ideal of inevitable progress advanced by Hegel and Comte; by now it should be clear that that ideal, which I examined near the end of Chapter 1, presupposes an Eleatic vantage point outside of time that would enable us to view history as an object—which amounts to the same thing as viewing human agency as an object. In place of this ideal, Ortega proposes notions of selfhood, agency, and understanding as fundamentally historical phenomena.

The core of Ortega's ideal of historical reason is the human individual. I suggested in passing earlier in this chapter that it is no accident that Rorty focuses on the individual, even though he purports to offer no actual argument in favor of liberal individualism. Ortega helps us to see why and in what sense that is the case. "The human individual," Ortega notes, "is not putting on humanity for the first time." On the contrary, anyone's humanity

> takes its point of departure from another, already developed, that has reached its culmination. . . . [Each individual] finds at birth a form of humanity, a mode of being a man, already forged, that he need not invent but may simply take over and set out from for his individual development.[122]

The Cartesian ideal depicts human agency in inverted form by characterizing thought as existing prior to concrete, lived experience rather than as necessarily operating on the data of lived experience. The phenomenologists attempted to shed light on this deficiency of the Eleatic ideal by arguing against Descartes that consciousness is never free-floating in this way but is always consciousness *of* something in the field of experience; Sartre refined this formula by further observing that even "consciousness of . . ." stops short of a full acknowledgment of the fact that consciousness is always consciousness *of something to be done*. Ortega sees the point of departure for all meaningful action to be the already-disclosed sphere of meaning into which each individual has been cast by the contingencies of birth: Selfhood is first of all inherited, and then the task for each individual is to take a stand on that inherited meaning.

Even if, as thinkers have increasingly argued in recent generations, the individual self is "socially constructed," it remains the case—and even Rorty expresses an awareness of this—that *choices and history are made by individual agents*, even when those agents are acting in concert with one another. Taking a stand—agency—in this connection requires historical awareness and reflection. The individual's task is to

> see, foreshortened, the whole of man's past still active and alive. . . . History is a system, the system of human experiences linked in a single, inexorable chain. Hence nothing can be truly clear in history until everything is clear.

Total clarity is, as I have noted, a regulative ideal whose value lies in affording us a sense of orientation and direction in our strivings. What Ortega envisions is a "systematization of *res gestae*" that "becomes reoperative and potent in history as *cognitio rerum gestarum*"—knowledge or understanding of things that have been achieved.[123] Naturally this endeavor to systematize depends on a principle of (i.e., some basis for) unification, as well as a shared sense of what counts as an historical achievement. It requires a sense of the "rational substance" of history and the "original, autochthonous reason" that accompanies it in virtue of being *"constituted by what has happened to man."*[124]

There is a kind of vision at work here: One has to look into the past and make discriminations based on established ways of seeing and valuing; but one equally has to look to the future and commit oneself to retaining, modifying, abandoning, and/or replacing prevailing commitments based on the goals one recognizes to be most appropriate under the current circumstances. Ortega's approach involves a commitment to a shared *ideal of humanity*, an ideal constantly subject to contestation and revision but nonetheless possessing an enduring authority that must be constantly revised and (re)affirmed. Anyone inured to an Eleatic conception of selfhood and meaning will have great difficulty seeing the power of Ortega's move to historical reason, because the Eleatic ideal brings with it the expectation that this "shared ideal of humanity" will be determinate, timeless, and straightforwardly explicable—which is precisely, for reasons offered by Ortega, what an historical ideal of reason can never be.

The Eleatic model may have borne fruit for early modern physics, Ortega notes, but "it has always failed where man was concerned."[125] The human being cannot be understood in terms of an enduring essence because "man has no nature; what he has is history," which is "fundamentally *mobility* and *change*. . . . Man cannot be *identi-fied*" inasmuch as "the real is the non-identical, it is pure event, mobility, flux."[126] Far from being aimless, this mobility and change exhibit an orientation or direction derived from a reflection on what we have been and what we have achieved, where the operative 'we' is the human community engaged in the project of "narrative [= historical] reason."[127] The central challenge for such an approach, as I have already noted, is to characterize the kind of vision that informs this narrative process such that it constitutes a progression. Ortega faces the same problem that Heidegger and Rorty face: In Heidegger's case, it is how to distinguish between choices with moral import that are legitimate from those that are illegitimate; for Rorty, the problem is how to substantiate a form of life such as liberalism (or, more specifically in Rorty's case, "liberal ironism") as one we ought to embrace. This "ought" is a problem for Rorty in spite of—indeed, its problematic nature is exacerbated by—his failure to acknowledge it.

Ortega's response to this challenge is to proceed from an ideal of humanity that began to emerge perhaps even before the advent of recorded history, and to characterize this ideal as one constantly subject to revision and refinement. The process of historical narrative is this process of revision and refinement. But refinement toward what? Only if we take an Eleatic approach, one taking the position that the self has a nature or essential identity, could we expect to receive an explicit, determinate answer to the question. In the absence of such a fixed identity, the self (both individual and collective) must forever contemplate a future that is inherently open. To embark on this open sea of possibility is not the same, however, for Ortega as it is for Nietzsche. Nietzsche presents an ideal of absolute autonomy and creative self-invention on the part of the so-called "higher" type of humanity, an ideal according to which the creative agency of the individual (the higher type, at any rate) must not be brought under the yoke of shared, prevailing social and moral conventions. For Nietzsche, to embark upon "the open sea" of possibility is to arrogate to oneself free rein to structure one's life in any way one sees fit.[128] Ortega, operating with a decidedly different vision of society, opts for a considerably more inclusive ideal of human community.

This ideal is based on a recognition of something shared by all individuals: the "immediate intuition" each of us has of our own life, an intuition in virtue of which autobiography is a key component in the process of historical reflection.[129] Inasmuch as we have no fixed nature and are essentially "inheritors," each of us has no choice but to reflect on our own past, which is embedded in a larger past shared by humanity, in order to ascertain and subject past commitments to critical scrutiny and employ our insights in confronting our current circumstances and exigencies. Ortega gives us good reason to see the modern liberal notion of individuality not as just some cultural invention that happens to have arisen at a certain point in time, but rather as a distinctive historical *achievement*. Consider, for example, the contrast between Aristotle's and, say, Locke's or Rawls's conceptions of citizenship: Aristotle restricts the prerogative of citizenship to free males, excluding women, enslaved persons, and children on the grounds that they are either *aloga* or insufficiently rational. By Locke's time, the scope of citizenship has broadened considerably, even if its promise would not be fulfilled for quite some time—in fact, we are still waiting for it to be fulfilled. That, however, is not a sign that the ideal itself is flawed, but rather that our acts of historical reflection have not yet caught up with this inchoate ideal. It may also mean, as I have already intimated, that reflection alone is not sufficient but stands in need of the "extra-rational source of meaning" disclosed by the affective dimension of our being—a dimension that Ortega affirms, at one point suggesting that "practical reason . . . must engage man's passions and blindspots" in order to come to grips with problems such as enmities between nations.[130] Ortega ultimately embraces an ideal exhibiting affinities with Habermas's ideal of communicative action and Dewey's ideal of "a genuine community of action" informed by a "concerted consensus."[131]

This is an ideal that, as I suggested earlier, is neither arbitrary nor insignificant. It is an ideal of full participation on the part of all individuals in the human community who possess the requisite agency, an ideal whose scope has broadened (in principle if not entirely in practice) in the course of history in response to humanity's increasing recognition of the fact that autonomy and agency are not the exclusive possession of those in power but in fact are *the way of being exhibited by beings like us*. For a culture long inured to a specifically Eleatic approach to understanding reality, this insight proves elusive because that approach posits beings as existing independently of us and as exhibiting certain objectively describable properties. The problem this approach encounters is that we are the beings who project or identify objects, and we ourselves on this view are never simply objects. How, then, might Ortega's ideal be defended? The answer is that this cannot happen outside the process of historical reflection and narrative, but must be undertaken from within the very operation of historical reason. Ortega's approach, in other words, is a textbook case of an immanent as opposed to total critique of reason.

Harry Frankfurt's analysis of Descartes's ideal of reason proves illuminating in this connection. In the *Meditations*, Descartes purports to call all his former beliefs into question, with an eye toward establishing a new and secure foundation for modern science. The First Meditation constitutes a retreat into the hinterlands of Eleatic interiority, and the subsequent five Meditations trace out the Meditator's attempt to overcome the hyperbolic doubts raised in the First. Many commentators have alleged that the Third Meditation proof of God's existence is circular in virtue of relying on clear and distinct insights whose validity presupposes God's existence as a non-deceiver. The crux of this charge of circularity is the fact that the Meditator, early in the Third Meditation, asserts the rule that "whatever I perceive very clearly and distinctly is true" but is able to do so only provisionally.[132] The Meditator's rationale for positing this provisional rule is that the appeal to clear and distinct insight has proved fruitful in the first two Meditations. Frankfurt offers an interpretation of the rule of evidence (the assumption that clear and distinct = true) that actually vindicates Descartes against the charge of circularity. Frankfurt's interpretation turns out to have important implications for Ortega's attempt at an immanent critique of reason.

Frankfurt proceeds from the proposition that Descartes's quest for certainty is not directed at establishing a correspondence between thought and external reality, but rather seeks simply to vindicate the internal consistency of reason.[133] The strategy is to employ reason in an essentially experimental fashion, in order to ascertain its potential and its limits. The entire notion of clear and distinct insight, which forms the basis for Descartes's quest for certainty, is "a rather straightforwardly logical matter of recognizing that no coherent grounds for doubting a proposition are conceivable, and of understanding the proposition enough to avoid confusion."[134] Frankfurt seizes upon the fact that a number of substantial cognitive achievements have been realized *prior* to the proof of God's existence in the Third Meditation, in particular that the *cogito* is logically unassailable, the essence of the mind is to think, and the essence of body is to be "extended." The most significant of these achievements is the '*sum*' (I am) that the Meditator recognizes to be logically inseparable from '*cogito*' (I think): "A person can never . . . have good reason to doubt his own existence," or, as Descartes frames the idea, my existence is logically "necessary" in any moment in which I am thinking.[135] These cognitive achievements, Frankfurt believes, *justify* the Meditator in proceeding with the construction of the edifice of wisdom that motivates Descartes's project.

Descartes operates with a pointedly Eleatic ideal of reason—the "logical" import of reason (for example, the *sum* or the essence of body) does not vary from person to person nor from culture to culture, but remains the same through time, as is evident in Descartes's invocation of mathematics as the paradigm case of clear and distinct insight. Nonetheless Frankfurt makes an intriguing observation that helps to bridge the gap between Descartes's Eleatic ideal of reason and Ortega's historical ideal. Frankfurt states that Descartes's "*sum* may be regarded as a kind of necessary statement despite the fact that it is logically contingent."[136] By acknowledging the contingency of Descartes's *sum*, Frankfurt implicitly situates even the Cartesian ego within a larger reality whose nature is at first obscure but becomes increasingly clear through the application of Cartesian method and the securing of enduring foundations for natural science. There is a trivial sense in which Descartes's *sum* (or, as it is more typically referred to, the *cogito*) is "contingent": I couldn't utter or contemplate the *cogito* unless I happen to have been born (born, that is, possessing the requisite cognitive apparatus). But there is a deeper sense in which the *cogito* is contingent: It required an historical agent, a Descartes, to articulate the *cogito* in the first place, on the basis of a reflection on prior thinking about selfhood that passed through a series of historical achievements such as Saint Augustine's much earlier anticipation of the *cogito* and Martin Luther's assertion a hundred years before Descartes that all devout individuals are equally endowed with the religious conscience needed to interpret Scripture. Far from concocting the *cogito* from out of thin air, Descartes arrived at it by refining the thought of prior historical agents in the light of exigencies prevailing in his own time, most particularly the revival of ancient skepticism undertaken in the late sixteenth century by Montaigne and the pressures Descartes experienced as an itinerant soldier in the Thirty Years' War.[137] In other words, while the *cogito* itself is internally coherent and unassailable, it is also hopelessly abstract and empty until we acknowledge the historical and cultural context out of which it arose but which Eleatic reason would prefer to forget. An essential part of that context is a looming sense of *in*security characteristic of the human condition.

An important lesson to be learned from Frankfurt's analysis of Cartesian certainty is that Eleatic reason *purports* to be absolutely autonomous (in the sense of producing truths that are permanent) but in fact is insuperably tied to the past and the future. Eleatic rational operations never occur in a timeless, free-floating manner in any disciplines other than mathematics and logic, those fields of inquiry that tell us about the unchanging features of human thought; and even in those disciplines, rational operations are performed by embodied historical agents, not by isolated Cartesian egos. Frankfurt's analysis also sheds light on the fact that *when we undertake a critique of reason, we must do so by employing the very resources that we seek to subject to critical scrutiny*. As I have suggested earlier in this chapter, I believe that this holds for Rorty in spite of his protestations to the contrary—that what he holds out as a total critique of reason is in fact simply an immanent one, one that he could have extended to the point of *defending* liberal ironism rather than simply offering it in the form of a hypothetical imperative.

Frankfurt's analysis also provides us with an occasion for reflecting on the limits of Cartesian certainty, that certainty that Descartes thought held the promise of rendering human beings "the masters and possessors of nature." Descartes is clear that his ultimate interest is in the use of scientific knowledge to control natural processes, not in abstract philosophical reflection. The quest to exercise mastery over nature is beset, as Ortega points out, by the fact that thinking is logical whereas reality is *il*logical.[138] But at least in the case of the endeavor to exert control over natural processes, our efforts have borne fruit in the four centuries since Descartes; the persistent deficiency of Eleatic logic is most apparent in the cultural and moral sphere, the sphere comprised not of passive objects but of historical agents. I have already noted that Descartes never fulfilled the promise of articulating "the highest and most perfect moral system, which presupposes a complete knowledge of the other sciences and is the ultimate level of wisdom."[139] The reason for this, I believe, is that he failed to recognize the specifically historical character of reason and its dialectical relationship with our affective (emotional) life.

Plato and Kant, too, failed to recognize the specifically historical character of reason as well as the essential role played by our affective life in moral judgment. Regarding reason, Plato treated the forms as permanent constituents of reality that form the basis for our perception and evaluation of sensible particulars. Kant criticized Plato for being a metaphysical realist rather than an idealist about the forms or ideas, but he retained the Eleatic notion of the Ideas or forms as timeless. "For Plato ideas are archetypes of the things themselves," which demonstrates that "he has not sufficiently determined his concept"; rather than corresponding to the things themselves, Kant argues, each archetype or "true original . . . is to be found only in our minds" and is "an indispensable foundation" for authoritative moral judgment.[140] Interpreted against the background of Kant's ideal of timeless moral truth and the categorical imperative, this passage makes it clear that Kant would retain the Eleatic premium on permanence while dispensing with the Platonic commitment to a correspondence between our ideas and reality.

A useful way to test Ortega's thesis about historical reason is to ask the question whether the move that Kant proposes constitutes an advance over Plato. While there is no definitive proof that it does—one can imagine Rorty suggesting that it is simply a departure from Plato rather than any kind of progression—Kant's insight at least has the virtue of acknowledging something that Ortega would later seize upon: that thought and reality do not have the same essential structure. That in itself constitutes an advance over the Eleatic assumption of a direct correspondence between thought and reality, for example, Aristotle's assumption that thought and reality share a predicative structure. What thinkers such as Plato and Aristotle failed to recognize is that the "reality" with which they sought correspondence was itself a product of the mind's Eleatic operations rather than a direct confrontation with what lies beyond us. Kant saw that the nature of a finite mind is such that its grasp of reality is always mediated, but he made the mistake of proceeding from the assumption that the mind is essentially separate from reality.

Ortega subjects this assumption to critical scrutiny and argues for the conclusion that the mind needs to be inscribed within history rather than being viewed as essentially separate from reality. Again, there is no "knock-down" argument that the move from Plato to Kant to Ortega constitutes a progression; when there is talk of a progression, there is always the question of the standard or measure in terms of which progress is being assessed. If there is anything like a "transcendent reality," it can be nothing more than "the dialectical series of [human] experiences."[141] Thus, if there is a measure for progress in history, it must emerge immanently within our experience and exercise its authority over us through the unfolding of time. Ortega says little about this immanent measure beyond suggesting that "to progress is to accumulate being, to store up reality."[142] But this brief remark tells us quite a bit: We can come to recognize progress only retrospectively, through a reflection on past human acts and with an eye toward confronting current and future exigencies, whereas an Eleatic conception of reason involves the assumption that such measures are prospective, i.e., somehow independent of both our experience and our judgment—prospective in the sense that they are taken as ideal endpoints of striving that trace out in advance the "proper" trajectory of that striving, much like the notion of an *eschaton* functioned as the gathering principle of Christian eschatology.[143]

Ortega recognizes that the move to historical reason requires an acceptance of our fundamental insecurity, both physical and cognitive. Like Husserl, Ortega proceeds from a recognition that reason is currently in a state of crisis. He goes so far as to state that the very idea of morality had "gone up in smoke" by the turn of the twentieth century, and he wonders whether and how humanity will respond to the fact that "we have lost all final courts of appeal that might have served to direct our lives."[144] This loss of all final courts of appeal is, on Ortega's view, directly attributable to the tradition's unremitting adherence to the Eleatic conception of reason, a conception according to which to recognize anything physical or cognitive is to do so in detached terms that abstract away from the lived character of phenomena, such as insecurity. Ortega's insight into the limitations of Eleatic reason shares a great deal with Heidegger's critique of language as *apophansis*: It is not the case that we are unable to articulate abstract notions detached from lived experience; on the contrary, we do so all the time. The problem arises when we treat those determinate predicative assertions as if they transparently conveyed facts about reality to us, when in fact they are always (with the arguable exception of propositions in mathematics and logic) products of a departure from lived experience and a narrowing down of our focus to some individual aspect(s) of that experience.[145]

Schopenhauer was trying to draw our attention to a route of access to lived experience that does not pass through the Principle of Sufficient Reason (the imposition of pure forms of space, time, and causality) but which affords us a direct encounter with reality. That route is the will; I believe that the solution to the crisis cited by Ortega will involve serious consideration of the will and most particularly our affective life, which is what connects us to nonhuman animals. As I have argued from the start of this book, I am concerned that this route has been neglected precisely because its embrace would demand an acceptance of Schopenhauer's insistence that human beings and nonhuman animals are essentially the same—an acceptance that would bring with it tremendous implications for the moral status and entitlements of nonhuman animals.

Notwithstanding his efforts to inscribe human beings within the larger context of a reality that unfolds through time, and notwithstanding his occasional acknowledgments that practical reason must engage our passion, Ortega retains the same anthropocentric orientation as all thinkers who make reason central. He cites criteria for distinguishing humans from nonhuman animals, including the idea that only human beings are inheritors and that law is what distinguishes the two.[146] And while I think it is probably fair (I can't be certain, as I have never managed to get inside the experience of a nonhuman animal) to say that living in a specifically historical manner (involving reflection on the past and attempts to refine and work toward the ideal of community) is unique to human beings, I do not consider those sorts of differences between human and nonhuman animals to have any implications for the moral worth of either. As I suggested previously, I believe that the only real implications of this difference pertain to the *obligations* that human beings have toward nonhuman animals and perhaps to the environment more generally.

The essential problem with the Eleatic approach as regards nonhuman animals is that it unavoidably represents nonhuman animals as objects with certain sorts of properties. It does so because of the centrality it ascribes to human experience, the only experience to which we have immediate access. By the time the notion of animality has passed through the lens of Eleaticism, it has taken the form of something essentially deficient or lacking in comparison with our experience of ourselves. From there it is a short step to treating nonhuman animals as resources, a status whose problematic nature is not ameliorated by the qualification that we ought to use nonhuman animals compassionately, nor by the fact that so many people enjoy the companionship of nonhuman animals. Is there not some way we can derive a more adequate sense of the moral status of nonhuman sentient creatures, perhaps by delving a bit deeper into the notion of historically situated agency? With this prospect in mind, I will conclude this chapter by examining some key ideas in the thought of John William Miller, who elevated the insights of thinkers such as James and Ortega into a full-fledged philosophy of actualism. I hope to show not only that Miller presents an important elaboration of Ortega's ideal, but also that by itself, even that elaboration falls short of accounting for the possibility of ethical principles or universals.

5. Miller's Actualism and the Problem of Universals

My central thesis in *Animals and the Limits of Postmodernism* was that postmodern thinkers, in virtue of their appeal to "the undecidability of meaning" (our inability, due to the inherently fluid nature of language, to arrive at enduring truth) are unable to arrive at any sort of stable moral principles, such as those governing our treatment of nonhuman sentient life. There I argued that

> the fatal limitation of postmodernism is . . . that it embraces two notions that are fundamentally incompatible with one another: a commitment to the indeterminacy of meaning and a sense of justice that presupposes the very access to a sense of determinacy that postmodern epistemology dismisses as illusory.[147]

Thus it is unclear *on what basis* postmodern thinkers, such as Derrida, decry the injustices we perpetrate against nonhuman animals: The ability to classify an action or policy as unjust presupposes some sense of justice that permits or informs the classification. In the absence of such criteria, postmodern lamentations of injustice reflect the light of arbitrariness: Why consider this action to be unjust but not that one? The postmodern thinker, as I have suggested already, is left in the position of simply staring at the young Nietzsche's "moveable host of metaphors" and making a choice that, in Schmitt's words, "emanates from nothingness."

It is in the face of this threat of arbitrariness that the tradition opted for the Eleatic conception of reason. In doing so, the tradition framed emotion as a potential threat to the integrity of judgment and, like Descartes, seized upon mathematics as the paradigm case of cognitive certainty. What the tradition failed to recognize is that the Eleatic model, oriented as it is on permanence, is ultimately ill-suited to the challenges of moral judgment, which by its nature (which is to say, by virtue of *our* nature as finite beings) is directed specifically at the temporal realm—as Ortega observed, the task is to try to make reason, which is "logical," match up with reality, which is decidedly "illogical." Hence the crisis in reason, and most particularly practical reason, to which Ortega draws our attention. I believe that a potentially fruitful resolution of this crisis could come through a rethinking of the nature of ethical principles or universals in the light of Ortega's turn to historical reason.

Eleaticism has traditionally conceived of universals as static structures. These structures have in some cases been taken to have metaphysical status, hence to be "real" universals, and in other cases to be linguistic-conceptual entities which, while having no necessary metaphysical status, nevertheless have an enduring character that remains identical through the flux of time and circumstance. It would be incumbent on anyone who takes Kant's transcendental turn seriously to reject or at least call radically into question "real" universals in the sense of notions that exhibit a direct correspondence with reality outside us; but Kant for his own part retains a commitment to a variety of conceptual universals, namely the Transcendental Ideas of Reason (God, freedom, and the immortality of the soul) that I examined in Chapter 2. For Kant, these postulates are timeless and guide our aspiration to realize an ideal kingdom of ends that resides completely outside of time and space, a kingdom that functions as the putative aspirational goal for human striving. Contemporary thinkers who advocate moral realism share Kant's commitment to the proposition that moral truths are permanent and unchanging products of pure practical reason.[148]

The virtue of the pragmatic turn in philosophical thought was to recognize *the contingency of historical situation* as a fundamental aspect of the conditions of experience and judgment. To acknowledge that the particularities of time and place are part of the conditions governing our understanding of ourselves and the world is to acknowledge that the Eleatic aspiration to timeless truths *possessing content* is as unfulfillable as it is understandable. It is to resign ourselves to the fact that Descartes's endeavor to extend the scope of absolute certainty beyond the field of pure mathematics was ultimately a fool's errand. With that resignation comes the additional acknowledgment that Descartes's aspiration to use clear and distinct insight to render human beings "the masters and possessors of nature" constituted a refusal to respect the meaning and limits of human finitude.

And yet for all that, to be human is to seek to exercise *control* over our experience, ourselves, and the world with which we are forced to come to terms. Philosophers who recognize the centrality of control in human life—or more precisely, the centrality of *acts* that seek to *exercise* control—are guided by a key insight: that the metaphysical tendencies of the tradition led philosophers to *reify* truth, selfhood, and reality. These fundamentally *active* phenomena were reduced to static entities, with the consequence that the essence of each was at best distorted and at worst completely lost.

Thinkers such as Ortega, Santayana, Dewey, and John William Miller all, each in his own way, stress the primacy and centrality of human efforts to *exercise control*. But their conception of what it means to exercise control is pointedly at odds with the Cartesian ideal.

Of the four thinkers I have just mentioned, Miller is relatively unknown; and yet his thinking constitutes an important development of themes explored by the other three philosophers. Miller seizes upon Ortega's notion of *res gestae*, historical accomplishments, and makes it central to his own philosophy of the act. Miller characterizes intelligibility as taking three different forms: the theoretical detachment of Eleaticism, the situated reflection of historical reason, and "the act," by which Miller means "*the noncognitive basis of cognition*," i.e., "the source and control of all cognition."[149] In identifying a level of experience that is prior to thought or cognition in the sense that it "controls" cognition, Miller is implicitly endorsing Heidegger's critique of "apophantic" *logos* as a "founded" way of relating to things: *There is an entire realm of intelligibility that is pre-theoretical and perhaps even pre-predicative, and it is from that level of experience that we gain a sense of orientation that in turn provides us with a sense of what needs to be done.*

On Miller's view, that pre-theoretical level of experience is the primary locus of what Miller calls "the act." The best way to think of this notion is to contrast it with the general account of perception inspired by Eleatic reasoning: On the traditional view, which is the view of everyday common sense, there exists an environment in which various sorts of entities appear, one kind being human beings; these beings employ their sensory apparatus to become aware of this environment and its contents, and on the basis of this acquaintance we can endeavor to act. Like a number of thinkers I have discussed, Miller recognizes the philosophical naivete of this view and opts instead for a conception very much like that of the phenomenologists—that instead of there simply *being* an environment containing entities of various sorts, we must take as our point of departure *the active encounter* we have with whatever lies beyond us. This encounter is the origin of meaning, one consequence of which is that there is no such thing as an environment existing independently of human agents. There simply is "no cognition of . . . absolute objects," but only what shows up in our active engagement with things. "The act *declares* the environment and articulates it. The act is *unenvironed*."[150]

From the standpoint of Eleatic rationality, this statement makes no sense: How can a being that gains its entire sense of purpose from contingencies in its environment act in the absence of an environment? Miller's response to this question is essentially the same as the early Heidegger's, namely, the Protagorean response that things are not simply "there," waiting to be discovered, but instead show up within the field of human experience, a field of meaning that itself constitutes anything like an "environment." Things are "there" precisely to the extent that we *experience* them as *being* there. This is what Heidegger was getting at when he advanced the theses that human "Dasein is world-forming," whereas nonhuman animals are comparatively "world-poor": To "form a world" is to *generate* (as well as modify) an environment, and on the view of thinkers such as Miller and Heidegger this act is unique to possessors of language.

Miller expresses this idea by stating that "the real world . . . [is] a world resulting from what [is] *not* real at all but actual."[151] "The real" in this connection refers to the world or environment that we have constructed on the basis of "the actual," which is the locus of our *active encounter* with what lies beyond us. Another way to contrast the real with the actual is to recall Kant's lesson that form is what we human agents project onto a field of indeterminacy. "The actual" in this sense, which Miller christens "the midworld," is a locus of action in which we *constantly process* our encounter with what lies beyond the bounds of sense. The "real world," in contrast, is a product of our ongoing efforts to render determinate what first shows up as indeterminate.

In taking this approach, Miller seeks to elevate the concrete content of our experience where the tradition sought either to forget it or to subsume it under abstract generalizations that hold the unfulfillable promise of absolute certainty. Where, for example, the Eleatic approach proceeds from a set of concrete experiences of space to the assertion that space exists as "a cognitive object," which for Eleaticism means that the nature of space is independent of whether or not anyone has a tangible experience of it, Miller argues that all such cognitive notions are mediated by acts of what he calls "local control." In contrast with Descartes, who employs an Eleatic conception of reason with an eye toward systematic mastery of all of nature, Miller proceeds from Ortega's observation that "human existence is essentially problematic" and characterizes human efforts at control as being particular or "local." These efforts are mediated by language or symbolic activity. When we confer a name on something in the course of our experience, we confer on it "a unity lacking in passive perception" and thereby contribute to our efforts at control.[152] To name a set of observations a "disease," to call them a "symptom," or to identify an action as "a cure for the disease" is to act in such a way as to give shape to reality. I *use* a stethoscope, I *determine* "an elevated heart rate," I *apply* the notion of causality in a more or less Kantian manner, I *ascertain* that the cause was most likely excessive consumption of caffeine, and I *recommend* measures calculated to alleviate the problem—and it is through *acts* of this kind that we *create, maintain, and revise* the environment within which particular things can show up *as* the things that they are.[153]

The core feature of Miller's thought is *action*. We are so accustomed to think of nature as simply "being there," regardless of whether we have an encounter with it, that it is difficult to see what Miller means when he states that even linguistic signs are not "things" in the traditional sense but instead are themselves *events*: "Every event perceived is perceived via a sign, yet the sign too is an event." Like Saussure and Heidegger, Miller sees "a sign" as "part of a total situation in virtue of which it is a sign at all."[154] That "total situation" is one that we grasp (i.e., render at least provisionally determinate) by employing tools and language to confer meaning on what we first encounter as unstructured and meaningless—precisely because it is through language and the use of instrumentalities that we render things coherent for ourselves. Where on the Eleatic approach terms such as 'space', 'time', and 'cause' reflect (i.e., correspond to) entities and relations "in the real world," Rorty and Miller alike recognize that language does not afford us a transparent window onto the inner nature of things but instead is a human device whose precise relationship to things beyond our ken must remain forever obscure and therefore always remains to be revisited. I noted earlier that thinkers such as Herder sought to dispossess us of the traditional view that language has a divine origin, and to argue for a conception of language as having a "natural" origin. Thinkers such as Rorty and Miller (and, of course, Heidegger and Derrida) radicalize this insight and argue that rather than having a "natural" origin (which presupposes a naïve conception of nature as simply "given"), language has a specifically human one. And while this might lead one to conclude that there is ultimately no difference between, say, Rorty and Miller regarding the prospects for overcoming the threat of arbitrariness in judgment, Miller (like the Pragmatists and Ortega) argues that human beings possess cognitive resources adequate to counter the threat of arbitrariness, even if those resources fall far short of fulfilling the promise of rendering us the masters and possessors of nature.

Where Descartes sought comprehensive, systematic control of the entirety of nature, Miller tailors his aspirations for humanity to the modest proportions of our experiential endowments. No longer viewing the human being as created in God's image and as having at least limited access to divinely ordained (permanent) truth,

Miller acknowledges what many thinkers since Nietzsche, Rorty included, have acknowledged: that there are few if any "permanent" truths, that even these (like *all* truth) originate in human experience, and that the limitations of our experiential capacities render us incapable of determining definitively whether and to what extent any of these "truths" hold for anything outside of our own concrete encounters with particular phenomena. Truth itself, in fact, is an "event" rather than a fixed relation, and this means that we must approach truth case by case or particular by particular, rather than approaching it from the standpoint of a "true world" that is supposed to afford us an Archimedean point somewhere outside the scope of our own experience. Control, in other words, must be *local* rather than global, and it is in principle open to constant rethinking and revision; as time progresses, we gain a better sense of the landscape of history, and we find ourselves confronted with new exigencies.

On Miller's view, our acts of local control occur by means of "functioning objects." This term is potentially misleading, inasmuch as a functioning object is not really best understood in terms of "objectness." A functioning object is not a thing, a *res*, in the traditional sense but instead is a *means* to acquiring and/or enhancing local control. On Miller's historicized conception of reason, a thing is not a static entity but rather "a function."[155] It is not itself the real but instead "prescribes the conditions for the real."[156] Here again the parallel with Heidegger is instructive: Functioning objects (or "things" in this specific sense) are symbolic *activities* that set the parameters for what can show up in our experience and only subsequently become objects of representation. Miller gives as examples of functioning objects linguistic signs and measuring devices such as clocks and yardsticks, and he argues that only once an environment has been disclosed through activities such as describing and measuring can we endeavor to represent the environment as if it were independent of human perception and action.[157] It is only by dint of narrowing our focus to specific features of the environment (an environment that itself is due to human action) that we can contemplate mute Eleatic objects—which, once we have released them from their practical contexts, can appear to exist entirely apart from the context of ongoing practical activity out of which they became abstracted in the first place. Where Heidegger characterizes "equipment" in terms of its "in-order-to" (e.g., we hammer "in order to" drive nails, which we drive "in order to" bind materials, which we bind "in order to" construct an edifice, which we build towards ends that are ultimately for the sake of human beings), Miller characterizes functioning objects in terms of the ways in which the environment becomes disclosed to us (and, as in Heidegger, modified by us) little by little in the course of carrying out our projects, through the employment of linguistic symbols and activities such as measuring. For both thinkers, detached objectivity, to whatever extent it is possible, presupposes a prior acquaintance with things, an acquaintance we obtain not by staring at mute entities but rather by engaging in ongoing practical activities.

This emphasis on engaged practice informs Miller's critique of Kant, for whom "looking at nature one finds only objects. *One finds no subject*, no thinker, no thinking. All that has been omitted."[158] Heidegger saw thinkers such as Descartes and Kant to be key exponents of the notion of "the world as picture," the idea that the human subject is not actually located in nature at all but contemplates it from a standpoint of detached, objective repose with an eye toward proceeding to interact with and exercise systematic control over it.[159] Miller's aim, like Ortega's (and Rorty's, for that matter) is to reinscribe human agency within the flux of time and circumstance. But unlike Derrida and Rorty, Miller recognizes the need to confer *authority* on our assertions and choices, rather than simply assuming that there can be no such demonstration of authority inasmuch as all our choices are ultimately arbitrary.

The twin sources of inspiration for Miller's actualism are the recognition that we are irretrievably tied to history and a commitment to democracy as the highest achievement of human action. This orientation on the past and Miller's aspiration for the future find their nexus in the ideal of the free individual, an ideal that as I have already suggested can be seen as emerging progressively from ancient Greece to the present. It should be clear by now that I consider it far from accidental that the notion of the individual as a center of self-determination and dignity arose in the course of history. Nor do I consider it a particularly trenchant criticism that this ideal arose from within a very specific historical and cultural tradition, the criticism being that this local origin of the ideal shows that it is arbitrary and culturally relative: The question is whether the ideal is one *worth defending*. Where Rorty will aver that any "defense" is ultimately arbitrary, I do consider it conspicuous that Rorty appeals to a notion of the dignified self-determining individual that shares so much with the ideal pursued by thinkers such as James, Dewey, Ortega, and Miller—conspicuous, and more than accidental.

Miller places a fundamental premium on the individual in his actualism, which is to say that the individual is the core agency at work in history. Consonant with his critique of Kant, Miller proposes that "the individual person . . . [is] a center apart from or independent of the real world."[160] As the center and source of actuality, the individual generates, preserves, and modifies the real by *conferring order* on unstructured perception and by reflecting on what has gone before, where "order is the form of functioning, of present active participles."[161] To confer and maintain order is, among other things, to unify diversities in appearance (e.g., X, Y, and Z are all democratic as opposed to totalitarian ideals, even though they may not outwardly appear to be so, etc.) and to revise our sense of what matters on the basis of new insights and unanticipated exigencies.[162] At the core of all this action is the functioning freedom of the individual that is "the source of all distinctions."[163]

But if, as even thinkers such as Heidegger and Miller are aware, the individual is "socially constructed," does this not mean that all of the individual's experiential resources (most particularly as they rely on the contingency of language) are simply effects of contingent processes and thus possess no authority that could endure beyond the current local circumstances? I noted earlier that Heidegger acknowledges this phenomenon of social construction but nonetheless argues for a notion of "authentic" individual choice, which is Heidegger's way of speaking of authoritative (more than arbitrary) decisions. There I argued that Heidegger is left with an insuperable problem of decisionism that recalls Kierkegaard's ideal of spiritual inwardness and Schmitt's notion of the political exception. This is where Miller and Heidegger part company—precisely where Miller opts for an immanent critique of reason where Heidegger demanded a total one. This also constitutes the fundamental point of contention between Rorty and Miller: Is reason just another "effect" or bodily drive, as Nietzsche insisted? Or does reason operate in a manner more along the lines of the way Frankfurt sees reason proceeding in the first several Meditations, such that we come to grasp the potential and achievements of reason not by surveying it from Eleatic distance (as if that were possible) but through our ongoing historical engagement with it? To provide a logically unassailable argument that the free individual is the core agency in history would require the kind of intellectual intuition that we simply do not possess. Thus the fact that there is no such argument is hardly a fruitful criticism of the notion of individual agency. What we ought to do instead, thinkers such as Ortega and Miller believe, is become students of history and start to gain an appreciation of the accomplishments of individual agents, accomplishments that are inherently local whether undertaken by isolated individuals or by groups acting in concert.

These accomplishments involve elevating our acquaintance with discrete particulars to the level of universal pronouncements, as when we assess different specific occurrences as all constituting a certain kind of injustice. A simple, current example is the contested phenomenon of "food deserts," typically economically underprivileged urban areas in which it is all but impossible for residents to obtain (let alone afford) nutritious food of the kind typically easily available in more affluent areas: We see a set of circumstances in a given location, and we see that those circumstances persist through time; we then extend our considerations and find that these phenomena persist in a great many places, we identify the key commonalities, we make determinations as to where those conditions do and do not prevail, and against the background of our already-established sense of justice and injustice we make a pronouncement about the meaning of these conditions that we intend to hold for all of the particular phenomena that we have just compared with one another. Depending on how we interpret the phenomena (a fundamental problem to which I will return in Chapter 5), one can imagine some of us calling the situations in question "food deserts" (with all the attendant meaning that that term has accumulated) and others contesting this interpretation. The "truth" of the situation now becomes an occasion for reflection, debate, and possible revision—a process whose legitimacy, Miller believes, is not undermined by the existence of disagreement but in fact is facilitated through good-faith efforts of members of society to work toward a shared ideal of community.

Central to Miller's actualism is the need to unify universal and particular, an aspiration embraced already in antiquity but in such a manner as to attribute absolute authority to the universal. As a result, Miller notes, the particular has been denigrated in a manner that prevents us from truly coming to grips with reality. Miller cites as an example of this failing on the part of the tradition Boethius's *Consolation of Philosophy*, a text whose author places "emphasis on a reconciling totality" but who does so at the expense of affirming "a world in which the actual has no constitutional place."[164] Concrete particulars, in other words, make no contribution to the endeavor to exercise control, with predictable consequences: Boethius winds up presenting an abstract ideal of spiritual integrity whose relevance to lived experience remains attenuated at best. None of this is changed by the fact that Boethius spends quite a bit of time in *The Consolation of Philosophy* complaining about the concrete injustices he has suffered (and which would lead to his execution). This is clearest in Boethius's Platonic commitment to the proposition that "the order of the universe implie[s] an ethical system for mankind," where "the universe" is not a product of human action but instead possesses a structure that is eternal and completely independent of human agency.[165]

On Miller's view, if there is an ethical order—and there is, although it is constantly evolving—then it must be a product of human action. To opt for a Boethian view would be to fail to recognize that *we ourselves are responsible for the conferring of all order*, moral order included; it would be to affirm that "true world" that Nietzsche dismissed as "a fable." The process of ascertaining or assessing value does not proceed through a reflection on "the divine dictates," but instead occurs in particular moments, each of which gains its meaning from the environed context against the background of which we experience it as well as from a reflection on the past out of which it emerged. Lacking an absolute standpoint for determining value (which would require an axiological standpoint similar to Kantian intellectual intuition), we have no choice but to start with concrete particulars and think from them toward their possible unification under a universal—where that universal itself is a product of actuality, not independent of actuality. To subsume a particular under a universal in this sense is to seek to confer a sense of necessity on what first shows up as contingent; with regard to our values, this involves reflection on the accidental character of "specific elements in motive or action," with an eye toward revising our conduct so as to bring it into accord with our evolving ideal. "Every specific act emerges from a matrix of commitment" that stands in need of constant re-evaluation and revision.[166] When it comes to "correcting" our values, "one must begin by asserting *some* value, and hence must act. But the value asserted, being accidental, discloses its partiality only in the partial failure of the acts that it provokes."[167]

Consider the food desert example I presented a moment ago, or an example of this sort: In 1896 the U.S. Supreme Court, in *Plessy v. Ferguson*, deemed the policy of "separate but equal" to do complete justice to the democratic ideal. By 1954, the Court had revised its view in the light of what it perceived to be "the partial failure of the acts that it provoke[d]," and in *Brown v. Board of Education* declared the policy of "separate but equal" to be "inherently unequal." On a cynical reading, one might conclude that all that changed from 1896 to 1954 was the identities and attendant sensibilities of the Justices populating the court. It goes far beyond the scope of this discussion for me to defend my view, but I find it considerably more persuasive to conclude that the nation's sensibilities, at least as expressed at the level of the Supreme Court, evolved significantly in the intervening fifty-eight years. Moreover, Miller's account of how values actually get revised helps us to understand how simply enacting laws is only part of the process of revising our values. What remains is the ongoing endeavor to bring particular judgments, sensibilities, and actions into accord with the "separate but equal is inherently unequal" principle. This is not simply a matter of subsuming particulars under a pre-given universal, but involves continual refinement of the universal—which itself arises out of acts of local control. As regards race relations in the United States today, this process of refinement is far from concluded, if indeed it can ever come to a fruitful conclusion.

This raging dispute regarding the status of race relations in the U.S. is not simply about what constitutes justice as opposed to injustice; it concerns our very estimation of what counts as a "fact." This reminds us that facts are not simply "there," waiting to be discovered, but instead are, in important part at least, products of the activity of free individuals. For Kant, freedom consists in the capacity to legislate and subject oneself to the moral law that prescribes respect for rational persons. Miller's conception of freedom bears a resemblance to Kant's in its emphasis on activity, but differs from Kant's in eschewing anything like Transcendental Ideas of Reason or Postulates of Pure Practical Reason. But due to our adherence to the Eleatic model of phenomena such as agency, "we are demoralized today because we proclaim liberty but no actuality as local control and as revelation."[168] We are demoralized, in other words, because the Eleatic conception of freedom as residing somewhere outside the world rather than emerging from within it proves to be hopelessly abstract and thus incapable of being brought into concrete contact with that world. (Recall how precious few specific moral duties Kant is able to enumerate in connection with the categorical imperative.) Approaches to freedom such as Kant's leave no room for the actual, which on Miller's view is the process of providing *warrant* for our utterances. Miller's departure from Kant is perhaps most evident in his assertion that "philosophy is the discourse without an '*a priori*'," by which Miller means that the condition of human finitude is such that it is only through utterance that we can provide warrant to utterance.[169]

Miller's assertion that the actual is a process of human interpretation rather than anything like an *a priori* established fact or notion has clear implications for the idea of universality. On Miller's view, to the extent that universals are formulations of human thought that arise in concrete circumstances and serve our efforts at "local control," universals are active rather than static. "In the absence of doing, of the actual," Miller states, "the universal has always collapsed."[170] "Passivity," in other words, "has no universal."[171] Thus universals are not cognitive structures or entities. "As items of knowledge, as *within* knowledge, universals have always been mysterious. I am saying that they are not cognitive at all, that they inform the act, as in going to the post office we become aware of what we are doing."[172] Acts of cognition, on

Miller's view, require universals (hence action and interpretation) as their necessary presupposition, and the ongoing process of action and interpretation involves a constant reflection on and revision of the universals we employ. This is what it means to say that it is only through utterance that we can provide warrant to utterance.

Regarding Kant's conception of universals, Miller offers the following verdict:

> Like all others Kant had to account in some way for the alleged universal. He said it was a priori. We had no hand in discovering it. Where I allow myself to modify that great man is in saying, as I have been saying, that we do have a hand in it—the hand that picks up a yardstick, the tongue that tells time by the clock, the act that keeps a tally by cutting notches in a stick . . . the universal is the form of the actual, generated, enlarged, conflicting as the act maintains itself in local control.[173]

To the objection that the Kantian subject does indeed "have a hand" in discovering universals, the response must be that in comparison with Miller's conception of universals as a fluid, active, historically conditioned process bound up with ongoing criticism, the Kantian subject functions more like an anonymous and inert mechanism that "discovers" truths that are more or less eternal. Does the Kantian subject really "have a hand" in determining or discovering that spatial relations are characterized *a priori* by Euclidean geometry, or that all and only beings who are rational in a specifically human sense count as "persons"? This conception of "having a hand" simply does not do justice to Miller's insight that "the individual is the locus of order, or, at least, a factor in all universality."[174] The Kantian subject is not so much a "factor" as a *consequence* of notions or rules (for after all, Kant characterizes concepts as rules) that are essentially prior to, and hence function as, the necessary condition for any activity, be it scientific or ethical, on the part of the subject.[175] Stated bluntly, the upshot of Miller's critique of Kant on the question of universals is that Kant actually *deprives* the subject of autonomy by treating the self as essentially an elaborate calculating device. (And to say that Kant does this "actually" has a Millerian significance.)

Miller's focus on the actual and his insistence that "any universal entails the verb" has important implications for universal claims in ethics.[176] In a discussion of the cosmopolitan ideal and the nature of ethical commitments, Miller states that "the universal is the form of limitation and in all its modes declares the order of critical finitude."[177] The notion of critical finitude distinguishes Miller clearly from Kant: Miller sees finitude as irretrievably inscribed within history, and he sees values not as static entities but rather as provisional commitments that stand in need of constant contestation, substantiation, and revision. In this connection he is worth quoting at length:

> A corollary of the possibility of history occurs in the necessity of asserting the absoluteness of some actual embodiment of values. If this is narrowness and

idolatry, it is also loyalty, and the condition of affection. That morality occurs in limit and only there, that it is not the reaching for a timeless value but for some present, incarnate, and imperfect good, may seem a strange doctrine. But in that way a philosophy of history is extended into the region of ethics. Ethical theory has suffered from an inability to settle for limitation and commitment, while history, and the men who make it, have been forced to treat every aim as called for by particular situations. Action proceeds from limit, and it arrives at no finality, but only at another defective limit.[178]

Like historical judgment, "every moral judgment rests on the base of a current concern, on what one now identifies oneself as doing."[179] Note how this differs from Kant's "categorical" conception of moral judgment: When we make a moral judgment, we must focus that judgment on a particular concern with a particular location. And while "no location is absolute . . . some location must be taken to be absolute in order to make any measurement of relative velocities. This is the factor of limit."[180]

Thus at any given time, there will be an established background of commitments in relation to which current concerns will be recognized, identified, considered, judged, and acted upon. The assessments, judgments, and actions we undertake about those current concerns influence and give new shape to the background against which old concerns and commitments will be revisited and new ones approached. Ethical judgment is an historical process. In the spirit of Miller's thought, we could say that we continually "re-actualize" our concerns, commitments, criteria, and ethico-historical trajectory. In contrast with this conception of morality,

> the Kantian morality appears defective. . . . There is no duty nor any rationality until the non-rational and existent moment gives leverage to the moral law. The question 'What ought I to do?' can get no answer, because it makes no sense, until one is already doing something which has for oneself an uncompromising value. This is the idea of the relatively static.[181]

These notions of the "existent moment" and "the relatively static" bear traces of Kierkegaard and Heidegger and, to my mind, signal a serious difficulty with the notion of universals and universal moral judgment as historically contingent. As I noted a short while ago, Miller, like Rorty and Derrida, faces the challenge of explaining how judgments that take their bearings from within a realm of contingency can attain genuine authority or "warrant." It seems entirely reasonable to say, along with Miller, that "only a pure actuality can propose a pure ideality"[182]: the inherently finite nature of human action renders the prospect of "a pure actuality" the kind of fanciful wish-fulfillment that should be dismissed as a romantic flight of fancy comparable to Heidegger's appeal to the silent call of Being. Such notions of purity deserve to be relegated to that "cloud-cuckoo land" so incisively derided by

Schopenhauer. For it is exceedingly difficult to see how we could ever arrive at the kind of "pure ideality" promised by Kant. There seems to be something quite right in Miller's suggestion that "one must, of course, discover the dark depths of the non-rational immediacy" in order to do justice to the inherent limits of action, criticism, and ethical choice.[183] But Miller himself expresses an awareness of a fundamental danger: that of "the demonic," which occurs when we give undue centrality or emphasis to one partial aspect of the actual, thereby obscuring or doing an injustice to others.[184] Miller attributes this notion of the demonic to Tillich; but Tillich for his part seems to be at least implicitly indebted to Kierkegaard, who in *Fear and Trembling* offers a diagnosis of the demonic as the assertion of the absolute without reference to the universal.[185] Kierkegaard anticipates Miller's concern, albeit in different jargon—the concern that some one aspect of the entire field of concern will exercise a distorting influence on our efforts to grasp the whole *as* a whole.

Miller confronts the danger of "this dark partiality of the demonic" by characterizing it as "the condition for testing both its own authority and the authority of other interests. . . . History rides on the vehicles of partial truth, but their demonry is the sole condition of discovering their force."[186] It is here that Miller makes a crucial step beyond Kierkegaard. Kierkegaard sees the aesthetic sphere of existence as the locus of the threat of the demonic, and he sees in the spheres of the universal and religiousness the promise of vanquishing that threat. The move from the aesthetic to the universal constitutes for Kierkegaard a step beyond aesthetic selfishness and the acceptance of one's membership in an ethical community built on mutual recognition and accountability to others. Where the aesthete (perhaps the demonic individual) can dwell in spontaneity and coincidence, one who occupies the universal, which by definition is the ethical, brooks neither "coincidence" nor "any esthetic illusion."[187] Thus Agamemnon, Kierkegaard's exemplar of the tragic hero dwelling in the ethical or universal, can and must explain himself to his community in rational terms that can be grasped and, presumably, evaluated by others who dwell along with him in the universal.[188]

But Kierkegaard's universal suffers from an inherent deficiency to which Miller's is not susceptible: Kierkegaard's universal loses any dimension of fine-grained particularity, which is to say that it, like Kant's Postulates of Pure Practical Reason, remains entirely formal and abstract, lacking any power to define the concrete individual. By the same token, Kierkegaard's universal lacks the crucial dimension of criticism emphasized by Miller, which means that Kierkegaard's universal-ethical sphere has no essential tie to history. The core problem here is that Kierkegaard's universal, like Kant's, has no essential grounding in what Miller calls "actuality." This is particularly clear in the way in which Kierkegaard characterizes the dilemma of the "tragic hero" who exemplifies the ethical or universal sphere: We are to picture Agamemnon struggling to choose between two entirely abstract duties, one held by a father in relation to his daughter and another held by a king in relation to his nation. These duties are presented as categorical rather than as proceeding from any concrete actuality.

Kierkegaard seeks to solve this problem of abstractness by subordinating the universal ethical sphere to that of religiousness, which in its highest manifestation is a higher immediacy than the immediacy of the aesthetic: Where the aesthete's spontaneity and uniqueness know nothing of the universal or ethical and thus are susceptible to the threat of the demonic, the knight of faith has been through the universal and remains dialectically informed by it even when "teleologically suspending" it for the sake of a duty to God that the knight of faith recognizes to stand above the universal. This duty to God confers concrete, unique content on the knight of faith, "who as the single individual stands in an absolute relation to the absolute."[189] But precisely in standing above the universal, this singular individual's essential choice cannot be made intelligible to others. It is liable to appear to others to be "madness," as when Abraham demonstrates his readiness to sacrifice Isaac on Moriah.[190] "Whether the single individual actually is undergoing a spiritual trial or is a knight of faith, only the single individual himself can decide," inasmuch as the single individual lacks "any connections and complications" that would tie him or her to the communal life of the universal, which is the locus of critical contestation of commitments.[191]

Precisely in virtue of standing above the universal, the singular knight of faith stands in clear isolation from the community. Even though this individual has gone through the ethical universal and thus has an appreciation of community, the knight of faith is fully prepared at any spontaneous moment to sacrifice the universal for a putatively higher immediacy in which the individual is answerable to no one but one's own private god. Miller wisely avoids the epic threat posed by the attempt to place the individual above the community, an attempt that is at the same time a removal of the individual from the particularities of history, its inherent "constitutional conflicts," and the demand to test one's convictions through a critical process that can be conducted only within a historically situated community. The Kierkegaardian aspiration to find security in the absolute is entirely understandable: we are, after all, clever beasts who have to die. But by seeking that security in a realm of transcendent purity that is impervious to criticism, Kierkegaard strives to evade a fact that Miller made central in his reflections on the human condition: not merely that "history rides on the vehicles of partial truth," but that "their demonry is the sole condition of discovering their force."[192] The very demonry that Kierkegaard seeks to vanquish, first through a turn to the universal and then through a turn to religiousness, is not only an insuperable threat faced by any being who lives historically but is in fact *part of the conditions of criticism itself.* "It is only a philosophy which endows limit with ontological status which can turn conflict into constructive use. The use of conflict is the disclosure of the necessary."[193] On Miller's view, there simply is no way to ascertain what is ethically "necessary" or incumbent at any given time, in the absence of conflict, struggle, and the constant reaffirmation of a fundamental "incompleteness in ourselves" that cannot be overcome by the flight from history and circumstance to any sort of refuge in transcendent purity.

But therein lies a paradox: If we are bereft of criteria for conducting the process of criticism that are anything more than "relatively stable," and if criticism is irretrievably tied to an "existent moment" that vividly recalls the Kierkegaardian-Heideggerian "moment of vision" that Ernst Tugendhat and Karl Löwith both recognized to be beset with decisionistic implications, how are we to save the process of criticism from the kind of fatal relativism that renders it indistinguishable from the very demonry that it seeks to overcome?[194] For that neither Miller, nor anyone else who takes the historical character of the ethical seriously, has as yet offered an answer. My own sense is that the solution to this problem will involve recourse to two key resources: the notion of a background ideal of living that I examined in Chapter 1, and a more open acknowledgment of our specifically *embodied* encounter with the world around us. It is to the notion of a background ideal of living that I think Santayana appeals implicitly when he suggests that "our criticism will be solid in proportion to the solidity of the unnamed convictions that inspire it," convictions that themselves reside in a shared moral world that is the necessary condition for communication and moral solidarity.[195] But what are the terms of this shared moral world, and in virtue of which sorts of criteria are we justified in embracing it? I believe that a complete answer to this question demands a rethinking of the relationship between thought and embodiment, one that finally recognizes that our affective life is every bit as vital to our moral existence as is thought. While Miller does not fully embrace this insight, he exhibits some clear indications of an appreciation of it, the most conspicuous being his verdict that "the basic functioning object is the body."[196]

6. A Concluding Thought

What I have tried to show in this chapter is that the acknowledgment of contingency and the abandonment of the Eleatic ideal of rationality need not bring with them an abandonment (or "total critique") of reason, but instead require a fundamental rethinking of the nature of rationality, its potential, and its limits. The fact that reason is incapable of fulfilling our Eleatic aspirations is not a sign that reason itself is bankrupt; Husserl was right to suggest that the endeavor to abandon altogether (or to "reject" it in the sense I have discussed early in this chapter) is itself a manifestation of "lazy reason." While there is no context-free logical demonstration of the power of historical reason, *human history exhibits the operation of reason.* That we do not have an absolute, transparent grasp of our own functioning as rational agents is not a sign that reason is impotent; instead it simply confirms Ortega's and Miller's insight that we come to know reason through our concrete employment of it. Miller's elaboration of Ortega consists in his preliminary exploration of a level of meaningful interaction that precedes and provides the basis for rational operations such as moral reflection. What remains inadequately explored in Miller's thought is the precise nature of this deeper level of engagement with the world. In the next chapter I will turn to the thought of Heidegger and Merleau-Ponty in an effort to provide a more fine-grained characterization of that level of engagement, with a special focus on its irreducibly *affective* (embodied, felt) character.

In this chapter I have said very little about the implications of the ideal of historical reason for the moral status of nonhuman animals. Here my goal has been to defend a revised conception of reason consonant with the process of immanent critique. In the next chapter I will return to the historical controversy regarding the status of the emotions that I discussed in Chapter 2; in Chapters 4 and 5, I will seek to defend a conception of the affective dimension of moral choice, one that addresses the concerns of the tradition regarding the pernicious tendencies of the emotions and places emphasis on the continuities rather than the putative differences between human and nonhuman embodied agents. I will propose a model of moral judgment according to which reason and emotion necessarily operate in dialectical relationship with one another; and I will argue for the conclusion that historical reason, when understood as functioning in dialectical relationship with emotion, points us in the direction of accomplishing two things: the vindication of the democratic ideal as considerably more than an arbitrary choice on the part of humanity, and the inscription of human community within the larger context of a cosmic community that includes more than simply human beings. This last aspiration is one that I recognize to be highly controversial and thus subject to ongoing contestation. As I hope to show in Chapter 5, it will not simply be a matter of offering logical arguments for including nonhuman animals as direct and full beneficiaries of moral concern, but instead will require a *felt recognition* that the terms of the anthropocentric background ideal of living are inherently self-serving and inadequate to the demands of cosmic justice.

Notes

1 William James, *Pragmatism*, p. 520.
2 *Discourse on Method*, Part 5, *The Philosophical Writings of Descartes*, vol. 1, pp. 131f., 139.
3 Martin Heidegger, *Nietzsche*, vol. 4, p. 8.
4 Friedrich Nietzsche, *Beyond Good and Evil*, "Our Virtues," sec. 228, Friedrich Nietzsche, *Basic Writings of Nietzsche*, p. 347; "On the Prejudices of the Philosophers," sec. 19, *Basic Writings of Nietzsche*, p. 215. See also my remarks on Nietzsche in *Animals and the Limits of Postmodernism*, pp. 10–22.
5 Friedrich Nietzsche, *On the Genealogy of Morals*, Second Essay, sec. 12, *Basic Writings of Nietzsche*, p. 514.
6 Nietzsche offers precious few specific historical instances of "the higher type"; when he does, he cites figures such as Caesar, Napoleon, and Cesare Borgia. See *Beyond Good and Evil*, "Natural History of Morals," sec. 201; *On the Genealogy of Morals*, First Essay, sec. 16; Friedrich Nietzsche, *The Will to Power*, Book Four, sec. 871. The implicit background ideal that motivates Nietzsche is hard to classify: is it anthropocentric or non-anthropocentric? It arguably exhibits features of both. But what is ultimately decisive is that it lacks the broad commitment to community that is evident in both of the background ideals of living that I examined in Chapter 1. (Needless to say, the non-anthropocentric ideal expresses a considerably broader such commitment than does the anthropocentric ideal. What is important here is that both ideals, unlike the one articulated by Nietzsche, at least purport to be committed to peace, mutual respect, and

a harmonious community; the key difference is the *scope* of community embraced by each.)

7 *The Will to Power*, Book Two, sec. 462, p. 255: "*Fundamental innovations*: In place of 'moral values', purely naturalistic values. Naturalization of morality." See also *Beyond Good and Evil*, "Our Virtues," sec. 230.

8 Richard Rorty, *Contingency, Irony, and Solidarity*, p. 33.

9 Those familiar with Rorty's views will undoubtedly rankle at this claim. What I hope to show is that even though Rorty outwardly rejects as essentially incoherent the claim that his liberal ironism is the "best" or "true" account of optimal human functioning, he is nonetheless implicitly committed to that claim.

10 See Charles Patterson, *Eternal Treblinka*. I attended a public lecture given by Rorty at Yale University in 2000 and asked him during the question-and-answer period which sorts of criteria should guide us in making moral judgments and distinctions. Rorty responded simply by stating that "you don't need criteria." That statement frightened me then, and it frightens me now—for reasons that should become clear by the end of this chapter. Of course, Rorty did not mean literally that "you don't need criteria"; what I think he meant was that, as he states in *Contingency, Irony, and Solidarity*, p. 80, "there is nothing beyond vocabularies which serves as a criterion of choice between them." If there are criteria—and I will show later in this chapter that Rorty himself appeals to them in an explicit and sustained manner—then, on Rorty's view, they are contingent matters and thus subject to endless revision.

11 Carl Schmitt, *Political Theology*, p. 15.

12 *Political Theology*, p. 31.

13 See my remarks on decisionism in *Animals and the Limits of Postmodernism*, pp. 32–41.

14 Ernst Tugendhat, *Der Wahrheitsbegriff bei Husserl und Heidegger*, p. 361. Martin Heidegger discusses Kierkegaard's notion of the *Augenblick* in *Being and Time*, p. 388n.iii.

15 *Being and Time*, pp. 376, 387f.; Jean-Paul Sartre, *Existentialism and Human Emotions*, p. 24.

16 *Existentialism and Human Emotions*, pp. 20–1, 38–9.

17 *Existentialism and Human Emotions*, pp. 40–1.

18 Think, for example, of Aristotle's ideal of a political community in which no women are ever citizens (active participants in legislation or the administration of justice) and enslavement is good for everyone (including the enslaved person).

19 In this connection, one need only compare Heidegger's declaration after World War II that the overcoming of nihilism "announces itself in German poetic thinking and singing" with Sartre's sustained efforts to oppose French colonial rule in Algeria. Martin Heidegger, "The Rectorate 1933/34: Facts and Thoughts," p. 29.

20 José Ortega y Gasset, *Historical Reason*, p. 105.

21 *Historical Reason*, p. 105.

22 *Historical Reason*, p. 115.

23 *Contingency, Irony, and Solidarity*, p. 51.

24 *Contingency, Irony, and Solidarity*, p. 53.

25 Jacques Derrida, *Of Grammatology*, p. 153. Elsewhere Derrida explains what he meant by this often misinterpreted expression: "When I say that there is nothing outside the text, I mean that there is nothing outside context, everything is determined." Derrida, "Hospitality, Justice and Responsibility," p. 79.

26 "On Truth and Lies in a Nonmoral Sense," p. 84.

27 "On Truth and Lies in a Nonmoral Sense," p. 84. Rorty explicitly endorses Nietzsche's approach to language and asserts that "the world does not provide us with any criterion of choice between alternative metaphors . . . we can only compare languages or metaphors with one another, not with something beyond language called 'fact'." *Contingency, Irony, and Solidarity*, pp. 17, 20.

28 Edmund Husserl, *The Crisis of the European Sciences and Transcendental Phenomenology*, p. 16.

29 Rorty, for his part, seeks to detach "philosophy" from the traditional conception of reason, arguing that one can "think of philosophy as *in the service* of democratic politics." *Contingency, Irony, and Solidarity*, p. 172. I cannot help but wonder whether Rorty, in making this move, is relying on reason a little more than he acknowledges.

30 Of the numerous examples of studies of the link between reason and violence that could be given here, Max Horkheimer and Theodor W. Adorno's *Dialectic of Enlightenment*, serves as an excellent representative.

31 *The World as Will and Representation*, vol. 1, fourth book, sec. 53, p. 273. Schopenhauer borrows the expression from Aristophanes, who employed it in "The Birds."

32 "On the Essence of Truth," p. 138.

33 See my discussion of Condillac and Herder on the origin of language in *Anthropocentrism and Its Discontents*, pp. 175–84.

34 *Animals and the Limits of Postmodernism*, Chapter 2. There I note that the bottom line for Derrida is that "the primacy of the trace means that there are no authoritative criteria for discriminating between discourses" (p. 72). As I have already suggested, I believe that this verdict holds for Rorty as well.

35 "Letter on 'Humanism'," pp. 239, 252. This is not the place to go into Derrida's spirited critique of and departure from Heidegger; here I will simply note that Derrida has no patience for Heidegger's surprisingly traditional (and often evangelical-sounding) call for a sense of reverence not for humanity or nature, but for "being" pure and simple.

36 *Being and Time*, p. 62; Martin Heidegger, "What are Poets For?," p. 129.

37 On Saussure's view, as well as on Derrida's criticism of Saussure for retaining faith in a "transcendental signified," see my remarks in *Animals and the Limits of Postmodernism*, pp. 56–7.

38 Immanuel Kant, "An Answer to the Question: What is Enlightenment?," pp. 55–6.

39 *Contingency, Irony, and Solidarity*, p. 67.

40 *Contingency, Irony, and Solidarity*, p. 173.

41 *Contingency, Irony, and Solidarity*, p. 75. See also pp. 97 (a liberal ironist is one "attempting autonomy" by "trying to get out from inherited contingencies and make his own contingencies") and 73 (a "final vocabulary" is "a set of words which [human beings] employ to justify their actions, their beliefs, and their lives").

42 *Contingency, Irony, and Solidarity*, p. 84. Here the influence of William James's pragmatist turn is apparent: For James, "ideas (which themselves are but parts of our experience) become true just in so far as they help us to get into satisfactory relation with other parts of our experience," i.e., an idea is true just in case it proves "helpful in life's practical struggles." *Pragmatism*, pp. 512, 520.

43 *Contingency, Irony, and Solidarity*, p. 47.

44 *Contingency, Irony, and Solidarity*, p. 65.

45 *Contingency, Irony, and Solidarity*, p. vii.

46 *Contingency, Irony, and Solidarity*, p. 53.

47 *Contingency, Irony, and Solidarity*, p. 67.

48 *Contingency, Irony, and Solidarity*, pp. 48, 58.

49 *Contingency, Irony, and Solidarity*, p. 20, 67.

50 *Contingency, Irony, and Solidarity*, p. 32.

51 *Contingency, Irony, and Solidarity*, p. 86.

52 *Lives of the Eminent Philosophers*, vol. 2, p. 65; cf. *Contingency, Irony, and Solidarity*, p. 190.

53 *Contingency, Irony, and Solidarity*, p. 80.

54 *Contingency, Irony, and Solidarity*, p. 174. Rorty cites writers such as Proust and Orwell as possessing such special talents.

55 *Contingency, Irony, and Solidarity*, p. 68.

56 Martin Heidegger, "Postscript to 'What is Metaphysics'?," p. 237. See also Martin Heidegger, *1. Nietzsches Metaphysik. 2. Einleitung in die Philosophie. Denken und Dichten*, pp. 136–60.

57 "Letter on 'Humanism'," p. 275. When Heidegger refers to "thinking" in this connection, he is explicit that he does not have the traditional Eleatic conception of reason in mind; he is at pains to distinguish between what he calls "calculative" and "contemplative" thought, the latter being open to the mystery of reality in a way that Eleatic reason precisely is not. Contemplative thinking is the thinking that "lets beings be." See "Letter on 'Humanism'," pp. 271–2.

58 In spite of this shared appeal to poetry, Heidegger and Rorty are at odds regarding the notion of truth, the later Heidegger casting it in terms of "the truth of being," where being, as I have already noted, is to be understood as "the *transcendens* pure and simple"—a notion that Rorty categorically rejects.

59 *Contingency, Irony, and Solidarity*, p. xv.

60 *Contingency, Irony, and Solidarity*, p. xv.

61 *Contingency, Irony, and Solidarity*, p. xv.

62 *Contingency, Irony, and Solidarity*, p. xv, 51, 65.

63 See, for example, *Personhood, Ethics, and Animal Cognition*, p. 134. Varner does not actually deny this capacity in nonhuman animals but instead suggests that we have "no good evidence" for it in them. Varner never considers the possibility that this lack reflects a limitation on the part of human beings (namely, our incapacity to understand the nonhuman, even where the evolutionary continuity and physiological similarities are significant) rather than on the part of nonhuman animals.

64 *Contingency, Irony, and Solidarity*, p. 92. Cf. p. 36, where Rorty suggests that the "faculty for creating metaphors" is unique to human beings.

65 *Contingency, Irony, and Solidarity*, pp. 92, 94, 177. Rorty, of all people, knows very well that *words matter*. Thus we should not be too quick to "excuse" as momentary lapses these denigrating references to nonhuman beings—nor should we forget that in the one instance in the book in which Rorty discusses a specific type of nonhuman animal, he writes of "the relative mindlessness of the monkey" (p. 15)—a statement that can arguably be defended, albeit only up to a point.

66 *Contingency, Irony, and Solidarity*, p. 45.

67 An excellent example of this sort of seeming overreaching is Marc Bekoff and Jessica Pierce's book *Wild Justice*, in which the authors attribute a highly sophisticated sense of justice and injustice to a variety of nonhuman animal species. As much love and respect as I have for nonhuman animals, I can't help but wonder whether the evidence cited by Bekoff and Pierce really supports their central thesis. But as I have argued here as well as in previous work, whether a given nonhuman animal possesses inherent moral worth does not depend on whether it possesses a sense of justice—any more than a human child's moral worth depends on whether it has such a sense.

68 I present my views on the experiential capacities of nonhuman animals in the first three chapters of *Animals and the Moral Community*. There I argue that many nonhuman animals appear to have rich subjective lives, but I also deem it unlikely that any (or many) nonhuman animals are capable of predication and conceptual abstraction.

69 Douglas Keith Candland, *Feral Children and Clever Animals*, p. 369.

70 C. Lloyd Morgan, *Animal Behavior*, p. 270.

71 *Introduction to the Philosophy of History*, p. 74.

72 G.W.F. Hegel, *The Science of Logic*, p. 41 (GW 21.48); Charles Taylor, *Hegel*, p. 85.

73 Hegel, *The Philosophy of Right*, sec. 270, p. 165.

74 Johann Gottfried von Herder, "Treatise on the Origin of Language," p. 155 (italics in original; translation altered).

75 Arthur Schopenhauer, *The Fourfold Root of the Principle of Sufficient Reason*, Ch. 5, sec. 27, p. 151 (italics in original); *The World as Will and Representation*, vol. 1, book 1, sec. 16, p. 84.

76 *The Animals Issue*, pp. 189–90. Cf. p. 183: Carruthers's reasoning is that for an event to count as "conscious," it must be "immediately available to a faculty of reflexive thinking."

77 Martin Heidegger, *The Fundamental Concepts of Metaphysics*, pp. 177, 342 (where 'benommen' is translated as 'captivated').
78 *Contingency, Irony, and Solidarity*, p. 40 (italics in original).
79 *Contingency, Irony, and Solidarity*, p. 45.
80 On the inward turn in the early Christian Church, see my discussion in Chapter 5 of *Anthropocentrism and Its Discontents*; on Augustine's early version of the *cogito*, see my remarks in *Descartes as a Moral Thinker*, p. 92.
81 *On the Soul*, book 3, Ch. 4, *The Complete Works of Aristotle*, vol. 2, p. 683.
82 Martin Heidegger, "The Age of the World Picture," Appendix 4, p. 140. Cf. p. 141 (Appendix 5), where Heidegger names Leibniz, Kant, Fichte, Hegel, and Schelling as leading exponents of the modern "subjectivist" notion of *Geist*.
83 *Critique of Pure Reason*, "Refutation of Idealism," p. 244 (at B274).
84 "The Age of the World Picture," Appendix 8, p. 145.
85 "The Age of the World Picture," Appendix 8, p. 143. See also Martin Heidegger, "Plato's Doctrine of Truth," and Paul Friedländer's critique of Heidegger in Chapter 11 ("Aletheia") of *Plato: An Introduction* (where Friedländer challenges Heidegger's claim that truth "originally" meant *aletheia* rather than *orthotes* or correspondence).
86 An influential recent example is Elaine Scarry's *The Body in Pain*, although her primary focus is less the inexpressibility of pain than the ways in which the infliction of pain is used to manipulate and humiliate. (See also *Contingency, Irony, and Solidarity*, p. 94: "Pain is nonlinguistic.")
87 "Dedicatory Letter to the Sorbonne," p. 4. I have already noted that in *Descartes as a Moral Thinker* I dispute the conventional wisdom that Descartes was a purely secular thinker, particularly in matters of morality.
88 *On Abstinence from Killing Animals*, book 3, sec. 3. For a closer examination of Porphyry's views (including his extensive reliance on anecdotes offered by Plutarch), see my remarks in *Anthropocentrism and Its Discontents*, pp. 103–11.
89 Aristotle is ultimately ambivalent on the question whether any nonhuman animals really possess understanding and ingenuity; on this question there is a tension between the ethological texts (in which Aristotle demonstrates an intimate familiarity with the behavior of a great many nonhuman animal kinds) and the ethical and political texts (in which, as I noted in Chapter 1, he denies *logos* in nonhuman animals and excludes them from community with human beings). See my discussion of Aristotle in Chapter 3 of *Anthropocentrism and Its Discontents*.
90 *On Abstinence from Killing Animals*, book 3, sec. 9, 11.
91 *The Fourfold Root of the Principle of Sufficient Reason*, sec. 20, pp. 70–1 (italics in original).
92 *The Fourfold Root of the Principle of Sufficient Reason*, sec. 21, pp. 110–1. I examine Plutarch's and Porphyry's views in depth in Chapter 4 of *Anthropocentrism and Its Discontents*.
93 *The Fourfold Root of the Principle of Sufficient Reason*, sec. 21, p. 119.
94 *The Fourfold Root of the Principle of Sufficient Reason*, sec. 21, p. 111.
95 *The Fourfold Root of the Principle of Sufficient Reason*, sec. 21, p. 120.
96 *The World as Will and Representation*, vol. 1, book 1, sec. 6, p. 21.
97 *The World as Will and Representation*, vol. 1, book 1, sec. 7, p. 30 (to possess eyes is to understand); sec. 6, p. 23 ("in judging the understanding of animals, we must guard against ascribing to it a manifestation of instinct, a quality that is entirely different from it").
98 See *Discourse on Method*, Part 5, *The Philosophical Writings of Descartes*, vol. 1, p. 139–41 and my examination of Descartes's views about animal mechanism in Chapter 6 of *Anthropocentrism and Its Discontents*.
99 *The World as Will and Representation*, vol. 1, book 1, sec. 6, pp. 21–3.
100 *The World as Will and Representation*, vol. 1, book 1, sec. 26, p. 131f.
101 Arthur Schopenhauer, *On the Basis of Morality*, sec. 17, p. 151.

102 *On the Basis of Morality*, sec. 17, p. 151.
103 *On the Basis of Morality*, sec. 19, pp. 175, 180.
104 *On the Basis of Morality*, sec. 8, pp. 95–6.
105 *The World as Will and Representation*, vol. 1, "Appendix: Criticism of the Kantian Philosophy," p. 421. Cf. p. xv: As Payne, Schopenhauer's translator, points out, "by identifying the Kantian thing-in-itself with the will in ourselves, [Schopenhauer] maintained that experience itself as a whole was capable of explanation." Here it is crucial to bear in mind that for Schopenhauer, the will is "in us" in the sense that its manifestation in individuals is simply a fleeting manifestation of the world will.
106 *The World as Will and Representation*, vol. 1, sec. 34, p. 181; Arthur Schopenhauer, "On the Doctrine of the Indestructibility of our True Nature by Death," vol. 2, p. 269.
107 *The World as Will and Representation*, vol. 1, sec. 33, p. 176.
108 Heidegger makes a convincing case for the proposition that modern science, which outwardly appears to be "neutral" and which historically predates the technological revolution, is actually technological *in its very essence*, which is to say that the nature and emergence of modern science were driven by a prior motive of control—a motive that, with the development of scientific thinking in early modernity, became a motive of systematic control over all of physical reality (recall Descartes's call to employ modern physics to render human beings "the masters and possessors of nature"). Martin Heidegger, "The Question Concerning Technology," p. 21.
109 *Discourse on Method*, Part 1, *The Philosophical Writings of Descartes*, vol. 1, p. 114; Part 2, p. 120. Descartes presents his ideal of a scientifically informed morality in the "Preface" to René Descartes, "The Principles of Philosophy," p. 186.
110 See Jacques Derrida, *La bête et le souverain*, vol. 2, pp. 69 (Heidegger's "obsession with orientation"); 78 (to pick a direction for investigation is to settle on a "dominant figure or trope"); 101 (Heidegger's choice to start with what lies closest to us is "forceful and arbitrary"); 137 (dislocation has priority in all textual events).
111 José Ortega y Gasset, "History as a System," p. 115.
112 José Ortega y Gasset, "Man as Technician," pp. 92–3. Here again, the affirmation of human freedom comes at the cost of denying freedom to nonhuman animals. In this respect, even thinkers such as Ortega and Miller retain the same pointedly anthropocentric commitment as the medieval philosopher John of Damascus, whose final verdict on nonhuman animals was that *non agunt sed magis aguntur*: they do not act but are rather acted upon. See my remarks about Aquinas's reliance on this assumption in *Anthropocentrism and Its Discontents*, p. 128.
113 *Historical Reason*, pp. 53, 65.
114 *Historical Reason*, pp. 83, 117.
115 *Pragmatism*, lecture 2, p. 520; lecture 8, p. 612.
116 *Pragmatism*, lecture 6, pp. 583–5.
117 *Pragmatism*, lecture 6, p. 574 (italics in original).
118 Pico Della Mirandola, *Oration on the Dignity of Man 1486*.
119 *Historical Reason*, p. 186.
120 *Historical Reason*, pp. 185. Hans Jonas makes a similar point: He states that the threat of nihilism with which we find ourselves confronted today is considerably "more radical and more desperate" than the threat of nihilism confronted by the Gnostics in antiquity: If the world was meaningless for the Gnostics in virtue of having been fashioned by the Demiurge rather than by God, at least they could look forward to the promise of eternal salvation—an event for which modern nihilism leaves no room. Hans Jonas, *The Gnostic Religion*, p. 339.
121 "History as a System," p. 217.
122 *Historical Reason*, p. 220.
123 *Historical Reason*, p. 221.
124 *Historical Reason*, p. 231.
125 *Historical Reason*, p. 116.

126 *Historical Reason*, pp. 118, 114 (italics in original). Cf. p. 113: "The true task begins when thinking attempts to adapt logic, which is *intelligence*, to the illogical thing that *reality* is" (italics in original).

127 *Historical Reason*, p. 118.

128 Friedrich Nietzsche, *The Gay Science*, sec. 343.

129 *Historical Reason*, p. 188.

130 *Historical Reason*, p. 205.

131 John Dewey, *Experience and Nature*, p. 145.

132 René Descartes, Third Meditation, *The Philosophical Writings of Descartes*, vol. 2, p. 24.

133 Harry R. Frankfurt, *Demons, Dreamers, and Madmen*, pp. 25–6, 177. Cf. p. 179, where Frankfurt cites a famous passage from the Second Replies (in which Descartes speculates that things may appear differently to God or an angel than to a human being) as his basis for arguing that Descartes is not really interested in correspondence. What Frankfurt neglects to note is that Descartes concludes that passage by explicitly *rejecting* the possibility he has just entertained. See Replies to the Second Set of Objections, *The Philosophical Writings of Descartes*, vol. 2, p. 103–4. A considerably more accurate reading of Descartes's intentions is provided by Bernard Williams, who argues that "we cannot understand Descartes if we break the connection between the search for certainty the search for truth, or the connection between knowledge and the correspondence of the ideas to reality." Bernard Williams, *Descartes*, p. 200.

134 *Demons, Dreamers, and Madmen*, p. 147.

135 *Demons, Dreamers, and Madmen*, p. 105; Second Meditation, *The Philosophical Writings of Descartes*, vol. 2, p. 17 ("This proposition, *I am, I exist* [*ego sum, ego existo*] is necessarily true whenever it is put forward by me or conceived in my mind"—italics in original). The more famous expression "*cogito ergo sum*" appears not in the *Meditations* but in Part 4 of the *Discourse*.

136 *Demons, Dreamers, and Madmen*, p. 107.

137 Thus, as Frankfurt notes, it is a mistake to characterize Descartes as a skeptic. *Demons, Dreamers, and Madmen*, p. 16. Edwin Curley makes an extended case for this conclusion in *Descartes Against the Skeptics*. Regarding Descartes's service in the war, commentators occasionally suggest that an interest in cannonball trajectories inspired Descartes to develop his geometry.

138 *Historical Reason*, p. 113.

139 "Preface" to "The Principles of Philosophy," p. 186.

140 *Critique of Pure Reason*, "The Ideas in General," pp. 310–1 (at A313–15/B370–72).

141 "History as a System," p. 230.

142 "History as a System," p. 219.

143 See my remarks on eschatology in *Descartes as a Moral Thinker*, Chapter 5, *passim*.

144 *Historical Reason*, pp. 177, 208.

145 I examine Heidegger's distinction between "apophantic" and "hermeneutical" *logos* in Chapter 4.

146 *Historical Reason*, pp. 72, 181.

147 *Animals and the Limits of Postmodernism*, p. 4.

148 Moral realists believe that moral truths are timeless and independent of both mind and individual perspectives. See Matthew H. Kramer, *Moral Realism as a Moral Doctrine*, pp. 42, 85, 288; Russ Shafer-Landau, *Moral Realism: A Defence*, p. 15. To this extent, moral realists present a textbook example of thinkers who demand the sort of Eleatic standpoint that seems unattainable in principle.

149 John William Miller, *The Midworld of Symbols and Functioning Objects*, p. 11 (italics in original).

150 *The Midworld of Symbols and Functioning Objects*, pp. 14–5 (italics in original).

151 *The Midworld of Symbols and Functioning Objects*, p. 13.

152 John William Miller, *The Paradox of Cause and Other Essays*, p. 121. Here it becomes clearer how Miller's critique of Eleatic reason holds for Kant: While Kant does not assert that space "exists" independently of agents of experience, he does argue that experience is *structured as a unity in advance*, which is to say that the parameters within which we are capable of experiencing space (and, *mutatis mutandis*, time) are by no means derived from concrete experience but instead are its presupposition. Compare Miller's view, according to which space is not something more or less determinate and structured in advance of concrete experience, but rather is a "function" or "present active participle." John William Miller, "The Owl," p. 225.

153 Here again the parallel with Heidegger is illuminating: Miller's distinction between the actual and the real roughly corresponds to Heidegger's distinction between the "ready-to-hand" and the "present-at-hand" in our everyday dealings.

154 John William Miller, *The Definition of the Thing with Some Notes on Language*, p. 79. An example that was vivid at the time of this writing is the expression "Covid-19." Much more than a label or a set of squiggles on paper, this expression *functions* today in a manner in which it will not be able to function once the crisis passes. Today "Covid-19" is a rallying cry for those on opposite sides of a divide (over social distancing etiquette, vaccination, and mask-wearing) that, as I will note shortly, must be confronted and resolved in the interest of pursuing Miller's democratic ideal.

155 *The Definition of the Thing with Some Notes on Language*, p. 103.

156 *The Definition of the Thing with Some Notes on Language*, p. 59.

157 See *The Paradox of Cause and Other Essays*, p. 61: Among symbols as a special case of functioning objects, Miller includes "yardsticks, clocks, balances . . . names and words, written numbers and logical notation, works of art and political constitutions."

158 *The Paradox of Cause and Other Essays*, p. 59.

159 See "The Age of the World Picture."

160 John William Miller, *In Defense of the Psychological*, p. 24.

161 *In Defense of the Psychological*, p. 74.

162 *In Defense of the Psychological*, p. 38; John William Miller, *The Philosophy of History with Reflections and Aphorisms*, p. 35.

163 *In Defense of the Psychological*, p. 73.

164 *In Defense of the Psychological*, p. 73.

165 Boethius, *The Consolation of Philosophy*, p. 11. Cf. p. 84: "O Lord, you govern the universe with your eternal order."

166 *The Philosophy of History with Reflections and Aphorisms*, p. 33.

167 *The Philosophy of History with Reflections and Aphorisms*, pp. 34–5.

168 *The Midworld of Symbols and Functioning Objects*, p. 191.

169 *The Midworld of Symbols and Functioning Objects*, p. 7.

170 *The Midworld of Symbols and Functioning Objects*, p. 190.

171 *The Midworld of Symbols and Functioning Objects*, p. 135.

172 *The Midworld of Symbols and Functioning Objects*, p. 109.

173 *The Midworld of Symbols and Functioning Objects*, p. 10f.

174 John William Miller, *In Defense of the Psychological*, p. 137.

175 On concepts as rules, see *Critique of Pure Reason* at A106.

176 *The Midworld of Symbols and Functioning Objects*, p. 189.

177 *The Paradox of Cause and Other Essays*, p. 88.

178 *The Paradox of Cause and Other Essays*.

179 *The Paradox of Cause and Other Essays*, p. 89.

180 *The Paradox of Cause and Other Essays*.

181 *The Paradox of Cause and Other Essays*.

182 *The Paradox of Cause and Other Essays*.

183 *The Paradox of Cause and Other Essays*, p. 89.

184 *The Paradox of Cause and Other Essays*, p. 89f.

185 Søren Kierkegaard, *Fear and Trembling*, pp. 97, 106f.
186 *The Paradox of Cause and Other Essays*, p. 90.
187 *Fear and Trembling*, p. 106: "The demonic, for which the individual himself has no guilt, has its beginning in his originally being set outside the universal by nature or by historical situation."
188 *Fear and Trembling*, p. 87. See also p. 92: "Ethics demands that he speak."
189 *Fear and Trembling*, p. 111.
190 *Fear and Trembling*, p. 77.
191 *Fear and Trembling*, p. 79.
192 *The Paradox of Cause and Other Essays*, p. 90.
193 *The Paradox of Cause and Other Essays*.
194 Karl Löwith, "The Occasional Decisionism of Martin Heidegger." I noted Tugendhat's charge of decisionism earlier in this chapter.
195 George Santayana, *The Life of Reason*, pp. 77, 79.
196 *The Midworld of Symbols and Functioning Objects*, p. 43.

4

THE AFFECTIVE DIMENSION OF MORAL COMMITMENT

In the destructive element immerse.

Joseph Conrad, *Lord Jim*

1. Background Ideals of Living and the Putative Autonomy of Reason

Many people working in moral philosophy today are firmly committed to the proposition that reason is autonomous—autonomous in the sense that it is the final (and possibly sole) arbiter of truth and value, and in the sense that it is capable of carrying out the immanent critique of reason to which I have appealed in previous chapters. By examining the notion of a background ideal of living in Chapter 1, I have sought to demonstrate that the ideal of reason as a "timeless" faculty is not only philosophically naïve, but in reality is the product of certain ulterior motives that proponents of the Eleatic conception of reason fail to acknowledge. I take Aristotle's characterization of reason (*nous*) as eternal and Kant's transcendental turn as textbook examples of a certain blindness that afflicts the Eleatic standpoint, a blindness that is evident in the concatenation of blithe anthropocentric assumptions that I examined near the end of Chapter 1: that we human beings are capable of authoritative (or at least generally reliable) pronouncements about the experiential capacities of ourselves and of nonhuman animals, that no nonhuman animals possess agency in the specifically human sense, and that this supposed lack of agency on the part of nonhuman animals renders them inferior to human agents in the moral scheme. These assumptions, which find a focal point in the oft-repeated claim of Western philosophers that human beings are the most divine creatures in existence, are precisely what Grimm seeks to call radically into question in his appeal to the

DOI: 10.4324/9781003425595-5

"animal-in-itself" and what Taureck seeks to counter with his "imperative to discontinue the use" of nonhuman animals.

Grimm and Taureck demonstrate considerably greater willingness than traditional philosophers to rethink the terms of an immanent critique of reason. Both recognize, at least implicitly, that while reason plays a key role in moral judgment as well as in other areas of life, a genuinely fruitful immanent critique of reason demands an openness to what lies beyond the bounds of reason. It is this kind of openness that Heidegger had in mind when he proposed that overcoming the traditional Eleatic conceptions of reason and selfhood will require "a step back" from accustomed ways of thinking and valuing so as to "let beings be." Only this kind of openness, as I hope to show in this chapter, can enable us to see the inherent limits of reason as well as its dialectical relationship with our affective or emotional life.

You may find it tempting to respond to this proposal with the observation that thinkers such as Aristotle and Kant are keenly aware of the limits of reason in human beings. And while this is true in a rather mundane sense—neither thinker ever suggests that we can attain the vantage on things that a divine being would enjoy—these thinkers, like many others, devote the vast majority of their attention to explaining what reason *can* do, *not* what it *cannot* do. I find it very attention-getting that these pronouncements about what reason can do are tied intimately to assumptions and claims about *what we are entitled to do* in our dealings in the world. As I have argued in the previous three chapters, this should not come as a surprise: the global commitments that give content and direction to our rational endeavors are not products of reason but rather are its fundamental presupposition. Ortega's and Miller's reflections on the specifically historical character of reason provide a needed corrective to the aspirations of the Eleatic ideal by reminding us that reason is an historically emergent phenomenon rather than a fixed faculty, and that historical reason is, to borrow a term from Heidegger, "always already" bound up with a constantly evolving global background of life commitments.

To this extent, an immanent critique of reason cannot take the form in which Kant undertook it. Kant, as already noted, was critical of Plato for treating the forms as constituents of reality rather than as archetypes in the mind.[1] For Kant, the task of immanent critique includes not only positing the thing-in-itself as the counter-concept to lived reality but also articulating the transcendental conditions for knowledge as well as notions such as the Transcendental Ideas of Reason. Kant, as I noted in Chapter 2, considers the entirety of immanent critique to be performed by reason alone. For reasons that I hope will become clear in this chapter, there is something profoundly naïve—if understandable—in this unremitting insistence on the autonomy of reason. Kant believed that he was fulfilling Descartes's aspiration to use natural science to render human beings the masters and possessors of nature; and he likewise believed that he had succeeded in extending the authority of reason so as to fulfill a second aspiration of Descartes's, namely, to advance a theory of morality based entirely in reason. If we accept the verdict that thinkers such as Kant were ultimately mistaken about the nature of rationality and its relationship to our affective life, then these twin aspirations—to master nature through applied science, and to master our own lives through the persistent application of practical reason—become threatened. We then have to face up to some ugly historical facts about ourselves, facts that cast serious doubt on our historical claim to be—or even to merit the title of—"the titular lords of nature."

An immanent critique of reason unfolds very differently when we shift from the Eleatic ideal to the ideal of historical reason. Here reason still plays a vital role, one that I will examine in this chapter, but the starting point of the critique is not some anterior commitment to vindicating the autonomy of reason. Instead the starting point is an assessment of our current circumstances and commitments, viewed in the light of the history out of which they have emerged and with an eye toward the aspirations that we have set for ourselves. All aspects of this complex of commitment—current circumstances, past commitments, and future aspirations—are subject to constant contestation and revision. Moreover, even though thinkers such as Ortega and Miller have little to say about the affective dimension of this process of historical reflection and striving, one of my central concerns in this chapter is to show that the openness characteristic of the historical approach to reason is precisely what is needed in order to arrive at a clear recognition that our reason is irretrievably bound up with our affective life, and that it is simply anthropocentric prejudice to suppose that emotion is simply a handmaiden to reason. By the end of this chapter, I hope you will understand why I find wildly implausible Kant's extensive efforts to acknowledge and vindicate moral love while denying it any foundational place in moral commitment. If anything, Kant may have gotten the relationship between moral love and the legislative autonomy of reason backwards.

To say all this is not to suggest that anyone who expresses the requisite openness will necessarily find their way to a specifically non-anthropocentric background ideal of living. There are simply too many unknowns of the kind that proponents of the Eleatic ideal tend to ignore or underplay, such as the notorious difficulty of determining when we are employing our reason in a truly impartial manner as opposed to pushing for self-serving ends while claiming that we are being impartial; moreover, thinkers such as Ortega and Miller openly embrace the ideal of historical reason but have little if anything to say about the experience and moral status of nonhuman animals. What this tells us is something rather obvious: that neither articulating nor purporting to embrace a particular moral position is any guarantee that you are actually living up the ideal you espouse.

In Chapter 5 I will examine this problem by focusing on the perennial tension between individual and collective values. Here my focus is the importance of recognizing several facts that a good number of people who are not academic philosophers seem to take as matters of common sense: that we are not the titular lords of nature, that reason by itself is not adequate for establishing moral commitments, and that we human beings have a lot more in common with nonhuman animals than traditional thinkers would have us believe. The more accustomed you are to an anthropocentric background ideal of living, the more far-fetched statements of this kind are likely to seem. Thus a basic demand of the kind of immanent critique of reason that I am advocating is that we suspend our prejudices and remain open to ideas that might at first blush strike us as fanciful. We also need to remain open to the possibility that those ideas strike us as fanciful *precisely because they are threatening to us*.[2]

Before proceeding to examine the role of the affects in the process of moral commitment—for reasons that will become clear by the end of this chapter, I believe that the expression 'judgment' is too narrow—I ask you to consider several critical responses to the Eleatic mindset that aptly capture my central concerns. Taken together, these responses shed light on the kind of openness that promises to yield positive rather than merely negative results for the process of immanent critique. Take first Michel de Montaigne, who is well known for his strong critique of human pride. Of all the challenges that Montaigne poses to human pride, none are of more direct relevance to the present discussion than his mockery of the human pretension to mastery:

> Is it possible to imagine anything so ridiculous as that this miserable and puny creature, who is not even master of himself, exposed to the attacks of all things, should call himself master and emperor of the universe, the least part of which it is not in his power to know, much less to command? And this privilege that he attributes to himself of being the only one in this great edifice who has the capacity to recognize its beauty and its parts . . . who has sealed him this privilege?

Montaigne notes a wide variety of experiential capacities in nonhuman animals (ingenuity, foresight, etc.) that account for the "superiority that animals have over us," and concludes that our inability to communicate with nonhuman animals signals a deficiency not on their part but on ours.[3]

Montaigne supplies us with one crucial dimension of the kind of immanent critique I am proposing: the readiness to acknowledge how pervasively our prejudices influence what we might otherwise take to be the sober, objective insights of reason. A second, related dimension of this form of critique is hinted at by Marcel Proust: that beyond the bounds of reason we encounter not simply an anonymous and unknowable thing-in-itself, but rather an entire, living world that is disclosed to us not by reason but by a "feeling which makes us not merely regard a thing as a spectacle, but believe in it as in a unique essence [*être sans équivalent*]."[4] To "regard a thing as a spectacle" is to depict it in the determinate terms to whose measure Eleatic reason is cut; I will elaborate on this notion shortly in connection with Heidegger's contrast between Protagoras and Descartes. Here it is important to note that Proust sees an inner connection between our affective life and the grasping of essences in the world, whereas adherents of the Eleatic tradition insist that essences are grasped only through rational inquiry. In doing so, Proust implicitly endorses the approach to reality counseled by Schopenhauer, who as I noted in Chapter 3 sees the will as an alternative route to contact with reality. Later in this chapter I will delve deeper into Schopenhauer's approach and relate it to his critique of Kantian ethics.

A third and final contribution to establishing the kind of openness needed in order to conduct immanent critique in a positive manner is made by Virginia Woolf, who radicalizes the critique of reason by questioning whether the reason-emotion dichotomy is itself a reductive abstraction.

> What reason or what emotion can make us hesitate to become members of a society whose aims we approve, to whose funds we have contributed? It may be neither reason nor emotion, but something more profound and fundamental than either. . . . reasons and emotions have their origin deep in the darkness of ancestral memory; they have grown together in some confusion; it is very difficult to untwist them in the light.[5]

Although Woolf's concern in making this statement is the ways in which gender differences can influence our values and judgments at a fundamental level, I believe there is something to be learned from it in the endeavor to rethink the rational and affective dimensions of moral commitment: that the very "split" between reason and affect may itself be a product not of a sober assessment of our own condition, but rather of Eleatic abstraction. Woolf's line of inquiry here implicitly relies on Aristotle's notion, noted in Chapter 3, that any two factors that bear a relationship to one another are bound together by "a precedent community of nature," i.e., a totality out of which individual factors can first emerge and be contemplated in isolation from the context out of which they have been distilled.

Montaigne, Proust, and Woolf provide us with serious food for thought. Woolf explores the mystery of "the darkness of ancestral memory" by directing her attention to the drive to dominate that has figured centrally in my remarks in this book. Woolf asks, "what possible satisfaction can dominance give to the dominator?" and offers the tentative diagnosis that the origin of this drive is some sort of "infantile fixation."[6] Of the reams that have been written on Aristotle, the Stoics, Descartes, Kant, and other thinkers proclaiming the lordship of human beings over the rest of creation, it is exceedingly difficult to find robust critiques of reason that give any sustained consideration to the drive to dominate that motivates Eleatic reason. Many but not all such critiques are presented by feminist thinkers, although I hope to show that one need not approach the question of the role of the emotions purely or primarily through a feminist lens. Woolf's concern is a sober-minded one: what can men reasonably expect a woman such as Woolf to say in response to the question how we are to prevent war? Even though my focus is different than Woolf's, I believe that her remarks shed important light on the notion of ulterior motives that I have associated with the anthropocentric background ideal of living.

One way to think about the significance of Woolf's passing thought that the drive to dominate may have its roots in some sort of infantile fixation is to consider the drive to dominate in the light of a metaphor employed by the German philosopher Walter Biemel. Biemel once suggested that the drive for security evident in Western thought might best be understood through a reflection on Kafka's story "Der Bau" ("The Burrow"), in which Kafka offers us an account of life from the standpoint of a subterranean creature whose entire existence is devoted to securing itself against external threats. Biemel's discussion invites the conclusion that philosophical commitments such as the Cartesian aspiration to render human beings the masters and possessors of nature might ultimately be nothing more than a desperate effort at fortification that looks a bit pathetic when viewed in the light of the underlying motivations at work.[7]

One need not take on a full-blown Freudian analysis—nor, for that matter, an ideology-based one—to see the profound significance of ulterior motives (typically selfish ones, as I will argue in the next chapter) in our assessment of the claims that people make. We simply need to acknowledge the mundane fact that the *tone or force* that people employ when they make claims in the name of reason often tells us a great deal about the nature of the claims themselves. Defenders of the Eleatic ideal maintain that they are simply appealing to "the force of the stronger reason," but I hope to show in the next section of this chapter that claims about the force of the stronger reason should not generally be taken at face value precisely because they depend on isolating a factor and considering it in complete detachment from the global context of meaning from which it derives its own significance. In the process of abstraction, a global context of meaning is broken down into what reason treats as separate constituent parts, thereby transforming the elements of experience and arguably *losing something essential*. This problem afflicts not only everyday situations such as people's pronouncements about their political convictions, but also the very construction of theories—particularly those purporting to make a logical link between degrees of cognitive sophistication and moral status.

There is a certain irony in the attempt to establish this link, given that it appears to violate the widely held proposition that an 'ought' cannot be derived from an 'is'.[8] Here I offer the admittedly tentative and risky suggestion that this proposition may ultimately be erroneous, the error being a product of Eleatic abstraction. The challenge, as I have already suggested, is to remain open to the prospect of an alternative route to an appreciation of the real, one that seeks to overcome the distorting influence of the specifically human perspective—a perspective that, as I have argued from the beginning of this book, makes putatively impartial pronouncements about the nature of reality when in fact, its very vision is shaped by prior anthropocentric commitments.

2. A Positive Path Beyond the Limits of Reason?

The rise of a Cartesian-Kantian conception of human selfhood in the modern age has greatly intensified this challenge. Contemporary views about the moral status of nonhuman animals continue to reflect deeply rooted anthropocentric prejudice. In the concluding pages of *Animals and the Limits of Postmodernism*, I argue that rational arguments by themselves are not sufficient to change people's core convictions about matters such as the moral status of nonhuman animals, and that what is ultimately needed is a shocking experience that globally alienates us from our received ways of valuing things.[9] It strikes me that people who are firmly committed to the proposition that nonhuman animals are things for us to use—remember that many such people also "love" nonhuman animals—are not liable to be moved by the consideration that nonhuman animals are free beings with lives of their own that do not generally (i.e., apart from cases of genuine need) require the guidance or help of human beings. For people of the anthropocentric persuasion, practices such as milking cows or killing chickens to prepare a festive holiday picnic are not only morally unobjectionable provided they are conducted "carefully," but there is essentially no argument that will change such people's minds. The ethics of dairy farming is the object of a great deal of debate nowadays; items in the press tend to exhibit a very strong split between those who believe that the practice essentially treats cows as milk delivery devices (their lives are short, they live their lives in confinement, and they end up at the slaughterhouse) and those who believe that milking is in principle not only acceptable but positively good (it doesn't really harm the cows, we treat them like family, domesticated cows would not exist in the absence of dairy farming, the industry provides needed jobs, milk is American as apple pie).[10] From the Eleatic standpoint, the dispute is a matter of purely rational considerations and one in which one view is correct and the other incorrect. The approach that I am proposing in this chapter appeals to considerations beyond (or, more precisely, *beneath*) the rational, considerations hinted at by Woolf that appeal to our affective constitution at least as much as to our rational nature.

The meat controversy unfolds along the same lines: there are those who believe that, because nonhuman animals are things for us to eat, there is little if any moral problem with killing nonhuman animals to consume their flesh; and there are others who, either by dint of rational arguments (this does sometimes occur) or due to a felt connection with nonhuman animals, consider the consumption of their flesh to be what Ovid chillingly characterizes as a "Thyestean banquet."[11] Ovid's characterization of meat eating as tantamount to cannibalism poses a stark contrast to the rosy picture and dismissive condescension offered by those who just can't wait for that summer barbeque: "These articles celebrating fried chicken and short ribs may arouse the ire of vegetarians and nutritionists. But I bet that I'm hardly alone in finding them a refreshing antidote to the rampant dietary puritanism that is blighting our times."[12] These two sentences do a remarkable job of distilling the essence of anthropocentric thinking about matters such as killing, cooking, and eating nonhuman animals: Meat consumption is utterly unproblematic from a moral standpoint, the practice gives many people joy (and, implicitly, this good trumps any goods that might be claimed to be proper to nonhuman animals), and any concerns that people such as me might harbor about meat eating are driven by the "blight" of "rampant dietary puritanism"—as if the reasons why people who are literally shocked by (and not just rationally aware of) the meat-industrial complex were entirely about their own spiritual purity and not about what caused the shock in the first place.[13] Given the way in which Eleatic reason can distort rather than clarify our grasp of reality by treating its constituents as intelligible in isolation from one another, there is probably no purely rational consideration that will do anything to move a writer of that sort of letter to rethink their anthropocentric convictions. For such a rethinking, as I argued in Chapter 1, is a threatening prospect, given that it opens us to the eventuality of having to make a global shift in our deeply rooted commitments. And we are nothing if not creatures of habit.

In Chapter 3 I examined Heidegger's contrast between two radically opposed ways of understanding human selfhood. Heidegger argues that Protagoras's statement "man is the measure of all things" takes on a new meaning when Descartes proclaims the autonomy of the Cartesian ego. Now no longer open to the fundamental mystery of existence, the self assigns itself the role of final arbiter of truth and accepts as true only those ideas that fit into the narrow confines of Cartesian clear and distinct insight. It is this move, which takes the standpoint of detached human rationality as the sole and definitive measure of the real, that accounts for the comprehensive treatment of nature, nonhuman animals included, as sheer mechanism. The fundamental shift that takes place from Protagoras to Descartes is from a standpoint of openness to a comparatively closed "subject-object" ontology.

Man *as subject* wants to be, and must be, the measure and center of "beings" (i.e., of "objects"). The human being is now no longer μέτρον in the sense of orienting one's hearing on the context of unconcealment, which is always oriented on the "I" and is what gives rise to presence. As subject, the human being is the "measure" to the extent that it arrogantly asserts itself as the foundational standard for all measures employed to determine what counts as "certain," i.e., as "true" and thus as "real."[14]

Heidegger's endeavor, both in *Being and Time* and in his essays on technology, is to diagnose what happens when human beings arrive at a self-understanding that not only asserts our superiority over the rest of nature, but pursues the dualistic path of viewing ourselves as detached "subjects" who technically do not reside in nature at all but instead contemplate nature from the external vantage point of a being whose essential constitution is fundamentally different than the essential constitution of nature. This is as true for Kant, who views the subject as the constructor of nature, as it is for Descartes, who maintains that our embodiment is not part of our true essence. Both thinkers view self and world as fundamentally different than one another, and both exhibit at least implicitly a certain contempt for the "merely" physical. (Recall in this connection Kant's distinction between *homo noumenon* and *homo phainomenon*.)

Heidegger's central concern is that by the time we have come to view ourselves as detached "subjects," we have committed ourselves to a global way of understanding ourselves and reality that has decisive and dire consequences. We are no longer a part of the world that we inhabit. As I noted in Chapter 3, when Heidegger associates the Protagorean approach with unconcealment, he is trying to retrieve a characterization of the real that became forever superseded when Descartes asserted the *ego cogito* and Kant undertook the transcendental turn. Unconcealment (*aletheia*) signifies precisely the kind of openness that I have argued is necessary in order to recognize the deep bond we share with nonhuman animals and, by extension, the rest of nature. The modern view of nature, sentient nonhuman life included, as "objects" can accept as true or real only those factors or considerations that are amenable to the kind of conceptual "packaging" of which we human beings are capable.

Heidegger notes that this process of packaging is most evident in specific scientific domains, which he refers to as the "positive" sciences. Disciplines such as geometry, physics, and even history are "positive" in the sense that each *posits* a field of study in advance, and this act of positing includes axiomatic expectations about which objects can be found in the domain being studied, the proper procedures for studying them, and which sorts of insights may ultimately be taken to be 'true'. To this extent, the positive sciences presuppose as much as they discover; they are not genuinely "open" but instead each one "cuts out particular spheres" by focusing narrowly on its own particular subject matter.[15] What unites the different positive sciences is a prior global subject-object ontology in which the subject's relationship to objects always takes the form of *predicatively-structured cognitive representations*.[16] On Heidegger's view, the modern age is "the age of the world picture" in the sense that the world has been *reduced* in acts of Cartesian-Kantian cognition from the way we actually experience it to an essentially determinate "picture" that we survey as if we were an observer staring at a scene wholly foreign to us.

In contrasting the approach of a pre-Socratic such as Protagoras with the distinctively modern approach that led Heidegger's student Biemel to equate the modern sensibility with that of a subterranean creature whose entire existence is devoted to fortifying itself against external threats, Heidegger was exploring the possibility of retrieving a deeper relationship to the real that he believed was at work in the intimations of certain pre-Socratic thinkers. From the standpoint of Eleatic rationality, such an endeavor is the product of mythical thinking that has failed to respect the wisdom of Kant's transcendental turn. But we have to bear in mind that Descartes and Kant alike were seeking absolute certainty in the foundations of knowledge, and it is no accident that both appeal to mathematics as a paradigm case of certainty. Descartes goes so far as to suggest that he will be able to attain the same kind of certainty in morals that the ancients had already discovered in mathematics, and moreover that we should not attempt to articulate a definitive morality before we have completed our study of both metaphysics *and positive sciences such as physics*.[17] Kant, for his part, sets as his goal apodeictic certainty, and he avers that morality can be legislated entirely on the basis of detached rational insight (pure practical reason).

Notwithstanding the incredulity with which Heidegger's Protagorean musings are certain to be met, I believe there is something fundamentally right in them—even if, from the standpoint of our Eleatic fortifications, we find it difficult to grasp exactly what is right in them and why they are right. I pointed out in the previous chapter that Heidegger, in virtue of his embracing a total rather than merely immanent critique of reason, is left with the problem of decisionism, a debilitating arbitrariness in the commitments we embrace. That is surely part of the reason why many philosophers have declined to pursue the path sketched by Heidegger. But another part of the reason is that Heidegger's call for openness and "letting beings be" demands that we human beings relinquish our historical pretension to being entitled to treat all nonhuman beings (and some human ones, for that matter) as objects to be used—even though, as I have noted, Heidegger himself appears to have had no real moral concern whatsoever for nonhuman animals. I have made it clear from the outset that one of my central aims in this book is to vindicate the power of reason while acknowledging its inherent limitations and ineluctable reliance on our emotional constitution. Ortega and Miller give us good reasons to embrace historical reason, and they shed important light on the positive contribution that reason can make to our efforts to come to grips with reality. It has always struck me that thinkers who embrace a total critique of reason do so because they have become despondent about the prospects for making a robust distinction, particularly in the moral and political spheres, between acts of reason that merit the appellation "authoritative" and those acts that purport to be authoritative but in reality are disguised efforts to consolidate one's power (through manipulation, oppression, etc.). This problem was one of the core motivations for my work in *Animals and the Limits of Postmodernism*, in which I argued for the conclusion that postmodern thinkers, in virtue of undertaking a total critique of reason, are left in the problematic position of not being able to articulate any sort of determinate principles for the guidance of our judgment and conduct.[18]

But the thinkers who recognize the dangers of a total critique of reason have tended to cling to an ideal of reason that, as I have argued in the preceding chapters, simply cannot be fulfilled. Rather than rejecting reason altogether, I believe that we need to see it as dialectically related to the affective states through which significance in the world is first disclosed to us. While he neglected to say much about the reciprocal influence of cognition and emotion on one another, Schopenhauer was keenly aware that cognition is not our only mode of access to the real and that *our embodiment affords us an entirely different kind of contact with the world.* Schopenhauer contrasts representation and the will as distinct modes of access to the real. Representation as Schopenhauer conceives it signifies something akin to the notion of representation that Heidegger associates with subject-object dualism, whereas the will signifies our unmediated, bodily encounter with the real. Schopenhauer models his notion of representation on Kant's faculty of understanding; to understand an entity or situation is to grasp its causal connections with other entities or situations. But Schopenhauer departs from Kant in maintaining that human beings are not alone in possessing the faculty of understanding. On Schopenhauer's view, any being possessing eyes also possesses knowledge, and this means that nonhuman animals are exactly like human beings in possessing understanding, the faculty of representation.[19]

This by itself constitutes a major step away from the anthropocentric ideal, even though as I noted in Chapter 3 Schopenhauer retains some highly traditional anthropocentric prejudices. In order to see just how major a step this is, recall Kant's insistence that insofar as we are *homo phaenomenon* we are just like any other animal, but insofar as we are *homo noumenon* we are categorically *un*like nonhuman animals. Forever excluded from the noumenal realm, Kant believes, nonhuman animals can be nothing more than mere "things."[20] The instability of Kant's position, and a sign of the instability of the Eleatic project, is the contradiction that arises when we try to make sense of what it would mean to have positive regard for a being that is by definition an instrumentality—e.g., the horse toward which Kant counsels us to exhibit "gratitude."

Notwithstanding his anthropocentric leanings—recall that he finds a way to rationalize practices such as vivisection—Schopenhauer's observation that nonhuman animals share in cognition and understanding (the faculty of knowledge) constitutes a significant departure from traditional denials that nonhuman animals possess "genuine" understanding. As I noted in Chapter 1, exponents of the anthropocentric ideal seek to account for the ingenuity and adaptability of nonhuman animals by arguing that nonhuman animals do not "really" possess any understanding but are driven entirely by impulse, instinct, or passion—all faculties that the tradition depicts as wholly passive in contrast with reason or understanding, which traditional thinkers assume without clear proof (and contrary to vast and increasing evidence of ingenuity in nonhuman animals) to be the exclusive possession of human beings. (Typically, when traditional thinkers claim this exclusivity, they exhibit no acknowledgment of just how much *human* behavior is precipitated by forces such as instinct.) There is a certain amount of sober common sense in Schopenhauer's attribution of intelligence (understanding, knowledge) to sentient creatures generally: To suppose that, in spite of the extensive evolutionary continuity and physiological similarity between human and nonhuman animals, the experiential capacities of the two differ so profoundly is to disregard considerations that we see and feel every day—or *could* see and feel, if we could attain the standpoint of openness that lies beyond the fortifications of Eleatic reason. (The more you read Aristotle's extensive texts on the capacities of nonhuman animals, the less pioneering a lot of contemporary ethological research begins to look.[21])

Schopenhauer accounts for the obvious differences between linguistic (human) and non-linguistic sentient creatures by attributing the faculty of reason to the former. Schopenhauer shares the traditional assumption, one that even I tentatively accept as accurate, that only those beings capable of predicative, concept-based language are capable of the general standpoint requisite for articulating scientific laws and moral principles. But even here Schopenhauer presents a comparatively earthbound conception of reason: Where for Kant, reason functions independently of the insights gained by the understanding (although reason must employ the categories in an "illicit" manner in order to think our experience as a whole), Schopenhauer characterizes the faculty of reason as a faculty of abstraction or generalization from the particular causal insights disclosed by sensibility and the understanding. Indeed on Schopenhauer's view, the sole function of reason is "the formation of the concept" from out of a prior experience of particular phenomena.[22]

Our conceptual ability, Schopenhauer suggests, enables us to do something that no nonhuman animal can do: We can live in the past and future as well as in the present, whereas beings bereft of reason "live in the present alone."[23] It is in virtue of our capacity to transcend the present that we can conduct scientific investigations, formulate general rules for conduct, and contemplate death. Nonhuman animals, on Schopenhauer's view, can do none of these things. And while on Schopenhauer's view this confers certain entitlements on human beings to use nonhuman animals, Schopenhauer nonetheless shows us a step beyond the Eleatic ideal by continually stressing *the essential sameness of human and nonhuman animals.*

> In all essential respects, the animal is absolutely identical with us. . . . The difference lies merely in the accident, the intellect, not in the substance which is the will. The world is not a piece of machinery and animals are not articles manufactured for our use.[24]

Where Kant and thinkers like him focus on the intellect, Schopenhauer places stress on *perception* rather than the imposition of a priori form as our primary route of access to knowledge about reality; he further argues that grasping things in terms of their causal relations (which is the business of the understanding in all sentient creatures) is simply one of two separate routes to contact with the real.

In this connection, Schopenhauer maintains that Kant erred in failing to recognize that the thing-in-itself *is* knowable, albeit not by means of the Principle of Sufficient Reason (namely, the representation of events as causally related to one another). Kant failed, in other words, to recognize that *the thing-in-itself is identical with the will.*[25] When Schopenhauer refers to "the substance" of sentient beings, he is acknowledging precisely what Kant purports to recognize but effectively retracts with his *homo phaenomenon/homo noumenon* distinction: that *human and nonhuman animals are essentially the same* in struggling to survive and facing the ultimate verdict of death. Those familiar with Schopenhauer's thought will recognize that his conception of the will differs significantly from the Cartesian-Kantian view of the will as a faculty possessed only by beings endowed with reason, i.e., by autonomous individuals. Schopenhauer proceeds from the axiom that all change and striving in the world are manifestations of one global "world will," out of which certain beings can emerge in the form of individuality. The world will constitutes "our real nature—a nature that is untouched by time, causality, and change," all of which are, as Kant suggested, projections of form employed by the understanding in our endeavor to secure ourselves.[26] As individuals, we are "phenomena" or fleeting manifestations of this world will, and as such our sense of autonomy is an illusion precipitated by our occupation of a spectatorial standpoint detached from the world will.

This sense of detachment is easily mistaken for metaphysical independence, and from that standpoint all sorts of doctrines such as the immortality of the soul have been asserted. What Schopenhauer wants to stress is that, when we see the fundamental distinction between representation and will, we should recognize the inherently "founded" and provisional character of individuality as well as the fictive character of many of the anthropocentric pronouncements that have guided us over the millennia. If we contemplate

> *human life in detail* . . . then the impression this now makes is like that of a drop
> of water, seen through a microscope and teeming with *infusoria*, or that of an
> otherwise visible little heap of cheese-mites whose strenuous activity and strife
> make us laugh.[27]

To accept this verdict is to accept Schopenhauer's lapidary conclusion that there is no separate noumenal realm, that the thing-in-itself is the very world in which we find ourselves rather than an unknowable limit on experience.

Schopenhauer states that it is not through representation that we establish direct contact with this reality, but rather through our *lived embodiment*. We can naturally relate to our own embodiment by means of representation, but Schopenhauer is at pains to observe that we are "conscious of this particular representation [of one's own body] not merely as such, but at the same time in a quite different way, namely as a will."[28] Corresponding to this distinction between representation and will is that between "abstract, discursive knowledge and intuitive knowledge," a distinction that Kant failed to make.[29] On Schopenhauer's view, Kant's failure to grasp the true nature of the will and the possibility of intuitive knowledge is responsible for Kant's having developed a moral theory that has the character of an "*a priori* building of houses of cards to whose results no man would turn in the storm and stress of life."[30] By seeking to depict lived experience *as it is actually lived*, Schopenhauer seeks to correct the dualistic error to which even Kant, notwithstanding his transcendental turn, succumbed: to treat our embodiment as the "inferior" part of our being, and to focus on the potential threat to the integrity of rational judgment posed by embodied states. Such a standpoint, Schopenhauer recognizes, consigns nonhuman animals to the status of "things" by failing or refusing to recognize that rational abstraction, when undertaken on the basis of anthropocentric commitments, practically guarantees that nonhuman animals will become reduced to this status. *The only way to arrive at an adequate sense of the lives of nonhuman animals is to proceed from our fundamental affinities, not from our putative differences.*

By taking this approach, Schopenhauer, notwithstanding his anthropocentric leanings, shows us a way to appreciate the pathos of openness that I have suggested is requisite to a rethinking of our own nature and limitations as well as of the nature and moral status of nonhuman animals. A broad array of thinkers have recognized the fundamental continuity between human and nonhuman animals, and a great many of these thinkers have further recognized that lived embodiment is the shared factor. The tradition of dualistic thinking in the West has strongly reinforced the notion that "mind" is superior to "mere" embodiment, and moreover that only beings who are rational in the specifically human sense actually possess mind. With this premium on detached cognition comes the denigration of the affective dimension of our lives, a dimension that many people consider to be the one major dimension of experience that we share with nonhuman animals. One conclusion toward which my entire discussion in this book has been building is that *it is our affective life that provides us with a sense of orientation* in the world, not simply in a spatial or cognitive sense but in the sense that our emotional experiences have the power to *disclose significance* to us. As such, our emotional life (which we have tried so hard to forget, because it reminds us of our inherent animality) constitutes an essential component in the ongoing process of forming, evaluating, and revising our moral commitments.

3. Reclaiming a Guiding Place for the Emotions

One contemporary thinker who has challenged the strict dualism of thought and feeling is Bennett Helm, who focuses the problem of moral commitment as one of forging a connection between deliberative choice and being moved to act.[31] What Helm acknowledges in doing so is that "deliberative choice" by itself does not and cannot move us to embrace a commitment or course of action. Once we acknowledge this limitation of reason, the problem with Kant's attempt to treat "moral love" as important but ancillary to our true moral motivation becomes clear: What Kant should have done, I believe, is acknowledge that moral love is an essential part of the *foundation* of moral commitment, not simply an adjunct to it.

Helm argues that the link between deliberative choice and being moved to act must involve an essential affective component. But rather than simply asserting the significance of our emotional constitution in the process of moral commitment, Helm makes the radical suggestion that we would do best to dispense with the cognitive-conative distinction (roughly, that between thought and motivation) altogether.[32] Neither thought alone nor emotion alone is adequate to move us to axiological commitment. The way in which Helm proceeds, however, makes it clear that what he has in mind is nothing like Woolf's gesture toward ways of being in the world that are essentially prior to the reason-emotion dichotomy. Helm is right to observe that affective states are essential to agency.[33] But Helm also maintains that emotional states are wholly passive; that judgment "normally" has priority over felt evaluations; and that "in every case, the kind of control we exercise over ourselves is rational control."[34] Helm thus starts out by presenting a potentially fruitful challenge to the traditional pretensions of reason, only to end up embracing precisely the rational ideal that his challenge to the cognitive-conative divide was supposed to call into question.

At the same time, Helm notes a very important dimension of axiological commitment that merits further examination. He states that things gain import for us by means of "an attunement of one's sensibilities."[35] This notion of *attunement* proves crucial to thinking through the central problem posed by Helm: how do we get from dispassionate judgments to actual, robust *commitments* on which we are prepared to act? Helm's answer is that our emotional responses take the form of wholly passive "felt evaluations" that can be rendered commensurable with reason. "Felt evaluations are defined by their place within a certain kind of rational structure."[36] The criteria or "standards of warrant" for both our evaluative judgments and our felt evaluations are "evaluatively thick properties" that "provide reasons for making certain evaluations."[37] Here I believe that Helm is employing a cognitive abstraction when he refers to the notion of attunement; it would have been highly productive to explore this notion in depth, the way that Heidegger does in Division I of *Being and Time*. Then it might have become clearer that the "dialectic" between conceptual understanding, judgment, and felt evaluations is one in which the affects arguably play a considerably more *active* role than the traditional defenders of rational autonomy have recognized.[38] It might also have become clearer just what Woolf has in mind when she gestures toward the possibility of an inner unity between reason and emotion. The exploration of the dialectic to which Helm refers also involves recourse to John William Miller's notion of local control, which certainly has a rational component but is a far cry from the Eleatic ideal of wholly rational control.

There is an inner connection between Heidegger's notion of attunement and Miller's notion of local control. Recall that for Miller, it is through our employment of "functioning objects" (measuring devices, etc.) that the world is first disclosed to us, and that *my own body is the primary functioning object*.[39] It is precisely through my embodiment that *all* encounters with the real occur, and on the basis of which it becomes possible to engage in the sorts of Eleatic abstraction that exponents of the tradition have wrongly supposed to be our primary mode of access to the real. To ascribe primacy to our embodied being is to take a decisive step away from the premium placed by modernity on Cartesian-Kantian detachment. It is to take seriously the prospects for us late-borns to make sense of a retrieval of Protagorean openness, and of arriving at a view of agency that places considerably more centrality on the affects.

More centrality, I want to say—not absolute centrality or priority. This is where I part company with Hume, who tends to be either lauded or condemned for his contention that "reason is, and ought only to be the slave of the passions, and can never pretend to any other office than to serve and obey them."[40] One penchant of the Eleatic mindset, one to which even Hume (notwithstanding his subordination of reason to passion) proved susceptible, is the tendency to depict things both as determinate and ordered in binary oppositions. You can see the footprint of this way of thinking about things in a related tendency: to invert established binary oppositions rather than plumbing their depths so as to explore the Aristotelian "precedent community of nature" out of which these two factors have been abstracted. Today, for example, there is a great deal of talk, in both academic and popular discourse, about *bodies* (particularly black and brown ones) being harmed or having injustices perpetrated against them, rather than talk of, say, *persons* (which have traditionally been associated specifically with mind rather than with body). This shift from persons to bodies constitutes an inversion of an old binary, that of mind commanding body (along with its affective life)—except that this inversion does not so much simply invert the relationship as subsume aspects of mind into aspects of embodiment.

This qualified inversion has much in common with Hume's call to see reason as the slave of the passions. Hume does not employ the traditional strict dualism of mind and body but instead derives both out of the field of impressions and ideas. Mind, for example, is not a metaphysical substance but instead is "nothing but a heap or collection of different impressions, united together by certain relations, and suppos'd, tho' falsely, to be endow'd with a perfect simplicity and identity."[41] Even though Hume acknowledges the role of reason in influencing our passions and at one point hints that the precise relationship between reason and passion is unknown to us, I believe that Hume underestimates the role played by reason in the establishment and revision of our axiological commitments.[42] As Ortega and Miller argue, reason plays a key role, not merely an ancillary one, in our reflections on our received values and current contingencies. Moreover, as I will discuss in Chapter 5, reason can help to give vital shape and direction to the evolution of our affective life, an evolution that is part of a larger endeavor at control—but control in a sense that must be contrasted with the traditional interpretation of control as lordship over reality. I will return to this idea of control after examining the contributions of Heidegger and several other philosophers to our understanding of the role of the affects, contributions that promise to rectify what I take to be some basic misunderstandings about the very nature of affective relations.

Heidegger for his part is notorious for having failed to do justice to the embodied dimension of our being in *Being and Time*. Nonetheless the early Heidegger provides us with important resources for rethinking both the Eleatic ideal and the nature and significance of our embodiment in matters of axiological commitment. Heidegger calls radically into question Hume's assumption that our first encounter with reality takes the form of receiving discrete bits of information that we then collect into a coherent whole by employing rules or what Hume calls principles for the association of ideas. Kant's view is similar, the key difference being that Kant wants to establish the prospect of apodeictic certainty in fields such as mathematics and natural science, where Hume treats most if not all of these matters as sheerly contingent; Kant builds his theory of knowledge on the idea that the faculty of receptivity provides us with a "manifold" or multiplicity (*Mannigfaltigkeit*) of sense data on which the understanding must impose *a priori* form if it is to achieve the apodeictic certainty that Kant saw in Newtonian physics.

Heidegger's view of experience is comparatively holistic. Consonant with the thinking of the Gestalt psychologists about perception and Saussure regarding the nature of language, Heidegger presents an account of the human condition in *Being and Time* according to which our experience is always structured *in advance* in terms of a global whole out of which we may subsequently distill or abstract individual features. Ortega and Miller proceed on the basis of the same commitment about meaning: By the time any individual enters the world, the world is already a meaningful whole; we do not invent meaning but instead inherit it and then have to evaluate that inherited meaning and decide whether and to what extent we are in accord with the values and commitments with which history has left us.

This character of world, which Heidegger calls "worldhood," merits comparison with the conception of world typically found in the Eleatic mindset. For many adherents of the Eleatic view, "the world" is an ordered physical system that we endeavor to know in order to master; it is an array of objects waiting to be scrutinized by subjects who are of a fundamentally different order than the world they endeavor to master. Heidegger recognizes a fundamental problem with this traditional conception of world, namely, that it elides altogether the role played by agents of experience in *disclosing* the world. Kant, too, recognized this role; but he nonetheless treated the world as an objective system, even if one "constructed" by subjects. He did this by assuming that what we can know about the world is crucially dependent on fixed *a priori* forms of experience rather than on anything truly novel, either in ourselves or in the world. Ortega and Miller show us precisely how both can surprise us in ways not possible in a Cartesian or Kantian framework. Heidegger provides us with an alternative account of world that facilitates a basic rethinking of what is involved in disclosing significance.

"The worldhood of the world" is Heidegger's characterization of the ways in which anything like "nature" as a scientific system can arise in the first place. What Heidegger is challenging is the Cartesian assumption that there are "certain seeds of truth which are naturally in our souls"—seeds that, as I have already noted, include the laws of nature—and the kindred Kantian definition of nature as "merely an aggregate of appearances, so many representations of the mind . . . that we can discover . . . only in the radical faculty of all our knowledge, namely in transcendental apperception."[43] In Chapter 3 I argued for the proposition that reason is not timeless after all, but instead is inherently historical in character. Heidegger's appeal to Protagorean openness poses a different but complementary kind of direct challenge to the dualistic aspirations of the tradition, which by the time of modernity proceed on the assumption that self and world are first separate and stand in need of unification. Heidegger's conception of world in *Being and Time* constitutes an effort to contemplate the idea of being in the world in comparatively "open" terms—terms that precisely do *not* proceed on the assumption that the world is divided metaphysically into subjects and objects. Rather than conceiving of world in terms of Cartesian *extensio* or in terms of Kantian a priori construction, Heidegger conceives of world in terms of a whole, meaningful environment in which we find ourselves and on the basis of which interpretive activity can take place. But he hastens to point out that "the 'around' ["Umherum"] which is constitutive for the environment does not have a primarily 'spatial' meaning. Instead, the spatial character which incontestably belongs to any environment can be clarified only in terms of the structure of worldhood."[44] Worldhood, in other words, is the condition for the possibility of experiencing things in spatial (and, by extension, a variety of other) terms. This has very significant implications for the ideal of scientific truth, chief among them the insight that representing things in scientific determinacy is not our primary mode of access to things, in spite of the fact that the subject-object ontology presents scientific knowledge as if it had primacy over other ways of relating to the world. This is what Heidegger has in mind when he notes the "positive" character of individual sciences and proposes that there is a level of experience that is prior to the positing of assumptions about what can and cannot be found in the field of experience. The notion of worldhood in the sense of an environing context of meaning constitutes Heidegger's attempt to describe a level of experience that is the condition for the possibility of acts of abstraction such as objectification. An additional feature of this phenomenon of worldhood is that it proves highly elusive to rationalist representation: Heidegger is attempting to characterize in the language of detachment a basic phenomenon that in key respects is not amenable to objectification, and whose character becomes reduced to a broad-strokes caricature when subjected to the criteria of rational demonstration. As will become clear shortly, any act of decontextualized description must, by its very nature, have recourse to the notion of representation, which narrows the focus of scrutiny exclusively to those features of observed phenomena that can be reduced to conceptual determinacy; problems arise when we attempt to describe in detached terms phenomena that do not themselves in the first instance take the form of objectness. After examining Heidegger's account of worldhood and his gesture toward the role of affect in the disclosure of the real, I will offer a few observations about the ways in which the Eleatic mindset distorts the affective dimension of experience by seeking to reify it.

Heidegger borrows the notion of environment from the ethologist Jakob Johann von Uxküll and develops an account of it according to which to be *situated* in a world is not in the first instance to be located in Cartesian space but instead is to inhabit what Heidegger calls "a wherein" [*Worin*] or locus of meaningful interpretation.[45] Human existence ("Dasein") is not initially an entity but rather is *a process of interpretation* out of which meanings such as 'entity' (with all its ramifications) can become abstracted. My undergraduate mentor, Hubert Dreyfus, was fond of saying that, on Heidegger's view, "Dasein is interpretation all the way down." Heidegger's own way of putting the point is that "Dasein is world-forming," i.e., our very encounter with things takes the form of interpretive activity that gains its content, direction, and force from what has gone before.[46] And once we have established our bearings in this "always already" fully established context of meaningful interrelationships, we must take a stand on the commitments and ways of being which we have inherited.

By ascribing this primacy to interpretive activity, and by suggesting that the kind of understanding of things we arrive at by means of detached scrutiny is not indicative of the way in which we become "primordially familiar" with them, Heidegger is arguing for the conclusion that knowledge of things in terms of their determinate or determinable qualities is in fact a "founded" (i.e., reductive) mode of experience.[47] Descartes implicitly acknowledged this founded character of metaphysical and scientific knowledge when he developed a method for superseding the "obscurity and confusion" of ordinary, lived experience toward the acquisition of "clear and distinct" cognitive insights so solid that they would never be subject to revision or alteration. Where Descartes took it for granted that detached scrutiny provides us with the most authoritative access to the real, Heidegger challenges this idea by arguing that its dogmatic assumption of a rigid subject-object dichotomy signifies a failure to appreciate the phenomenon of meaning or significance that makes possible metaphysical abstractions such as "thinghood" in the first place; one key observation that Heidegger makes in this connection is that Descartes failed adequately to interrogate the 'sum' in 'cogito ergo sum', just as an entire tradition of philosophers in the West had failed to plumb the depths of the term 'being'.[48]

I have argued in previous chapters that the ideal of reason that has guided Western thought for the past two millennia purports to be impartial but in fact is always informed and driven by *prior commitments*. Some of these are metaphysical, as when Aristotle proclaims that *nous* (reason) is eternal and even when (on a strict reading of Kant's own revised definition of 'metaphysics') Kant asserts a strict dualism of subject and object. Kant goes so far, on the basis of his embrace of the notion of subjectivity, to assert that "reason is, indeed, so perfect a unity that if its principle were insufficient for the solution of even a single one of the questions to which it itself gives birth we should have no alternative but to reject the principle."[49] By calling the tradition's dogmatic assumptions about the 'sum' into question, Heidegger is directly challenging this pretension of reason and exploring the possibility that the faculty of reason suffers precisely the inadequacy that Kant denies. By now it should be clear that I share Heidegger's misgivings about traditional claims to the absolute autonomy of reason; but it should also be clear that I believe Heidegger was wrong to undertake a total critique of reason. Heidegger's own notions of worldhood and interpretive activity hold the key, I think, to seeing how the affective dimension of our experience can and must make a fundamental contribution to the endeavor to answer Kant's three questions. An examination of Heidegger's notion of interpretive activity will prepare the way for rethinking, from a non-anthropocentric standpoint, potential answers to those questions. (E.g., should I hope for a world in which people treat certain nonhuman animals very kindly and then kill them for food, even though those nonhuman animals pose no threat and consuming their flesh is not a necessity but a luxury? Or: Can I know that nonhuman animals lack agency? And so on. My contention is that *reason alone is incapable of arriving at authoritative answers to these questions*.)

Heidegger seeks to shift the terms of reflection on the human condition away from rationalistic abstractions that represent human beings as certain sorts of entities that can proceed to engage in activity, and toward a view of humanity in terms of interpretive activity that is, as Heidegger states, "always already" immersed in a fully functioning world of meaning. It is one thing to suppose, as phenomenological thinkers have done, that consciousness is always "consciousness of . . ."—by which these thinkers mean to stress that consciousness is never free floating but is always oriented on objects (which need not be physical objects but may also be conceptual). This is the sense in which consciousness is said to be "intentional," i.e., directed an aim or an object.[50] But for Heidegger, this characterization views experience in overly objectified terms, by placing emphasis on detached staring as a precursor to taking action, and by representing choice as involving explicit acts of predication. Heidegger instead opts for an approach that would become clearer in Sartre's revision of the traditional notion of consciousness: For Sartre as for Heidegger, consciousness is always consciousness *of something to be done*.[51] The explicitly teleological structure of worldhood as Heidegger develops it in *Being and Time* is based on the recognition that action is primary to human existence. Things show up in experience first of all and most of the time not as objects to be studied, but as *means* to be employed in the furtherance of the ends we set for ourselves—hence Heidegger's emphasis on the "um-zu" or in-order-to and its subsumption under higher teleological ends, all of which are ultimately for-the-sake-of human Dasein.[52] Like many thinkers before him, Heidegger incorporates the category of understanding into his view of how we relate to things in the world. But for Heidegger, understanding is not a noetic grasping of things in our perceptual or conceptual field; instead he characterizes it in terms of projection, an active movement toward the future in which we seize upon things in terms of their suitability for certain purposes. Understanding in this sense is interpretive and exhibits the structure of the 'as': we "see" something first of all not as a mute object whose properties we can ascertain, but rather as (already) useful for this or that purpose. Things, in our first encounter with them, are already invested with meaning. Moreover, on Heidegger's view, seeing something as suitable for some purpose or other is the necessary precondition for seeing it in the mode of detached observation.[53] Consider how far removed this conception of understanding is from, say, that of Descartes, who characterizes the faculty of understanding as a passive receptor. (Remember that for Descartes, the activity of judgment is performed by the will, not by the understanding.)

But on Heidegger's view, action requires more than understanding in the sense of a *procursus* into the future through the choices one makes. It requires a *prior disclosure of a totality* of useful things, a disclosure for which *both* understanding *and* a sort of affective orientation are necessary.[54] Simply put, we don't generally go looking for things unless we have an anterior drive or need (orientation on the future), and we can't go looking for things if they have not already been disclosed to us. This holds even (and perhaps especially) for scientific investigation, notwithstanding the common refrain that many scientific inquirers seek knowledge simply "for its own sake." On Heidegger's view, "the possibilities of disclosure which belong to cognition reach far too short a way compared with the primordial disclosure belonging to moods, in which Dasein is brought before its being as 'there'."[55] Indeed, in moods "Dasein is disclosed to itself *prior to* all cognition and volition, and *beyond* their range of disclosure."[56] This is a view according to which, when we consider a statement such as "all men by nature desire to know," we ought to devote a good deal of thought to the *kind* of desire that is at work here.[57]

Heidegger's view, then, is that we never really engage in acts of understanding in a fully detached, neutral manner, but instead are always already situated in a totality of meaning in virtue of which we can pursue our ends in the first place. The term 'mood' is potentially misleading when describing how we find the world disclosed to us, because it carries too many associations with the psychological notion of mood. Mood or disposition in Heidegger's sense is not best conceived as a psychological state; while it should be understood as a *felt* rather than cognitive connection to the world, and while much of what Heidegger has to say about mood recalls the language of psychology, Heidegger is after something deeper in our experience than, say, an individual person's tendency to be thin-skinned or overly confident. He is addressing the very manner in which things can show up as significant to us in the first place. The terms that Heidegger employs to explore this notion of situatedness are 'Befindlichkeit' and 'Stimmung'. 'Befindlichkeit' is a difficult term to translate into English; Macquarrie and Robinson translate it as 'state-of-mind' but note that this expression really signifies a sense of where one stands with things; Macquarrie and Robinson suggest that a literal translation would be "the state in which one may be found," and Hubert Dreyfus used to say that it has the character of "where-you're-at-ness."[58]

In discussing moods Heidegger also uses the term 'Stimmung', which is typically translated as 'attunement' and signifies a sense of being "in tune" with things. Earlier I noted that Helm employs this same term, but there I suggested that his use of it requires fundamental revision. Helm is right that some kind of attunement is required in the process of axiological commitment. But by hewing too closely to a rationalist conception of self and world, and by treating attunements as individual psychological states, he stops clearly short of the radical view of *Stimmung* advanced by Heidegger. For Helm, it would appear that attunement functions as a necessary *adjunct* to activities such as judgment, whereas for Heidegger it functions as a necessary *precondition* for them.

But what *are* moods, understood phenomenologically? (Contrast the psychological notion of moods, which has been developed on the basis of certain "positive" assumptions about what moods are and which sorts of entities can experience them.

The psychological approach to moods *presupposes* the global sense of orientation signified by *Befindlichkeit* or *Stimmung* in the specifically phenomenological sense.) For Heidegger, the key lies in recognizing that moods *let things matter to us.*[59] In this sense they are the cornerstone of care, which is the all-encompassing category of *Being and Time*. But in letting things matter, moods or dispositions do not simply disclose and let some particular object of concern matter to us; instead, a mood in the existential sense *"has already disclosed, in every case, Being-in-the-world as a whole, and is what first makes it possible to direct oneself towards something."*[60] Having a mood in this sense is to be distinguished from an individual psychological state with a single object; *moods, existentially conceived, disclose the world of meaningful concern in a global way that informs our entire understanding of things.*

Heidegger offers only a few specific candidates for moods in this existential-ontological sense: boredom, fear, love, and anxiety. All except the last are instances of mood or disposition in the mode of Dasein's everydayness, which for Heidegger means that they tend to facilitate Dasein's evasion of itself and what Heidegger calls Dasein's "ownmost possibility." Thus there is a certain tension in Heidegger's view of moods, just as there is in his account of everydayness: On the one hand moods, as modes of everydayness, are the way in which we exist first of all and most of the time, and they are the necessary precondition for the possibility of "authentic" (freely chosen) existence. But on the other hand, on Heidegger's view all moods except for anxiety are vehicles for flight from the prospect of authentic existence. How this tension is to be resolved, Heidegger never explains—and this failure on Heidegger's part may be traceable to the fact that his existential analytic of Dasein is heavy on notions such as resoluteness, with its unmistakable political overtones, and light on (indeed, deficient in) any real appreciation of the embodied character of our existence. As I will argue in Chapter 5, a serious consideration of embodiment, understood in non-objective terms, may hold the key to redressing the problem of recalcitrant affective dispositions (as well as to providing the crucial sense of orientation in the world for which approaches such as Cartesian-Kantian representation leave no room). Heidegger characterizes such dispositions in terms of "falling," by which he means active avoidance of one's "ownmost possibility," i.e., that possibility of existence that the individual sees and seizes upon as most appropriate to the current circumstances. For my own part, I believe that the entire Eleatic tradition could be charged with a sort of cultural falling, in this case a falling away from the prospect that guides my entire discussion in this book: the prospect of a moral community that is grounded in felt kinship and has no place for the pathos of domination that has characterized Western philosophy virtually from its beginnings.

The one everyday mood to which Heidegger dedicates a separate numbered paragraph in *Being and Time* is fear. In doing so, Heidegger expresses a recognition of a simple fact of human existence—that the world is populated, among other things, with threats to our well-being. These threats are typically identifiable as something particular. In this respect, a mood such as fear seems indistinguishable from the rationalistic conception of moods as predicatively structured intentional states with explicit noesis-noema structures. But Heidegger stresses that in a state such as fear, "we do not first identify a future evil (malum futurum) and then fear it." Nor "does fearing first ascertain what is drawing close; it discovers it beforehand in its fearsomeness." Only then "can fear look at the fearsome explicitly, and 'make it clear' to itself."[61] We are so conditioned to think of fear as a psychological state involving a fearing subject and a fearsome object that it is difficult to think past that conditioning and see what Heidegger is trying to say: that a fear-inducing object always emerges from a *totality* of things, activities, and meaningful inter-relationships, indeed in such a way that any such object implicitly gestures toward the totality from which it emerged and from which it draws its essential meaning.

Moreover, a state such as fear is never "simply" a mood or a way we happen to feel, but instead is part of a complex of historically conditioned, action-oriented ways of being. A given threat is an obstacle to my current endeavor, and by extension to my sense of personal integrity and perhaps to the integrity of the community of which I am essentially a part. By the same token, moods such as boredom and love disclose globally, even when they outwardly appear to be about a single object. In boredom, for example, it may appear to be "only this book, or that play, that business or this idleness that bores us," but in fact "deep boredom, drifting here and there in the abysses of our existence like a muffled fog, transports all things and human beings and oneself along with them into a remarkable indifference. This boredom discloses beings as a whole."[62] Similarly, in love it is not "simply an individual person" but rather their "presence" (*Gegenwart*) that induces joy—not their presence-at-hand (sheer determinacy) as an object in a field of objects, but rather the entire coming together of past meaning and future possibility that facilitates an encounter between these two individuals in this current time and place.[63]

Mood or attunement in the phenomenological sense, then, takes the form of a global as opposed to particular disclosure of things. When an individual (or perhaps a culture) encounters the world through the lens of boredom, everything appears in its light—not simply entities but commitments and relationships. It is not difficult to imagine an entire society gripped by boredom (just think of all the Heideggerian "idle talk" that occupies people on social media nowadays), by a globally distracting form of idle (as opposed to goal-driven) "curiosity," or even by a mood of conquest. Indeed, I find the proposition that the Eleatic mindset has been informed in advance by this last particular mood to explain quite a lot about both history and prevailing contemporary values—and not just values about nonhuman animals.

In the face of the sorts of moods that typically arise in our everyday dealings, Heidegger seizes upon one distinctive form of moodedness or orientation that holds the promise of shaking us out of our everyday presuppositions and enabling us to see our condition (namely, the basic character of our involvement with the world) in a comparatively undistorted fashion. Heidegger's most extensive remarks about mood are to be found in his discussion of the dislocating mood of anxiety, a mood or affective state in which beings as a whole "slip away."[64] This slipping away is not the disappearance of beings, but instead the kind of distance from things and all possibilities of action that discloses at the most basic level *that* human existence is thrown possibility. With the notion of thrown possibility, Heidegger brings together two notions that play a central role in Ortega's and Miller's conception of historical reason: that we never simply invent or construct an understanding of reality from scratch but instead have always already inherited an entire meaningful world ("thrownness"), and that our essential freedom enables us (indeed it makes it incumbent on us) to choose our own path going forward ("possibility"). Heidegger has quite a lot to say about anxiety, particularly in *Being and Time* and the essay "What is Metaphysics?" and this is not the place to go into an exhaustive discussion of it. In the context of a reflection on moods as existentially and phenomenologically distinctive modes of disclosure, what is of particular importance is that anxiety is distinct from all other moods in being characterized by a sense of "uncanniness" that Heidegger links to the prospect of taking up one's past in a distinctive or "authentic" manner.[65]

What is at stake in the question whether Heidegger makes a compelling case for anxiety as a distinctive human experience is whether we are capable of mastering our lives *affectively* rather than "rationally." Anxiety in the phenomenological sense is not a cognitive state but rather a mood or disposition that forces us to take a critical stand on the world and the choices we have made. Heidegger links the mood of anxiety to a notion that he adapts from Kierkegaard, namely, the notion of the *Augenblick* or "moment of vision." For Heidegger this signifies a privileged form of vision into the historical situation in which I find myself; what I see when I experience this vision is my "ownmost possibility," that possibility of action that I recognize to be most incumbent upon me here and now.[66] This is a possibility that the tradition has rarely if ever taken seriously, and it is a possibility that thinkers such as Heidegger urge on us, if only by way of subtle gestures. I have already noted that Heidegger neglected the significance of embodiment. At the same time, his notion of disposition or mood is highly compatible the reflections on embodiment undertaken by thinkers such as Schopenhauer and a good many contemporary feminist thinkers: If we stop thinking of Heideggerian mood in the rarefied terms in which Heidegger presents it and think of it specifically in terms of embodiment, then two important conclusions should become apparent: that *it is through our embodiment, which includes the pervasiveness of affective dispositions, that we encounter the world at the "primordial" level that is Heidegger's persistent concern*; and that *it is only by way of a "reduction" or abstraction away from that primordial field of encounter that we are able to arrive at the sorts of detached insights that the Eleatic tradition has (wrongly, I believe) considered privileged.* This is the core significance of Heidegger's overarching theme in *Being and Time*: that "in care is grounded the full disclosedness of the 'there'."[67] Caring is the foundation for every way we have of relating to the world and ourselves, and as such it is the necessary precursor to detached knowing.

One final dimension of Heidegger's analysis of human existence in *Being and Time* is needed to see the logic behind his claim about the "founded" character of detached knowing, and that is Heidegger's notion of the "as-structure" of interpretation. Where the tradition has tended overwhelmingly to characterize experience as if one more or less determinate entity (the human being) receives sense-impressions from an array of other more or less determinate entities in the external world and then, provided that it possesses predicative language, proceeds to describe the determinate features of those entities (this is the approach that would reach modern fruition in the subject-object ontology), Heidegger argues that this sort of characterization of self and world skips over a more original level of experience altogether, thereby providing what is essentially a caricature of the ways in which we actually encounter and experience things. Heidegger starts from the proposition that we are always already involved in a fully functioning context of meaningful interrelationships. What is conspicuous about the various versions of the Eleatic approach, which includes not only obvious examples such as the Cartesian-Kantian subject-object ontology but even Hume's resolution of the field of lived experience into what are essentially pixels of sense-data in need of unification by the mind, is that they all treat what Kant called the "manifold" of experience as discrete bits that bear no relationship to one another until the mind supplies the needed principles for relating those bits to one another.[68]

Heidegger's recourse to the notion of the 'as' is designed to help us recognize the "founded" character of all such approaches to both the nature of world and the human-world relationship. Understanding, on Heidegger's view, is not simply an inventory of things existing independently of us that we can subsequently incorporate into our endeavors. Instead it is "a disclosure," a future-oriented one in which things become disclosed not in their mute determinacy but instead in terms of their place in a larger context of *meaning*.[69] The world, in other words, is always already meaningful, and that context of meaning is the ground from which we abstract in our efforts to represent the world in determinate ways designed to facilitate manipulation and control. We always understand a situation *as* having a certain character; things show up *as* useful (or useless) for this or that purpose, *as* bearing certain relations to other things in the environment, or *as* having a certain significance for our lives.

The 'as' is at work in acts of detached cognition, in which we seek to ascertain the explicit features of something in the abstract (i.e., removed from all meaningful interrelationships). An excellent example of this functioning of the 'as' in detached cognition is the Cartesian meditator's conclusion that the essence of the wax is the abstract, mathematically describable property of "extension"; another is the meditator's proclamation of the *cogito*, which strips from the self all connections with lived experience apart from the sheer inward staring of Cartesian introspection.[70]

Heidegger characterizes these sorts of cognitive insights as "founded" in the sense that they are products of a deliberate looking *away* from essential features of human existence and the world, and toward those features of both that appear to hold the promise of facilitating our efforts at control. There are all sorts of significations at work in our engaged dealings with things, but only those that are of direct relevance to our efforts at control survive the process of abstraction. Anticipating his later characterization of the Cartesian as opposed to Protagorean mindset, Heidegger characterizes this formal level of discourse in terms of the "apophantic" 'as', where 'apophansis' signifies "letting something be seen by pointing it out."[71] Naturally there are various forms of letting things be seen by pointing them out that fall far short of the specific form of *apophansis* that takes place in scientific cognition or philosophical theorizing; one can simply say, for example, "hand me the big Phillips head screwdriver over there." Heidegger's point is that, by the time we find ourselves in a position to make even a statement of that kind, we have gone through an entire dimension of meaningful encounters with things such as screwdrivers and the sorts of projects in which they are designed to be employed. There are levels of removal from fully engaged context, and Heidegger intends the notion of apophantic *logos* to encompass them.[72] He further intends to make a case for the proposition that the sorts of statements that comprise scientific or philosophical theories necessarily take the most rarefied (i.e., decontextualized) form of *apophansis*, and this means that all such statements run the serious risk of leaving key considerations out of the picture. As I have argued from the beginning of this book, I believe that some foundational facts about ourselves and our relation to the world around us have been deliberately if unwittingly (I believe both are possible simultaneously—think of Woolf's notion of an "infantile fixation") omitted in the service of certain ulterior motives.

Heidegger observes that the tradition has proceeded on the assumption that "thinghood" in the sense of sheer determinacy (Heidegger calls it "presence-at-hand") is our primary mode of access to reality. In doing so, the tradition has employed a form of dogmatism according to which there are first mute entities, and only subsequently the projection of values onto them. Moreover, according to this dogmatism our most genuine encounters with things, those that yield (potentially "apodeictic") knowledge, are those that take the form of predicatively structured, conceptually informed *judgments*. One of Heidegger's central contentions is that this is simply not the way in which we encounter things in our ordinary lived experience; in fact, this characterization confirms that the tradition has missed the phenomenological conception of world altogether, having opted for an "objective" characterization of the world—which as we have already seen leaves us with the problem of how we become related to the world in the first place.[73] This problem does not arise if we proceed from the axiom that we are first of all and most of the time fully immersed in a meaningful context of action, and only subsequently view ourselves through the lens of conceptual abstraction as separate from the world.

Here the "founded" character of apophantic discourse becomes clear:

> The statement, as indicatively showing something [*aufweisendes Sehenlassen*], is possible only on the basis of our already being with the subject matter to be

shown. . . . To that degree, it entails that the plurality of meanings in the unity of a sentence is possible only on the basis of and in the medium of meaning.[74]

Here, when Heidegger goes on to explain that "the 'as' is the basic structure whereby we understand and have access to anything," he is now referring to *a different form of the 'as'*, one that is the condition for the possibility of employing the apophantic 'as'. Even our everyday, engaged dealings with the world involve seeing things "as"; but in the case of engaged practical dealings we see things "as" suitable or unsuitable for certain purposes that we are pursuing, rather than "as" mute objects or processes with determinate properties. By the time we have engaged in acts of predication, be those scientific statements or "give me that hammer over there," we have already become acquainted with a whole context of meaning that is precisely *pre*-predicative. Heidegger calls the "as" that is at work here the "hermeneutical" as opposed to apophantic 'as'. It signifies full immersion in the context of meaning, which is "an *existentiale* of Dasein, not a property attaching to entities."[75] It is on this basis that Heidegger urges the conclusion that "the aforementioned basis of metaphysics and its orientation toward propositional truth is indeed necessary in a certain respect, yet it is not originary [*ursprünglich*]."[76]

This characterization of our everyday dealings and the sight that guides them in terms of the hermeneutical 'as' affords an entirely different way of thinking about our relationship to and place within the world than the mode of access and characterization of vision dominant in the philosophical tradition.[77] It is entirely possible, and in many instances highly advisable, to conceive of phenomena such as vision in detached, objectified terms—e.g., when we are seeking to theorize about a condition such as macular degeneration and are attempting to find a cure. This is presumably what Heidegger has in mind when he acknowledges that the detached standpoint of metaphysics (and, by extension, natural science) is "indeed necessary in a certain respect." It is a simple fact of life that we humans are highly vulnerable to threats in the environment, and this means that acts designed to enhance control can be of great benefit to us. But acts of control are not the only sorts of acts in which we engage, even if they are crucial to our existence.

On Heidegger's view, *all* human conduct takes the form of care. That includes metaphysical and scientific theorizing. The specific nature of and driving force behind the care specific to theorizing should be clear by now: The fact that we sometimes seek truth purely "for its own sake" does not mean that we should take anything like detached staring as the model for how we guide our conduct. The historical record, as I have tried to show in previous chapters, demonstrates that the pathos of detached staring is not as neutral and innocent as its major exponents have proposed. It, too, occurs within "the care structure" of human existence; and its ethos, its sense of where human beings fit into the cosmic scheme, is one according to which human beings reign supreme and everything else is essentially a resource (even if it might be a resource toward which we ought to behave kindly).[78]

At the end of Chapter 1, I introduced Heidegger's notion of "letting beings be," which poses a stark contrast to the comparatively "closed" approach to reality expressed through the Eleatic mindset. It is this pathos of openness that I believe holds the key to a fundamental rethinking both of ourselves and our relationship to the world around us, most particularly our relationship to and with nonhuman animals. I have noted several times that Heidegger himself exhibited no particular interest in nonhuman animals and went so far as to characterize their sense of being in the world as impoverished in comparison with us "world-forming" human beings. If only against his own intention, Heidegger points the way toward the prospect of reevaluating our received characterizations of nonhuman animals, and in turn a reevaluating of prevailing sensibilities about their moral status.

Heidegger's appeal to an essential pre-predicative level of experience poses a very strong challenge to the historical assumption that propositional truth has priority over all other forms of or approaches to truth. I think it is obvious that many of our moral commitments take shape without any recourse to judgment, although of course we can always review our commitments at the level of rational abstraction and seek to reduce them to determinate judgments. Why, then, the unremitting rationalist insistence that moral commitment fundamentally takes the form of judgment? As I have argued from the start, at least one of the motivations seems obvious. Alongside the entirely reasonable interest in overcoming the distorting influence of selfishness and partiality in the formation and living out of our commitments—I will return to this problem in Chapter 5—we must acknowledge the drive to dominate that has led the tradition to turn away from the prospect of "letting beings be" and toward artificial constructions that both facilitate domination *and* purport to justify it.

Here is where Heidegger's appeal to the hermeneutical 'as' and an entire pre-predicative dimension of experience enters the picture. Recall my discussion of Peter Carruthers early in Chapter 1. He is one of a number of thinkers who argue (or rather: claim) that few if any nonhuman animals possess conceptual abilities and thus anything like predicative language. Carruthers goes so far as to suggest that because nonhuman animals cannot reflect on their experiences of pain in acts of detached cognition, there is a sense in which nonhuman animals are not really conscious of their pain at all and therefore their experiences of pain "make no real claims on our sympathy."[79] Carruthers aptly captures a key commitment of the Eleatic pathos: that something is not true or real unless you can prove it according to the axioms and proof procedures admitted by that pathos. Contrast this with a comparatively "open" approach, one that can entertain the possibility that *not just human beings but many nonhuman animals inhabit the world in meaningful ways*, even if nonhuman animals cannot step back from those ways of being in order to represent and think about them in conceptual-linguistic terms.

4. Pre-Predicative Meaning and Affective Engagement

To consider the prospect that the lives of nonhuman animals are meaningful *to them* is to entertain the possibility that (many) nonhuman animals make use of the hermeneutic 'as' in their dealings in the world, i.e., that they do not simply "behave" in a blind reactive manner but actively conduct their lives in ways that in many respects closely resemble our own.[80] The Carruthers-style response is: prove it! That strikes me as a very convenient response for anyone who seeks to justify the regime of animal use that has prevailed for so long. Contrast that sort of response with the approach taken by the phenomenological thinker Merleau-Ponty, who proceeds from an open acknowledgment that *human meaning is a special case of the phenomenon of meaning as it is experienced by a wide variety of sentient creatures.* Like Heidegger, Merleau-Ponty takes his cues about inhabiting an environment (*Umwelt*) from the ethologist Uxküll. But unlike Heidegger, Merleau-Ponty explicitly acknowledges that some nonhuman animals live life meaningfully. Merleau-Ponty makes a distinction between "lower" and "higher" nonhuman animals, i.e., between those that can and cannot issue "a reply to the exterior world" that goes beyond mere mechanistic reaction.[81] How "open" is Merleau-Ponty's approach to the proposition that human beings are not unique in participating in meaning? Just open enough to entertain the possibility that the "higher" sort of nonhuman animal does not simply react but actually "interprets its situation" in the course of navigating its environment.[82]

In affirming this capacity of many nonhuman animals (you can easily include at the very least Tom Regan's "subjects-of-a-life" here) to engage in acts of interpretation, Merleau-Ponty is giving one of a number of indications of the absolute primacy he places on embodiment in the process of meaningful engagement with the world. "There is a Logos of the natural esthetic world," he writes, "on which the Logos of language relies."[83] The term 'esthetic' recalls Kant's treatment of sense-experience in his Transcendental Aesthetic, which is devoted to an examination of "aisthesis," the Greek term for sense-experience. But where Kant presents a view of our sensory encounter with things as an unformed and essentially meaningless "manifold" that requires unification according to *a priori* forms of intuition employed by the mind, Merleau-Ponty takes a comparatively open approach that does justice to Proust's insight that our embodied encounter with things moves us beyond the "spectacle" of detached staring and provides us with intimate access to essences in the world. Any sentient creature capable of interpretation has an active relationship with its *Umwelt* which, as in Heidegger, is never an array of discrete units standing in need of unification by a subject but instead is *always already a meaningful whole.*[84]

Before looking more closely at Merleau-Ponty's conception of lived embodiment, it is worth dwelling for a moment on the notion of the "spectacle" to which Proust alludes. A spectacle in this sense is the reduction of lived experience to an image or set of images, the rendering static of what is essentially an active, lived engagement with things. Guy Debord, writing in a different context, suggests that the production of the spectacle is motivated by, and designed to serve, interests of power. Debord's concern is the way in which the phenomenon of the spectacle serves the capitalist mania for the production and consumption of commodities, thereby cutting society off from a deeper, more genuine connection with things. But his analysis of the spectacle has important implications for the present discussion. The reduction of lived experience to spectacle entails an "obvious degradation of being" and serves the "totalitarian management of the conditions of existence."[85] In the context of his critique of capitalism, Debord concludes that emancipation from the hegemony of the spectacle requires the raising of Marxian class consciousness.[86] There is a clear parallel here, I think, with the concerns I have been addressing: The reduction of lived experience to the epistemological spectacle that culminates in the modern complex of subject-object-representation entails a critical loss or diminution of the affective dimension of sentient life, i.e., an "obvious degradation of being"; the anthropocentric background ideal of living from which it proceeds serves a certain sort of "totalitarian management of the conditions of existence" by projecting a comprehensive map of the real that locates human beings at the apex of existence; and the prospect of overcoming the self-serving distortions of the anthropocentric background ideal demands a new kind of consciousness, one that can both see the ideological character of anthropocentrism and anticipate its overcoming toward a more inclusive and just sense of community.

The endeavor of Merleau-Ponty and a number of other recent figures to rethink the place of embodiment in human experience is a direct response to a long tradition of treating embodied states as potential threats to our well-being. As I have noted already, many traditional thinkers acknowledge the role of our affective life in helping to provide us with a sense of orientation in the world; but almost without exception they consign the affects to the status of potentially unruly reactions that require management from above, or they characterize the affects as being forms of rationality. On the dominant view in the history of Western thought, reason is active and our affective life is comparatively "passive," as the term 'passion' reinforces. But what if it turns out that "the emotional act, like any act of consciousness, is an attempt to *do* something," i.e., what if we have misunderstood the true nature of our affective lives in our haste to proclaim the authority of the one experiential faculty (namely, linguistic rationality) that we consider to be uniquely human?[87] What if our affective life is not passive after all, but is in fact just as active as our rationality? (A related question: What if Descartes was right that in a certain sense, reason by itself is actually passive? Recall that in the Fourth Meditation, Descartes is clear that the activity of judgment is performed not by the intellect but by the will.) This is where the significance of Heidegger's notion of attunement or disposition as a fundamental dimension of "understanding" becomes clear: Merleau-Ponty is inscribing Heidegger's notion of attunement within the context of lived experience as an irretrievably bodily condition, in an effort to remind us of something that found no place in Kant's subject-object ontology: that *already in our embodied sensory encounter with the world, meaning is disclosed*. And as I hope to show by the end of this chapter, this has important implications for the endeavor to rethink our place in the cosmic scheme.

Like Heidegger, Merleau-Ponty takes Descartes's reductive ontology as his point of departure for reconceiving the human encounter with the real. Merleau-Ponty directs our attention away from the Descartes "after the order of reasons" (i.e., subsequent to the cognitive reduction undertaken in the *Meditations*) and toward the Descartes *before* the order of reasons: Here we find the "Descartes of the Cogito before the Cogito, who always knew that he thought, with a knowing that is ultimate and has no need of elucidation."[88] This *cogito* prior to the order of reasons lies "behind the spoken *cogito*" and "does not constitute the world, it divines the world's presence surrounding it as a field that it has not given itself." This "tacit *cogito* . . . is anterior to any philosophy"; its grasp of the world is "inarticulate" in comparison with the *cogito* that is informed by the order of reasons and bound up with language as its primary means of relating to things.[89] This tacit *cogito* does not show up thematically for itself and does not relate to the world as to a static, transparent, mute object. "The sentient subject does not posit [qualities of things] as objects, but sympathizes with them [*sympathise avec elles*], makes them his own and finds in them his momentary law."[90] In characterizing our primary, pre-thematic relation to things as *sympathetic*, Merleau-Ponty redirects attention away from cognitive detachment denuded of any affective dimension and toward a felt, embodied relation to what Merleau-Ponty called the "thickness" (*l'épaisseur*) shared by self and world.[91]

This sense of thickness as opposed to transparency reminds us of precisely what the ontology of subjectivity was designed to evade: the fact that our relation to things, indeed to our own perceptual activity, cannot ultimately be mastered through apophantic theorizing and cannot be adequately grasped if we see embodiment as incidental to our being. The primacy of embodied experience is such that at the most basic level,

> the perception of our own body and the perception of external things provide an example of *non-thetic* consciousness, that is, of consciousness that does not possess its objects in full determinacy, that of a *logic lived through* which does not account for itself, and that of an *immanent meaning* which is not clear to itself.[92]

The corollary of this form of perception is a thinking that "feels itself rather than sees itself, which searches after clarity rather than possesses it, and which creates truth rather than finds it."[93] Where Heidegger has only a vague intimation of the importance of embodiment in seeking to characterize the unifying principles of worldly experience, Merleau-Ponty straightforwardly asserts the irreducibly affective, bodily dimension of lived experience and the essential basis it provides for our global sense of orientation and our struggle for truth.

Heidegger and Merleau-Ponty make fundamental contributions to the endeavor to arrive at a more unified conception of human experience and its continuity with the world, a conception that avoids the pitfalls of characterizing feeling in either overly physiological or overly intellectualized terms. Contemporary neurophysiological approaches to emotion are not alone in characterizing the body as what Merleau-Ponty aptly called "a transparent object," thereby failing to see the body as "an expressive unity which we can learn to know only by engaging it [*en l'assumant*]."[94] Approaches such as the James-Lange theory similarly characterize emotions as physiological reactions to external events, thereby losing the phenomenological sense of thickness exhibited by self and world alike.[95] At the other extreme lie views of the emotions as cognitive judgments. This view has been embraced by thinkers from the Stoics to contemporary thinkers such as Martha Nussbaum, who proposes a view according to which emotions are "ways of seeing" that are "cognitive, evaluative, and eudaimonist" and "always involve some sort of combination or predication."[96] On Nussbaum's view, emotions are essentially Stoic *lekta* in connection with which the affective dimension appears to be no more than an afterthought.[97]

What is irretrievably lost in these approaches, which seek to characterize both reason and affect from a standpoint of detached repose, is the sense of *being immersed* in a world that ultimately exceeds our grasp and that in the first instance (and perhaps in the last as well) presents itself as a mystery to us. Our primary relation to the world is not one of detached knowing, because we are not primarily detached knowers and the world is not primarily an object. Nor is our primary relation to the world passive, as if our primary function were simply to contemplate the world and we only subsequently endeavored to take a stand on it. On the contrary, our relation to the world is fundamentally an *active* one, one of projects, needs, and the imperative to transform the world and ourselves in response to threats, uncertainty, and other exigencies. This, indeed, was the basis of Sartre's critique of Husserl's conception of the transcendental ego—that consciousness is not simply consciousness of something, but "consciousness of something *to be done*."[98]

A primary mistake of conceptions of consciousness in the wake of Descartes and Kant has been to retain these thinkers' primary focus on cognition and their corresponding relegation of emotion to an object of suspicion. Descartes, for example, characterizes the passions primarily as threats to our integrity, arguing that the will must "arm" itself against the passions by forming "firm and determinate judgments bearing upon the knowledge of good and evil."[99] Kant, as we have seen, dismisses feeling as a "pathological" influence on moral judgment. At the same time, both thinkers make a place for "positive" affects, Descartes giving pride of place to the passion of generosity and Kant arguing for the importance of "practical [i.e., moral] love."[100] But these appeals to the positive role played by the affects in human life constitute departures from the greater tenor of both thinkers' remarks about the passions, a tenor that is modulated by an unremitting emphasis on the primacy of detached reason over engaged practice.

5. The Moral Community is Neither Exclusively Nor Primarily Human

Heidegger, as I have noted, remains cryptic and indirect about the role and primacy of the affects in the active life because he appears to have had some serious hesitation about ascribing centrality to the role of embodiment in human existence. Merleau-Ponty makes a decisive step beyond Heidegger when he states that

> it is [the body] and it alone . . . that can bring us to the things themselves, which are not themselves flat beings but rather beings in depth, inaccessible to a subject who would survey them [*un sujet de survol*] and open only to those, if such a thing be possible, who coexist in the same world with those beings.

Detached contemplation, in other words, is predicated on a crucial *loss* of reality, whereas at the level of the "cogito prior to the cogito" that I noted earlier, the body presents itself not as a determinate object but as possessing a "constitutive paradox" that is inseparable from "the presentation of a certain absence" characteristic of the world at its most primordial level of disclosure.[101] At this level, being is "carnal" rather than objective, although it can be rendered objective, if only inadequately, in acts of noetic reconstruction. In this connection, the failure of philosophy in its attempt to understand embodiment has been its endeavor to exercise a "total and active grasp, intellectual possession" of what can be grasped only as "a dispossession [*dépossession*]." This is what Merleau-Ponty refers to as the idea of "chiasm": that "every relation with being is *simultaneously* a taking and a being taken, the hold is held, it is *inscribed* and inscribed in the same being that it takes hold of."[102]

Thus for Merleau-Ponty, it is an *a priori* that embodiment is the vehicle of our encounter with ourselves and the world, and that this encounter, indeed the body itself, is irreducibly *a paradox*. Both the encounter and our embodiment are, as Schopenhauer had recognized, essentially prior to any conceptualization that would resolve them into distinct entities with determinate features. At this level, the body is

> a truce of metaphors. It is not a surveying [*un survol*] of the body and of the world by a consciousness, but rather is my body as interposed between what is in front of me and behind me . . . in a circuit with the world, an *Einfühlung* with the world, with the things, with the animals, with other bodies . . . made comprehensible by this theory of the flesh,

which itself is "*Urpräsentierbarkeit* of the *Nichturpräsentierten* as such, the visibility of the invisible"—the original presentability of what is not originally presented.[103] Here, in his lectures on nature as in the *Phenomenology of Perception*, Merleau-Ponty stresses *Einfühlung*—our sympathetic relation with the world, thereby stressing that our most fundamental mode of engagement not only is not a detached surveying of things but is *a felt immersion in and with them*, replete with an open sense of their fundamental mystery, their essential withdrawal and their refusal to subject themselves to our efforts to master them. This is what Merleau-Ponty seeks to underscore when he characterizes the body as "libidinal" and our intercorporeality in terms of *Einfühlung*.[104]

In making a central place for *Einfühlung*—sympathy or empathy, an affective sense of interrelatedness with other beings—Merleau-Ponty is showing the influence of Husserl but also distancing himself from Husserlian phenomenology. For in *Ideas, Volume II* Husserl identifies *Einfühlung* as a fundamental mode of relating to other subjects that first makes intersubjective objectivity possible, whereas Merleau-Ponty is considerably more invested in giving prominence to the irreducibly subjective aspects of our experience and the ultimately imperceptible transition between ourselves and the rest of nature.[105]

But Merleau-Ponty is ultimately ambivalent about the amount of distance he ultimately wishes to take from Husserl. In his lectures on nature, Merleau-Ponty acknowledges the gradual emergence of human subjectivity from animal life, but he also states categorically that a human being is not to be understood as "the mere addition of reason to the animal (body) . . . The animal-human relation will not be a simple hierarchy founded on an addition: there will already be a different manner of being a body in humans."[106] In what does this different manner of being a body consist? For the Merleau-Ponty of the lectures on nature, it does not consist merely in the interpretation of symbols, as Merleau-Ponty freely acknowledges that an "architecture of symbols" is part of the experience that many animals have of their environments. And yet even here Merleau-Ponty denies that there is "a break between the planned animal, the animal that plans, and the animal without plan."[107] Nonetheless Merleau-Ponty remains fairly squarely within the tradition when he asserts that the distinctiveness of the human consists in a certain "structural freedom," not "a freedom in the Kantian sense" but nonetheless presumably informed by a linguistic capacity distinctive of beings capable of theorizing.[108] This structural freedom is, on Merleau-Ponty's view, what first makes possible "the junction of Φυσις and λογος," which in turn makes history possible.[109]

Nonetheless the human retains a certain continuity with the nonhuman animal, and this continuity is to be seen first of all in our affective rather than our cognitive modes of access to things. Merleau-Ponty elaborates on the link between intercorporeality and affectivity by suggesting that having access to a world that we share with others depends on a certain "identification between corporal schemas" that situates me in a world together with other experiencing subjects.[110] In characterizing these schemata and their identification as "libidinal," Merleau-Ponty stresses the fundamentally affective nature of our primary access to things. Thus it is no surprise that he appeals not to cognition but rather to *desire* as "the common framework of my world as carnal and of the world of the other."[111] Desire in this connection is not to be thought of as "mechanical functioning" but rather as "an opening to an *Umwelt* of fellow creatures" that shares fundamental affinities with nonhuman animal desire.[112] Merleau-Ponty calls this approach to the rudiments of human and animal experience an "esthesiology," a logic of sensible experience. Much of what Merleau-Ponty writes about sensible experience is focused on sight and touch.[113] The link between these forms of sense experience and desire is an intimate one, a link most evident when we think of sight and touch as active, directed, engaged processes rather than as physiological responses to external stimuli. This is what Merleau-Ponty evidently has in mind when he calls for "a new conception of intentionality," one not oriented on "the experience of the world as a pure act of constituting consciousness."[114]

In appealing to esthesiology rather than to pure acts of the understanding, Merleau-Ponty (like Schopenhauer) is calling radically into question Kant's famous dictum that "intuitions without concepts are blind."[115] Certainly thoughts without content remain, as Kant averred, empty; but the proposition that our intuitive grasp of things is bereft unless and until it becomes ordered in terms of concepts makes no more sense to Merleau-Ponty than the proposition that only apophantic *logos* counts as "meaningful" made to Heidegger. Merleau-Ponty capitalizes on Heidegger's insight into the hermeneutic "as" and makes it his own by moving past Heidegger's focus on meaning in general and arguing that meaning and communication are to be understood as inseparable from bodily desire—and not just desire as we think of it as a possibility for beings who exist "toward" death, as Heidegger had suggested, but desire in a sense that we share with nonhuman animals. Merleau-Ponty states that just "as esthesiology emerges from the relation to an *Umwelt*, human desire emerges from animal desire."[116] He further states that this *Umwelt* is one "of fellow creatures" and "communication," and in his lectures on nature he is clear that meaning and communication—indeed entire *Umwelten*—are not the exclusive possession of human beings but are lived by many nonhuman animals.[117]

Merleau-Ponty's reference to our "fellow creatures" anticipates contemporary thinkers such as Cora Diamond and Christine Korsgaard, both of whom employ this notion in an effort to rethink the nature and limits of community. Neither Diamond nor Korsgaard, however, takes an entirely "open" approach to the prospect opened up by the recognition of a deeper level of engagement than that afforded by apophantic *logos*. Korsgaard, as I discussed in Chapter 1, does present some important insights into the dignity that nonhuman animals possess in virtue of being ends-in-themselves; but she remains wedded to a view according to which human beings retain the prerogative to exercise dominion over nonhuman sentient life, at least in the sense of considering ourselves entitled to determine what does and what does not constitute the entitlements and proper treatment of nonhuman animals. Diamond acknowledges that nonhuman animals are our "fellow creatures," and she observes that "without [loving] attention . . . there is no perceiving the evil of genuine injustice."[118] But she also suggests that in our endeavors to establish a felt rather than a merely abstract appreciation of the suffering of other sentient beings, "images of fellow [nonhuman] creatures are naturally much less compelling ones than images of 'fellow human beings' can be."[119] Diamond does not explain what she means by 'natural' in this connection, and the fact that she takes much of her inspiration from Wittgenstein might lead one to wonder where this sense of naturalness is coming from if not from historically stabilized prejudices about the nonhuman. What Diamond does state is that in asserting the need for a sense of human solidarity, she is *not* relying on any specific capacities such as intellect or "moral personality."[120]

That is a step in the right direction; it constitutes an acknowledgment that our sense of moral worth is not ultimately based on capacities such as cognitive prowess, but rather on a prior sense of essential connectedness with others. But rather than explore this sense of essential connectedness in a spirit of open inquiry, as Schopenhauer sought to do, Diamond takes recourse to a surprisingly traditional conception of kinship: She identifies "human fate" as the basis for moral connection and concern. She adds that this central focus on the human "is not any kind of attempt to determine the limits of moral concern."[121] But history demonstrates very clearly that notions such as "the limits of moral concern" are notoriously slippery, and moreover that simply holding back from excluding a possibility (in this case, that nonhuman animals merit full and direct moral concern of a kind never seen before) is *not* the same thing as taking that possibility deadly seriously. I take this to be an excellent example of what Diamond means by "the difficulty of reality."[122]

In this connection, Grimm's and Taureck's insights hold great promise for our efforts to confront the difficulty of reality with which the historical treatment of nonhuman animals confronts us. Both thinkers take the view that we do not and cannot know what nonhuman animals are really like and what they genuinely merit while we are actively engaged in representing them as determinate entities possessing certain experiential capacities and lacking others, and while we are actively employing them as objects of use. As Taureck observes, what is needed is precisely the kind of "step back" that I discussed late in Chapter 1, a step back from willing. More than anything else, Taureck calls on us to proceed with an abiding sense of *humility* that is frankly difficult to reconcile with the repeated claim of Western humanity to be "the lords of nature." Rather than simply "rethinking" our received prejudices about nonhuman animals, we must prepare the way for such a rethinking by suspending the will—exactly as Schopenhauer proposes.

A truly open perspective on our place in the world is one that subjects our most deeply held assumptions to extreme critical scrutiny, a degree of scrutiny that helps to think past what I take to be the anthropocentric limitations of thinkers such as Korsgaard and Diamond. There is a straightforward sense in which human fate is unlike the "fate" to which non-linguistic beings might be said to be subject: We humans appear to be the only sentient creatures whose self-understanding is specifically historical in character; this has a variety of consequences, including the fact that we devote a lot of time and energy to thinking, writing, singing, and preparing elaborate rituals related to death. People often tell me that death is more significant for us than for nonhuman animals, their reasoning typically being that we experience a lot of grief due to our relation to death (fear of my own death, grief at the loss of loved ones, etc.) *and* that human beings are unique in having this sort of experience. This is the general line of thought that appears to have led Mary Anne Warren to assert that human lives "have greater intrinsic value [than those of nonhuman animals], because they are worth more *to their possessors*."[123] This strikes me as a textbook example of anthropocentric thinking: Warren appears simply to be assuming that there are objective metrics regarding matters such as intrinsic worth; that human beings are able without further ado (i.e., without checking their own prejudices, or after "correcting" them) to perceive and employ these metrics in undistorted fashion; and that such employment clearly (perhaps clearly and distinctly) confirms that human lives matter more to human beings than any nonhuman animal's life can possibly matter to it. This last proposition assumes that anthropocentrically-minded thinkers are open enough to admit that *things matter to nonhuman animals*, and that this sense of significance is not dependent on acts of detached reflection and predication.

How, exactly, does Warren or anyone else *know* that, say, my life matters more to me than my cat Pindar's life mattered to him? Think about it this way: Absent any ability to read the minds and felt evaluations of nonhuman animals (as I have suggested from the start, I take this to be a deficiency in human beings), what must we assume in order to come to the conclusion urged on us by Warren and shared by so many people? The short answer is: all the essential elements of the anthropocentric background ideal of living, including the rash assumption that nonhuman animals live in "an eternal present" and that they therefore are incapable in principle of anything like a sense of self or truly active (i.e., forward-looking) dealings in the world. The comparatively open approach that I have been pursuing is one that acknowledges the inherent limitations and self-serving character of subject-object representation and other forms of cognitive abstraction, as well as the fact that our embodiment affords us an appreciation of features of reality that are not available to detached reflection. One such insight, which I fully acknowledge not to be a matter that I can "prove" to a robust possessor of the anthropocentric mindset, is that "my cat Pindar's life [was], at the pre-predicative level, worth every bit as much to him as my life is worth to me."[124]

This, I believe, is the crucial juncture at which embodied, felt experience must be acknowledged to play a key role not only in our experience but in our sense of place in the world—not our objectively describable sense of physical location, but our sense of "where we stand" with ourselves, our fellow humans, nonhuman sentient life, and the rest of nature as a whole. Late in Chapter 1 I invoked this sense of place by focusing on Heidegger's etymological analysis of the term 'ethos' and his proposal that we conceive of ethics not first of all in terms of axioms and propositions, but rather in terms of dwelling in the world. Ethics, in this sense, "ponders the abode [*Aufenthalt*] of the human being."[125] Where Heidegger ties this prospect to a contemplative form of *thinking* that reaches deeper than the resources of apophantic *logos*, thinkers such as Schopenhauer and Merleau-Ponty stress the role of our felt, embodied connection to reality as a crucial complement to our efforts to navigate the world by using reason. And yet Heidegger is absolutely right to stress the possibility and significance of a "more primordial" way of relating ourselves to the world by means of thinking. That possibility and significance are explored by Ortega and Miller, with some inspiration from the Pragmatists. A key mistake to avoid here, I believe, is to move from one extreme to the other, by embracing a total "pathocentrism" of the kind recommended by Hume and some contemporary thinkers.[126] In Chapter 5 I will return to Woolf's tentative suggestion that there is some common origin to reason and emotion, and I will argue for a specifically dialectical relationship between the two faculties in the process of moral commitment in which neither can be said to have ultimate priority over the other.

A truly robust affirmation of the centrality of embodiment in our own experience provides us with the basis for expanding our sense of the moral community beyond the strictly human. I noted a short while ago that Diamond makes an important contribution to our thinking about the moral community when she states that inclusion in this community is not ultimately a matter of possessing certain experiential capacities. As an example to illustrate this principle, Diamond presents the case of "the rape of a girl lacking speech and understanding, lacking what we think of as moral personality and the capacity for autonomous choice, and incapable of finding the event humiliating and the memory painful as a normal woman might." The outrage we feel when contemplating such a scenario is not the kind of outrage we would feel "as Kantians," which Diamond equates at one point with the point of view of "rational Martians."[127] Instead, our outrage is based on something else—on a sense of *shared fate* that Diamond takes to constitute a "natural" dividing line between human and nonhuman animals. What Diamond says precious little about is what the experience of nonhuman animals may actually be like; instead she appears simply to assume, like a long line of thinkers, that if we can't detect it in a nonhuman animal then it simply isn't there.

Rather than stressing the differences between human and nonhuman animals, the spirit of openness demands a reaffirmation of our own inherent animality and the fact—I take it to be a fact—that *all sentient creatures share the same fate*, namely, the struggle for survival and meaning and the ultimate confrontation with mortality. This is precisely what Schopenhauer meant when he observed that human beings and nonhuman animals share a common essence, and it is precisely what a reflection on Aristotle's notion of "a precedent community of nature" ought to illuminate. It is remarkable how many thinkers have acknowledged a variety of experiential capacities in nonhuman animals that have traditionally been assumed to be the exclusive possession of human beings, only to hold back from an open consideration of the implications of that acknowledgment. Scheler, to take just one example, explicitly states that nonhuman animals employ the 'as' and are capable of "ekstasis," the ability to step back from engagement with an object so as to recognize its usefulness, but he retains the traditional prejudice that the capacity for objectivity renders humans "superior beings."[128] As I have argued from the outset, this confident assertion of superiority is predicated on a global background of anthropocentric assumptions rather than on anything like sober insight.

Thinkers such as Schopenhauer, Heidegger, and Merleau-Ponty make important contributions to the effort to rethink these traditional assumptions. Schopenhauer reminds us of the essential sameness of human and nonhuman animals, comparing our efforts to rise above our natural condition to the frenzied movement of a heap of cheese mites and proposing that any being possessing eyes necessarily also possesses understanding. Heidegger develops the notion of understanding by stressing the disclosive power of pre-predicative meaning, thereby further emancipating nonhuman animals from the charge that they can understand nothing because they lack human language. Merleau-Ponty makes a decisive step beyond Heidegger in stressing the fundamentally embodied character of affectivity, sometimes going so far as to suggest that "we can speak of a *logos* of the natural world" of which human "language is a resumption."[129] This notion of a *logos* of the natural world is far from new, having been advanced in antiquity but subsequently relegated to the dustbin of mythopoetic fancy.[130] To embrace fully such a conception of *logos* requires an overcoming of the traditional characterization of reason and affect as a strict either-or, i.e., a rethinking of *logos* itself as equiprimordially affective and cognitive. In the writings of Heidegger and Merleau-Ponty such a rethinking ultimately remains no more than a promissory note because both thinkers remained, if only against their own intention, too much within a tradition whose cognitivist limits they both recognized.

But perhaps this is not to be seen as a deficiency on the part of Heidegger and Merleau-Ponty so much as a reflection of the inherent limitations of conceptual language and an index of the power of long-standing anthropocentric prejudice. Perhaps Heidegger was right when he hinted that it is to poets rather than to thinkers that we must turn if we wish to retrieve a sense of the wholeness of being.[131] For it is no philosopher but Proust's protagonist who reminds us, as I noted earlier, that it is "feeling which makes us not merely regard a thing as a spectacle, but believe in it as in something unique [*être sans équivalent*]."[132]

That sense of uniqueness, that sense of a commanding presence that has not been flattened by cognition into the comparatively one-dimensional pigeonhole of objectification, is something that the Eleatic tradition has actively subjected to forgetting—due, I think, to the fact that acknowledging that sense of uniqueness and irreducible otherness would require a demonstration of humility that has been all but absent from the dominant voices in philosophical thought. Writing in a different context—his central concern was conflicting political ideologies—Herbert Marcuse makes an observation with crucial implications for the endeavor to think past the one-dimensionality of cognitive representation. He states that in a framework of understanding that has become so totalizing as to suppress any genuine moments of negativity (where negativity signifies what in the current parlance might be called a critical intervention from outside the dominant discourse), philosophy finds itself confronted with the possibility of taking on a new task, that of becoming "therapeutic." The realization of this prospect requires not simply new ideas, but an entirely new way of thinking:

> Philosophy approaches this goal to the degree to which it frees thought from its enslavement by the established universe of discourse and behavior, elucidates the Negativity of the establishment (its positive aspects are abundantly publicized anyway) and projects its alternatives. To be sure, philosophy contradicts and projects in thought only. It is ideology, and this ideological character is the very fate of philosophy which no scientism and positivism can overcome. Still, its ideological effort may be truly therapeutic—to show reality as that which it really is, and to show that which this reality prevents from being.[133]

There are some essential parallels between Marcuse's critique of prevailing political ideology and my critique of the anthropocentric background ideal of living. Each proclaims or presupposes a set of global axiological commitments. Each actively seeks to suppress or deny (Cavell and Diamond would say: deflect) "negative" moments that challenge the illusion that reality is exactly as the guiding commitments indicate. And the underlying motivation for these acts of resistance is an interest in dominance.

From the standpoint of efforts to dominate, acknowledging the uniqueness of affectively disclosed essences holds no allure because it does not hold the promise of enhancing our endeavor to master reality. The more we acknowledge the inherent ambiguity that characterizes our encounters with reality, the less persuasive are the confident assurances of Descartes to develop "a highest and most perfect ethics" whose foundation includes positive sciences such as physics. Ethics is not what Heidegger would characterize as a "positive" science in the sense intended by Descartes, precisely because it proceeds from our immersion in an element or dimension of being that resists reduction to neat conceptualization. This is Stein's counsel to Lord Jim: "In the destructive element immerse!"[134] That element is Schopenhauer's world will, and we treat ourselves as being independent of it at our own peril. To acknowledge it openly is to admit our essential commonality with all of sentient life, and in turn to shift the terms of ethical discourse away from its rationalist moorings and embrace the insight that "dependence, not autonomy, obligates us."[135]

Naturally this leaves us with an obvious problem, namely, that a great many people claim to care deeply about nonhuman animals and purport already to include them as direct beneficiaries of moral concern. The historical record of human conduct, however, tells a very different story. Almost without exception, those who advance philosophical views purporting to classify nonhuman animals as subjects of direct moral concern adhere to a distinctly hierarchical conception of membership in the moral community. I have tried to make a case for the proposition that this commitment to hierarchy is based on some questionable assumptions about the power of rational abstraction as well as on the self-serving prejudice that human beings are "the titular lords of nature." Thinkers such as Schopenhauer seek to look behind the veil of Maya, and when they do so some of them come away with the insight that Schopenhauer found but ultimately set aside: that we human beings are ultimately nothing more what the young Nietzsche called "clever beasts [who have] to die."[136]

To face up to this verdict is to confront a difficulty of reality, perhaps the single most such difficulty. In the next chapter I will examine this difficulty from the standpoint of the perennial tension between selfish or local affiliations on the one hand, and altruistic or global ones on the other. I will seek to explore—not resolve, but explore—this tension from the standpoint of the respective contributions and dialectical relationship between historical reason and the affective disclosure of meaning. Where the philosophical tradition has tended overwhelmingly to proceed on the assumption that there is some determinate, timeless set of truths that ought to govern our conduct (recall, for example, my discussion of moral realism), the turn to historical reason involves a confirmation that truth is not timeless but instead is inherently timebound. Where exponents of the Eleatic tradition see this as a capitulation to the threat of contingency, Ortega and Miller make a convincing case for the conclusion that contingency is part of our fate—that it is "the destructive element" in which we must immerse.[137] What remains to be seen in more precise detail is the way in which historical reason and our affective engagement with reality mutually inform one another. My hope—and it is admittedly a speculative one, given the staunch resistance with which so many people appear to greet it—is that a more subtle sense of the way in which this dialectical relationship functions will lead to *new* insights in the course of human history that will help us to arrive at a lived relationship with nonhuman animals that is *at once more just and more compassionate* than the regime of domination that has prevailed for thousands of years.

Notes

1 *Critique of Pure Reason*, p. 310 (A313–4/B370).
2 In my translator's preface to Dominic Lestel's *Eat This Book: A Carnivore's Manifesto*, I suggest that "the best way to test one's own convictions is to open oneself completely to the challenge posed by one's most strenuous critics or opponents—to confront doubt rather than extinguish it and to attempt to dwell in the space of irony" (p. xv).
3 Michel de Montaigne, "Apology for Raymond Seybond," pp. 399, 404, 417. Anyone who brushes aside Montaigne's charge of human impotence on the grounds that he offered it prior to the fruition of early modern science would do well to contemplate the epic damage humanity has done to nonhuman animals and the rest of the environment in the wake of (and, to a very large extent, thanks to) the scientific revolution.
4 Marcel Proust, *Swann's Way*, vol. 1, p. 90.
5 Viriginia Woolf, *Three Guineas*, pp. 103–4. I am indebted to my colleague Professor Erica Delsandro for introducing me to this text.
6 *Three Guineas*, pp. 129–30.
7 Walter Biemel, *Philosophische Analysen zur Kunst der Gegenwart*, pp. 66–140.
8 See David Hume, *A Treatise of Human Nature*, book 3, Part 1, sec. 1, p. 469.

9 See *Animals and the Limits of Postmodernism*, pp. 231–2. There I state that my own such shocking experience was reading Upton Sinclair's *The Jungle* when I was fifteen years old.

10 See, for example, Andrew Jacobs, "Is Dairy Farming Cruel to Cows?"

11 Ovid, *Metamorphoses Books 9–15*, book 15, line 462, p. 397. Upon learning of his brother's infidelity, Atreus took revenge by killing Thyestes's sons and serving them to him for dinner.

12 Felicia Nimue Ackerman, "Fried Chicken Heaven."

13 Not everyone who eschews the consumption of nonhuman animal products does so out of concern for nonhuman animals. Going back to antiquity, many people in many places have been motivated by an *ascetic* impulse: to purify their soul, thereby preparing it for greater enlightenment, by minimizing their reliance on material things. Even Porphyry, who as I have already noted is one of the few early thinkers to present a robust defense of nonhuman animals, offers ascetic considerations as one of his reasons for advocating vegetarianism. See my remarks in *Anthropocentrism and Its Discontents*, pp. 109–11.

14 Martin Heidegger, *Vorträge. Teil 2: 1935 bis 1967*, p. 773 (italics in original).

15 Martin Heidegger, *The Basic Problems of Phenomenology*, p. 13.

16 *The Basic Problems of Phenomenology*, p. 155.

17 *Discourse on Method*, Part One, *The Philosophical Writings of Descartes*, vol. 1, p. 114; "Preface" to "The Principles of Philosophy," p. 186; letter to Princess Elizabeth, May 1646 and letter to Chanut, February 26, 1649, *The Philosophical Writings of Descartes*, vol. 3, pp. 289, 368.

18 On the decisionistic consequences of the postmodern total critique of reason, see *Animals and the Limits of Postmodernism*, Chapters 1 and 2, particularly pp. 7–10, 32–41, 67–76.

19 *The World as Will and Representation*, vol. 1, sec. 7, p. 30; "Appendix: Criticism of the Kantian Philosophy," p. 439.

20 I have already noted that Kant, too, believes that nonhuman animals experience representations; that was the crux of his rejection of Descartes's wholly mechanistic account of nonhuman animals. Where Schopenhauer really differs from Kant in this connection is in being willing to attribute *knowledge* to nonhuman animals, and this constitutes a decisive difference between the two thinkers. One would do well to contemplate what "representation" can mean in a being that (Kant assumes) lacks linguistic and conceptual ability. I cannot escape the conclusion that such a view effectively denies to nonhuman animals the exact kind of freedom that I believe is characteristic of a sentient being. See my review of *Can Animals Be Moral?* Kant's entire justification for employing nonhuman animals as instrumentalities relies on the proposition that they lack freedom; and Kant, as I have argued in previous chapters, conceives of freedom in extraordinarily narrow, anthropocentric terms.

21 See my remarks on these texts in *Anthropocentrism and Its Discontents*, pp. 69–76.

22 *The World as Will and Representation*, vol. 1, sec. 8, p. 37; "Appendix: Criticism of the Kantian Philosophy," p. 439 ("This is the whole activity of reason (*Vernunft*), which nevertheless has the whole content of its thinking only from the perception that precedes thinking").

23 *The World as Will and Representation*, vol. 1, sec. 8, p. 36; see also sec. 16, p. 84.

24 "On Religion," p. 375. Here Schopenhauer implicitly rejects Descartes's conviction, presented in Part Five of the *Discourse on Method*, that nonhuman animals are biological machines and nothing more.

25 *The World as Will and Representation*, vol. 1, "Appendix: Criticism of the Kantian Philosophy," pp. 421–2.

26 "On the Doctrine of the Indestructibility of our True Nature by Death," p. 267.

27 "On the Doctrine of the Indestructibility of our True Nature by Death," p. 290 (italics in original).

28 *The World as Will and Representation*, vol. 1, sec. 19, p. 103.

29 *The World as Will and Representation*, vol. 1, "Appendix: Criticism of the Kantian Philosophy," p. 473.
30 *On the Basis of Morality*, sec. 6, p. 75.
31 Bennett W. Helm, *Emotional Reason*, p. 125.
32 *Emotional Reason*, p. 97.
33 Bennett W. Helm, "The Significance of Emotions."
34 *Emotional Reason*, pp. 116, 143, 178, 197.
35 *Emotional Reason*, p. 72; see also pp. 79, 88, 109.
36 *Emotional Reason*, p. 214.
37 *Emotional Reason*, pp. 232, 244.
38 *Emotional Reason*, p. 230.
39 *The Midworld of Symbols and Functioning Objects*, p. 45.
40 *A Treatise of Human Nature*, book 2, Part 3, sec. 3, p. 415.
41 *A Treatise of Human Nature*, book 1, Part 4, sec. 2, p. 207; see also book 1, Part 4, sec. 6, p. 252 and "Appendix," p. 634.
42 *A Treatise of Human Nature*, book 2, Part 2, sec. 2, p. 340; book 2, Part 3, sec. 8, p. 438.
43 *Discourse on Method*, Part 6, p. 144; *Critique of Pure Reason*, p. 140 (at A114).
44 *Being and Time*, Division One, Ch. 3, par. 14, p. 94.
45 *Being and Time*, Division One, Ch. 3, par. 18, p. 119.
46 *The Fundamental Concepts of Metaphysics*, par. 42, p. 177. Here Heidegger goes on to state that "the animal is world-poor" and "the stone is worldless." In a moment I will return to the question whether Heidegger was right to cast doubt on the interpretive capacities of nonhuman animals.
47 *Being and Time*, Division One, Ch. 3, par. 18, p. 119; Ch. 5, par. 33, p. 200.
48 *Being and Time*, Division One, Ch. 3, par. 20, p. 132; Ch. 6, par. 44, p. 254.
49 *Critique of Pure Reason*, "Preface to First Edition," p. 10 (at Axiii).
50 On the nature of intentionality in this sense, see my *Animals and the Moral Community*, Ch. 1 and 2.
51 See Joseph P. Fell's excellent discussion in *Emotion in the Thought of Sartre*, pp. 40, 80f.
52 *Being and Time*, Division One, Ch. 3, par. 18, p. 116.
53 *Being and Time*, Division One, Ch. 5, par. 32, p. 190.
54 *Being and Time*, Division One, Ch. 5, par. 28, p. 171f.
55 *Being and Time*, Division One, Ch. 5, par. 29, p. 173.
56 *Being and Time*, Division One, Ch. 5, par. 29, p. 175.
57 Aristotle, *Metaphysics*, 1.1, *The Complete Works of Aristotle*, vol. 2, p. 1552 (979b23).
58 *Being and Time*, Division One, Ch. 5, par. 29, p. 172n1. The German expression "wie befinden Sie sich?" is typically translated as "how are you doing?" and has some resonance with English expressions such as "I find myself in a (e.g., difficult) position."
59 *Being and Time*, Division One, par. 29, p. 176. Helm acknowledges this, but I believe that he does not draw out its full implications.
60 *Being and Time*, Division One, Ch. 5, par. 29, p. 176 (italics in original).
61 *Being and Time*, Division One, Ch. 5, par. 30, p. 180. See also Martin Heidegger, "What is Metaphysics?," p. 88.
62 "What is Metaphysics?," p. 87. Heidegger discusses boredom at great length in *The Fundamental Concepts of Metaphysics*, pp. 74–167.
63 "What is Metaphysics?," p. 87.
64 "What is Metaphysics?," p. 90.
65 *Being and Time*, Division Two, Ch. 1, par. 50, p. 294f.
66 Heidegger states that anxiety is not itself the moment of vision, but that it makes the moment of vision possible. *Being and Time*, Division Two, Ch. 4, par. 68(b), p. 394. See also pp. 376, 387–8. Heidegger contrasts the privileged form of sight in the moment of vision from the comparatively everyday form of seeing that he calls "circumspection" (*Umsicht*). Division One, Ch. 3, par. 15, p. 98.
67 *Being and Time*, Division Two, Ch. 4, par. 69, p. 402.

68 Aristotle is more difficult to assess on this score, inasmuch as he assumes a teleology in nature that could account for the interrelationships between natural things. In *Anthropocentrism and Its Discontents*, I argue that Aristotle stopped short of proclaiming an overarching, unifying telos of nature, and that it would remain for the Stoic philosophers to advance this additional claim. See Ch. 3, especially p. 77. Heidegger makes an interesting case for the proposition that Aristotle sees truth in terms of a binding and separating of discrete units (*synthesis* and *diairesis*) in *The Fundamental Concepts of Metaphysics*, par. 72(c), pp. 313–4.

69 *Being and Time*, Division 1, Ch. 5, par. 31, p. 184. At the same time, Aristotle is the obvious inspiration for Heidegger's characterization of the care structure as teleological.

70 Descartes, Second Meditation.

71 *Being and Time*, Division One, "Introduction II," par. 7(B), p. 56.

72 In the interest of maintaining my central focus, I will not go into detail in examining Heidegger's account of how and why such acts of removal from context take place. The core of his answer is the notion of "equipmental breakdown." See *Being and Time*, Division One, Ch. 3, par. 16.

73 *Being and Time*, Division One, Ch. 3, par. 20, pp. 132–3. The problem is this: if human beings and everything else in the world are essentially different, as the tradition has averred, then in virtue of which gathering principle or factor can they come into a relationship with one another? Heidegger implicitly accepts Aristotle's notion of a "precedent community of nature" and thereby avoids this problem altogether, arguing that "Dasein *is* its world existingly." Division Two, Ch. 4, par. 69(c), p. 416 (italics in original). In other words, there is first a totality of meaning in which I am immersed, and only then can I endeavor by means of apophantic abstraction to represent myself *as* separate and the world *as* a collection of "things."

74 Martin Heidegger, *Logic: The Question of Truth*, par. 12, p. 129 (translation altered).

75 *Being and Time*, Division One, Ch. 5, par. 32, p. 189, 193; par. 33, p. 201.

76 *The Fundamental Concepts of Metaphysics*, par 69(b), p. 290.

77 Martin Jay notes that Heidegger's retreat from the tradition's ideal of detached staring involves a turn to a more fundamental way of seeing things that "refuses to stare aggressively at its objects." "Sartre, Merleau-Ponty, and the Search for a New Ontology of Sight," p. 148.

78 As noted late in Chapter 1, I borrow this notion of ethos as a sense of proper place from Heidegger.

79 *The Animals Issue*, p. 190. I shared this suggestion with my longtime veterinarian some years ago, and her immediate response was, "spend five minutes in any veterinary ER and you will know that animals are aware of their pain." Clearly the kind of "knowledge" my veterinarian had in mind is not the kind that the Eleatic tradition would acknowledge to constitute knowledge.

80 Thus even if thinkers such as Mark Rowlands are right that nonhuman animals cannot not evaluate their actions in the detached, reflective ways in which human beings typically can, it is a mistake to assume without further ado that nonhuman animals are essentially bereft of *freedom*. See my review of *Can Animals Be Moral?*

81 Maurice Merleau-Ponty, *Nature*, Second Course, Ch. 2, p. 167.

82 *Nature*, Second Course, Ch. 2, p. 174. Here Merleau-Ponty contrasts the mechanistic reactions of a tick with the interpretive response issued by the presumably "higher" nonhuman animal bitten by the tick.

83 *Nature*, Third Course, Second Sketch, p. 212.

84 *Nature*, Second Course, Ch. 2, p. 175. See also p. 177: To the extent that humans are not alone in having an *Umwelt*, we need to abandon the reductive notion that nonhuman animals function "entirely under the dependence of physicochemical conditions. . . . The animal regulates, makes detours."

85 Guy Debord, *Society of the Spectacle*, sec. 17, 24.

86 *Society of the Spectacle*, sec. 221.

87 *Emotion in the Thought of Sartre*, p. 130 (italics in original).
88 Maurice Merleau-Ponty, *The Visible and the Invisible*, p. 273. On the order of reasons in Descartes, see Martial Gueroult's masterful *Descartes's Philosophy Interpreted According to the Order of Reasons*.
89 Maurice Merleau-Ponty, *Phenomenology of Perception*, pp. 403–4.
90 *Phenomenology of Perception*, pp. 214, 398.
91 *Phenomenology of Perception*, pp. 204, 398.
92 *Phenomenology of Perception*, p. 49.
93 Maurice Merleau-Ponty, "The Primacy of Perception and Its Philosophical Consequences," p. 22.
94 *Phenomenology of Perception*, p. 206.
95 See for example William James, *The Principles of Psychology*, vol. 2, p. 449.
96 Martha C. Nussbaum, *Upheavals of Thought*, p. 125f.
97 On the Stoic view, see Richard Sorabji, *Animal Minds and Human Morals,* pp. 40–4, 58f. and my *Anthropocentrism and Its Discontents*, pp. 77–92.
98 *Emotion in the Thought of Sartre*, pp. 40, 80f.
99 *The Passions of the Soul*, art. 48, *The Philosophical Writings of Descartes*, vol. 1, p. 347.
100 *The Passions of the Soul*, art. 160–64, *The Philosophical Writings of Descartes*, vol. 1, pp. 386–9. On Kant's appeal to moral love, see Chapter 2.
101 *The Visible and the Invisible*, p. 136.
102 *The Visible and the Invisible*, p. 266. Here Merleau-Ponty implicitly relies on Heidegger's notion of truth as *aletheia* or "unconcealment."
103 *Nature*, Third Course, First Sketch, p. 209.
104 *Nature*, Third Course, Second Sketch, p. 218. In a very similar spirit, Scheler employs the term 'Einsfühlung' (identification) to describe this primary sense of affective immersion in the world, arguing not only that it is essentially prior to states such as *Einfühlung* (empathy), *Nachfühlung* (having a sense of the other), and *Verstehen* (understanding), but moreover that *Einsfühlung* is positively excluded by these founded experiential states—that the move to states of these sorts entails a crucial loss of intimate connection with the way in which our being in the world is primarily disclosed to us. Max Scheler, *Wesen und Formen der Sympathie*, pp. 40, 44. See also p. 42, where Scheler observes that the "hypertrophy of understanding" in human beings has brought with it a "complete loss" of the "*Einsfühlungsfähigkeit* [capacity for identification] and other instincts possessed by animals."
105 Edmund Husserl, *Ideen zu einer reinen Phänomenologie und phänomenologischen Philosophie*, pp. 166ff.
106 *Nature*, Third Course, First Sketch, p. 214.
107 *Nature*, Second Course, Ch. 2, p. 176.
108 *Nature*, Second Course, Ch. 2, p. 178.
109 *Nature*, Second Course, Ch. 2, p. 199. (*Physis* and *logos* refer to nature and thought/reason/language, respectively.)
110 *Nature*, Third Course, Third Sketch, p. 225.
111 *Nature*, Third Course, Third Sketch, p. 225.
112 *Nature*, Third Course, Third Sketch, p. 225.
113 See for example *Phenomenology of Perception*, pp. 315ff., 351.
114 *Phenomenology of Perception*, p. 243.
115 *Critique of Pure Reason* at A51/B75.
116 *Nature*, Third Course, Third Sketch, p. 225. On Heidegger's claim that nonhuman animals do not "die" but merely "perish," see my remarks in *Animals and the Limits of Postmodernism*, Ch. 3.
117 See *Nature*, Second Course, Ch. 2, pp. 167ff. (following Uxküll). Contemporary debates about the possibility of non-conceptual content address some of the same issues that I am addressing here. See my remarks in *Animals and the Moral Community*, pp. 73–88.

118   Cora Diamond, "Injustice and Animals," p. 139.
119   Cora Diamond, "Eating Meat and Eating People," pp. 105–6.
120   Cora Diamond, "The Importance of Being Human," p. 61.
121   "The Importance of Being Human," pp. 58–9.
122   A "difficulty of reality" arises when the mind experiences "not being able to encompass something which it encounters . . . we take something in reality to be resistant to our thinking it, or possibly to be painful in its inexplicability." Diamond follows Stanley Cavell in noting that a typical response to such a difficulty is "deflection." Cora Diamond, "The Difficulty of Reality and the Difficulty of Philosophy," pp. 44–6, 57. In this connection, consider the difficulty many people today have in the endeavor to grasp how being "*anti*-racist" differs from simply "not being racist": The former commitment is considerably more *active* than the latter, in the sense that it expresses *a sense of responsibility to address and respond to the other* in a way that simply considering oneself "not to be racist" does not.
123   "The Rights of the Nonhuman World," p. 192.
124   *Animals and the Moral Community*, p. 126.
125   "Letter on 'Humanism'," p. 271 (translation altered).
126   See for example, Elisabeth de Fontenay, "Pourquoi les animaux n'auraient-ils pas le droit à un droit des animaux?"
127   "The Importance of Being Human," pp. 55–6.
128   Max Scheler, *Die Stellung des Menschen im Kosmos*, pp. 35, 39–40, 48. Scheler borrows the term 'ekstasis' from Heidegger, who uses the term to signify the human ability to move between different temporal horizons. See *Being and Time*, Division II, Ch. 3, par. 65 and 68. Elsewhere Heidegger equates Dasein's "ecstatic" nature with the possibility of humans "standing outside their being as such and within the truth of being." "What is Metaphysics?," p. 248. See also "On the Essence of Truth," p. 144, where Heidegger ties our "ek-static" nature to openness and "letting beings be."
129   *Nature*, Third Course, Second Sketch, p. 219.
130   See my remarks in *Anthropocentrism and Its Discontents*, pp. 83–5 (where I discuss Cleanthes's *Hymn to Zeus*).
131   "The thinker says being. The poet names the holy." "Postscript to 'What is Metaphysics'?," p. 237.
132   *Swann's Way*, p. 90 (translation altered).
133   Herbert Marcuse, *One-Dimensional Man*, p. 199. See also pp. 130–1, where Marcuse follows Heidegger in tracing the problem of one-dimensional thought to the prominence of apophantic *logos* and the "divided reality" that it assumes.
134   Joseph Conrad, *Lord Jim*, Ch. 35, p. 255.
135   Martin Huth, "Reflexionen zu einer Ethik des vulnerablen Leibes," p. 300.
136   "On Truth and Lies in a Nonmoral Sense," p. 79.
137   See *The Paradox of Cause and Other Essays*, pp. 136, 187, 191.

5

FELT KINSHIP

The Essential Tension Between Local and Global Commitments

> It is not by a true judgment, but by foolish pride and stubbornness, that we set
> ourselves before the other animals and sequester ourselves from their condition
> and society.
>
> Montaigne, *Apology for Raymond Sebond*

1. The Power and Essential Limits of Reason

In Chapter 4 I explored the prospects for including nonhuman animals as direct
beneficiaries of moral concern, by seizing upon embodiment rather than linguistic
rationality as the point of departure for thinking about membership in the moral
community. Where the tradition has placed almost its entire focus on the ways in
which human beings purportedly differ from nonhuman animals, I am suggesting
that we focus instead on the extensive continuities and commonalities between
human and nonhuman animals. The endeavor to acknowledge the direct moral status of nonhuman animals has been met with fierce resistance, in many cases even
by people who purport to care very deeply about nonhuman animals. What are
we to say to a person who staunchly insists that they really do care about nonhuman animals, but who also avers that all we really need to do in order to redress
the harms we have inflicted on nonhuman animals is simply to treat them better
while we prepare them for the slaughterhouse? What I am suggesting is that, in
effect, anyone who takes that position—or, for that matter, the position that pet
ownership is in principle unproblematic—has decided in advance, if only beneath
the level of explicit, reflective awareness, that nonhuman animals count less in the
moral scheme than do human beings and that human beings enjoy the prerogative to make definitive determinations about what does and does not constitute the
"proper" treatment of nonhuman animals. And that, following the lead of Michel
de Montaigne, is exactly what I would like to call into question.

DOI: 10.4324/9781003425595-6

One might well take the view that yes, nonhuman animals share embodiment with human beings, but that no, this does not mean that nonhuman animals share any sort of moral parity with human beings. I have argued from the outset that our insight into the nature and moral status of nonhuman animals has long been overshadowed by a pointedly anthropocentric background ideal of living, indeed to such an extent that taking seriously the notion of full and direct moral status for nonhuman animals strikes many not merely as absurd or incoherent but as *threatening*. I suggested in the introduction that one of the grounding motivations of philosophy is fear. That fact sheds light on the entire endeavor to appeal to "purely rational" considerations when assessing the relative moral worth of human and nonhuman animals: It is at least as much our affective responses as our "logical" ones that inform our values about nonhuman animals (and, for that matter, about matters of value generally).

This, in turn, means that we might be able to appeal to people's felt sense of connection with the rest of the community of sentient beings, rather than to (or in addition to) rational considerations by themselves. Here the same problem prevails as in the case of rational arguments: *the failure or refusal to acknowledge* something that ought to be plain to sober insight. And, just as in the case of assessing rational arguments, our assessment of our own affective sensibilities is beset with the problem of ideology. This is worth bearing in mind when a philosopher responds to the call for a sense of reverence for nature with the insistence that they "do not possess any of [the] three special senses" (viz., mystical, religious, or moral) to which people typically appeal when they argue for the conclusion that nature possesses inherent worth.[1] How are we to convince such a person that "the claim that humans by their very nature are superior to other species is a groundless claim," that it is nothing more than "an irrational bias in our own favor"?[2]

As I suggested in *Animals and the Limits of Postmodernism*, I believe that there is no knock-down argument in favor of nonhuman animal rights. There is also no knock-down feeling, so to speak, that will immediately part the Red Sea and clear a path to some non-anthropocentric Kingdom of Ends. At the end of *Animals and the Limits of Postmodernism*, as I noted earlier, I suggested that what is really needed to change someone's entire orientation on the nonhuman animal question is a shocking experience, such as my own experience of reading Sinclair's *The Jungle* as an adolescent. The problem is that these sorts of experiences do not come with a guarantee of enduring impact; it is quite often the case that we recoil very quickly from such dislocating experiences in an effort to return to the relative tranquility of what Heidegger calls "everydayness." What is needed is not simply a particular idea or a particular feeling, but rather *something that will forever remain a regulative ideal*, namely, the kind of "healthy consciousness" that can see through "the system of delusion" and aspire to the kind of "reconciliation" of perception and reality that has long eluded us.[3]

If this does not all come down to arriving at compelling arguments or felt experiences with the power to disclose meaning in a fixed, inalterable manner (i.e., if this is not all a matter of Eleatic insight or vicissitude-free feeling), then it has got to come down to a process of reflecting on our cultural inheritance with an eye toward getting clear about our historical background commitments and taking a critical stand on them here and now. Ortega's and Miller's insights into the historical character of reason make it clear that the notion of moral realism is the product of an understandable if ill-conceived Eleatic attempt to secure ourselves against the destructive power of time. There is something profoundly elusive in the endeavor to attain the sort of "healthy consciousness" that promises to move us past the delusions of the past and present. To the extent that we can begin to attain it, it is my hope that gestures such as Richard Posner's blithe dismissal of nonhuman animal rights as "weird" and "insane" will begin to appear in the sad anthropocentric light in which they deserve to appear.[4]

Even many who stop far short of Posner's harsh assessment of the idea of rights for nonhuman animals adhere to all the essential features of the anthropocentric background ideal of living. Anyone who argues without hesitation for the permissibility of using nonhuman animals to satisfy human needs and desires is, if only implicitly, expressing deeply anthropocentric convictions. I have suggested throughout my work on nonhuman animal ethics that I find it impossible to consider using and killing a sentient being to be essentially unproblematic, even if we use and kill that being carefully, gently, quickly, or in any other way intended to signal that we "really care" about the nonhuman animal we are using and killing. Depriving a sentient being of its ability to direct its own life, whether or not we kill that being in the course of using it, constitutes a violation of what I think of as "cosmic" as opposed to "social" justice. I have already noted this distinction earlier in this book, and I will return to it later in this chapter. Treating nonhuman animals as "replaceable resources" rather than as beings with full and direct moral status is to project a status onto them that just so happens to fit perfectly with the anthropocentric ideal of total control over nature.

Too often, human beings have taken the position that practices such as nonhuman animal husbandry actually benefit nonhuman animals, the implicit idea being that we are actually doing nonhuman animals a favor by what amounts to depriving them of their freedom. In this connection, two observations are in order. First, as I have suggested from the outset, humanity has left a rather abysmal historical record when it comes to caring for nonhuman animals and the environment. Human-induced climate change, human-caused extinction, and the epic and ever-growing scale of nonhuman animal slaughter all testify to the unreliability of human beings when it comes to the task of stewardship. (Our record in the case of human-human relations is arguably no better.) And second, as I have also suggested from the start, we ought to pause and ask ourselves whether nonhuman animals really need our help in the first place. Derrida characterizes the life force of sentient beings as "autotelic," by which he means that sentient creatures are so constituted that they can identify and pursue ends ('-telic' refers to the Aristotelian *telos* or point of fruition) for themselves ('auto-').[5] In this Derrida strikes me as one hundred percent correct. In fact, given how much better a job virtually every species of nonhuman animal appears to have done than human beings of relating to its habitat, a very strong case could be made for the proposition that we human beings are not only in key respects inferior to nonhuman animals, but that we exacerbate our own inferiority by engaging in a kind of self-delusion that seems possible only for beings capable of apophantic *logos*. In other words, one would not be out of order for wondering whether the very capacity upon which we have historically seized as a clear sign of our superiority is actually the root cause of a great deal of misery—our own, and that of an incalculable number of nonhuman sentient creatures.

The core of the problem is the way in which ideology provides the motivation for an ideal of reason that purports to be impartial and objective. As thinkers from Marx to Marcuse have argued, ideology has a sort of magical ability to conceal its influence under the cloak of an ideal of reason that purports to be independent of what Kant called "inclination." In Chapter 4 I presented considerations in support of the conclusion that *reason by itself does not give rise to meaning or value*, and to this extent reason is not really autonomous in the sense intended by thinkers such as Descartes and Kant. I hope to have shown by now that the comprehensive ideal of reason proffered by the tradition, inscribed as it is within a project of mastery or domination, serves as a demonstration of this fact. I take Kant's epic (and, I think, unsuccessful) struggle to consign moral feeling to the status of a handmaiden to autonomous reason to be an unwitting demonstration of the necessary reliance of detached rationality on the prior disclosure of an entire context of meaningful interrelationships. And as I have suggested, this prior disclosure is not itself "rational" but rather affective—it takes the form of active, embodied, meaningful inter*action* with others and the environment.

I noted a moment ago that Horkheimer and Adorno focus on the prospects for achieving a reconciliation of perception and reality in the endeavor to pierce the veil of ideology. By now it should be clear that the anthropocentric background ideal of living that I have examined in earlier chapters has many affinities with the notion of a political ideology. In particular, it provides a comprehensive map or plan of reality as a scene of value, it does so in advance, and it functions in such a way as to serve interests that tend to remain concealed under the veil of impartiality even though it turns out not to be all that difficult to identify those underlying interests if you put your mind to it (which is to say, if you have the will to do so). What, then, would be needed in order to move toward the kind of "healthy consciousness" that would constitute a recognition of the historical injustices we have inflicted on nonhuman sentient life?

In Chapters 3 and 4 I argued that if we are to establish a more adequate relationship to nonhuman reality, we will need to retreat from the historical pretensions of reason and acknowledge our own inherent animality. A recurring theme in this book has been Heidegger's contrast between Protagoras and Descartes, a contrast intended to remind us of the distortions and excesses of Eleatic reason. Heidegger's appeal to a specifically Protagorean sense of the dictum that "man is the measure of all things" is likely to strike many anthropocentrically-minded individuals as a fanciful appeal to the kind of undistorted experience that only a divine being could have (perhaps in a moment of Kantian "intellectual intuition"). How, after all, can we have an encounter with reality without the benefit of *a priori* forms of sensible intuition, concepts, and the unifying power of apperception?

From the standpoint of the anthropocentric ideal, the only possible answer seems to be some illegitimate appeal to mystical experience—as if the only alternative to detached cognition in matters of meaning and value were something like magic. The extensive philosophical work that has been done, in fields from phenomenology to feminism, to restore centrality to embodiment in human experience affords us an alternative explanation for the genesis of meaning, an explanation roundly rejected by exponents of the Eleatic ideal. In Chapter 4 I examined the key commitments that motivate this acknowledgment of the primacy of embodied action. One of these commitments is that consciousness is not initially spectatorial, but instead is actively engaged with concrete exigencies in the environment.[6] These exigencies show up *as meaningful*, they are faced consciously, and they are faced not simply by humans but by all sentient beings—after all, sentience *is* consciousness. It is only by dint of the capacity for abstraction possessed by some sentient beings that there can arise a sense of reality as somehow initially devoid of meaning and only subsequently invested with it. When we encounter the world—and again, this 'we' means *all* sentient creatures—we encounter it as already meaningful, indeed as already meaningfully structured. The fact that we humans, seeing things as we tend to do from the standpoint of Eleatic detachment, cannot fathom what it is like for a non-linguistic being (assuming that nonhuman animals are non-linguistic) to encounter a meaningfully structured world does not signify that nonhuman animals are incapable of having a meaningful encounter with things. It simply signifies that we human beings have encountered the limits of our ability to comprehend the other.

Thinkers such as Aristotle and Augustine mistake this limitation on the part of humans for an indication that human beings and nonhuman animals really have nothing "in common" with one another.[7] What I have attempted to highlight in previous chapters is *the extensive commonalities between human and nonhuman animals*, rather than their apparent differences. Both face what Freud called *ananke*, the demands placed on us by material reality. Both are frail, mortal, and typically a little lost (although nonhuman animals appear to be considerably less lost than humans) in what William James aptly called the "blooming buzzing confusion" of material existence. And both have their "most primordial" encounter with reality not by means of concepts but by means of active, embodied negotiation (interpretation) of concrete needs and challenges.

Moreover, Heidegger's and Merleau-Ponty's reflections on the nature of our encounter with meaning make it clear that experience, as Dewey noted, has a fundamentally social character, which is to say that it is not initially attached to the

'I' that plays such a central role in the Eleatic conception of reason.[8] This is not to say that the social world is not comprised of individuals, but rather—and this was Heidegger's and Merleau-Ponty's point—that the *kind* of 'I' or self that is engaged with meaning at the most basic level is *not* the detached 'I' of Cartesian-Kantian apperception. This is perhaps the greatest single sticking point when it comes to arriving at the core prejudice of anthropocentric thinking and valuing: There is, as I have already observed, no "knock-down" proof that a great many nonhuman beings possess some kind of concrete, active, engaged individual selfhood. But by now it should be clear *why* there is no such knock-down proof. Anthropocentrically-minded people are deeply invested in seeing human beings as unique in possessing anything like selfhood. At the same time, Heidegger's and Merleau-Ponty's critiques of linguistic rationality give us good reasons to accept the proposition that quite a bit of socially structured encounters with meaning have taken place *before* anything like a Cartesian-Kantian subject can become advanced as the putative core of human (and only human) experience.

The insights of thinkers such as Schopenhauer, Heidegger, Merleau-Ponty, and Dewey offer us a route of approach to the elusive Protagorean ideal of unmediated contact with reality. Schopenhauer establishes his distinction between representation and will as a way of exploring the Protagorean ideal. Schopenhauer takes Kant's notion of transcendental conditions for knowledge to its logical extreme, observing that understanding the world in terms of causal interactions is necessary for survival, and further observing that all sentient creatures (all with eyes, at any rate) possess understanding—which is to say that a wide variety of sentient creatures have an intimate awareness of causal relations, even if they cannot depict those relations in terms of universal laws. For Schopenhauer as for Dewey after him, consciousness is first of all directed at the exigencies of material existence; and for Schopenhauer, this means that human and nonhuman animal existence are in many respects entirely alike.[9]

And while it might not be impossible in principle to arrive at this insight by means of detached rationality, it is important to bear in mind that reason is not the eternal *nous* of Aristotle, but instead is a faculty always already informed by antecedent value commitments—the most influential of which is that to human uniqueness and superiority—in such a way that those prior commitments exhibit a very strong tendency to distort rather than facilitate insight into the nature of things, most particularly in questions of value. Recall that the motive force of reason, as it has been understood by thinkers from Aristotle to Miller, is the endeavor to exercise control. And to the extent that a full acknowledgment of the interpretive capacities of many nonhuman animals would dethrone us from our traditional place of pride in the cosmic scheme, it should come as no surprise that so many defenders of the anthropocentric orientation demand irrefutable proof of matters that tend to be much less obscure when viewed from the standpoint of the will rather than from the standpoint of detached rationality. In other words, the more we acknowledge the truth disclosed by our embodied encounter with meaning, the more urgent will become our sense of the need to rethink our anthropocentric commitments and begin to supplant them with new ones. The problem, of course, is that of acknowledgment.

In Chapter 3 I examined Ortega's and Miller's reflections on the essentially historical character of reason, and I suggested that even exponents of historical reason face the problem of warrant: How do we ensure that the axiological (moral, political, social) commitments that we hold most dear are commitments that we *ought* to embrace? My hypothesis from the start has been that there can be no context-free, "objective" procedure for addressing the problem of warrant, for the simple reason that reason does not operate in a vacuum but instead necessarily presupposes a value-orientation against the background of which it can operate. This is the sense in which moral commitment requires an appeal to what I have called an "extra-rational" source of meaning. Pathocentrists, those who believe that the foundation of morality is to be found entirely in feeling or emotion, propose that the problem of warrant can be solved through an appeal to our affective constitution. I have already noted that I find this approach to be what Marcuse would consider to be a simple binary inversion of the traditional picture of the reason-emotion relationship: if it isn't all a matter of reason, then surely it must all be a matter of our emotional proclivities. Late in Chapter 3, I noted that even Miller acknowledges "the dark partiality of the demonic," which is essentially an acknowledgment of the ways in which affect can mislead us—just as Kant and other exponents of reason have long observed.

To suppose that our affective sensibilities by themselves could provide us with everything we need to establish an authoritative ethos is to neglect the role of reason in human experience. Simon Blackburn is right to point out that "without emotions the will is rudderless."[10] But to suppose that a simple appeal to "basic sympathies and concerns" will be sufficient by itself to underwrite a sense of moral solidarity is to think of our own experience a little too much along the lines of the way in which many people think of nonhuman animal experience—as if we were essentially programmed with "basic" commitments and acted more or less mechanistically on the basis of them. Blackburn never goes to this extreme, instead arguing for the importance of deliberation in the process of forming, revising, and testing our moral convictions. He recognizes that our sense of rightness is guided by a process of "deploying values as we go."[11]

It is here that the problem of the interaction of reason and affect finds its focus. Life is full of examples of people acting on "basic sympathies and concerns" that later prove to be morally and/or politically pernicious. Our rational capacity affords us the ability to reflect on our basic affinities and enmities in an effort to determine whether they are warranted. By now it should be clear that this process of scrutiny does not occur in the safe space of Eleatic detachment, but instead always proceeds against the background of some prevailing set of global commitments. It is not difficult to imagine how greatly the notion of "basic sympathies and concerns" will differ when contemplated from the anthropocentric and non-anthropocentric standpoints, not to mention when comparing the views of different individuals who inhabit the same global ethos. Indeed, in the end our entire stand on the question whether we share common cause with nonhuman animals will come down to the question of just how ready we are to be deeply self-critical and unselfish. And needless to say, the capacity to be self-critical presupposes precisely the rational capacity whose centrality in the process of human commitment proponents of pathocentrism seek to exclude or at least marginalize.

Late in Chapter 4, I noted Cora Diamond's suggestion that to respond as a Kantian would to the sexual violation of an insensate young girl would amount to adopting the standpoint of "a rational Martian." Diamond's assessment of such a situation is that the appropriate response would be to seize upon our sense of *shared fate* with the girl rather than on some reckoning of her cognitive abilities. But I have also noted that on Diamond's view, images of fellow human creatures are "naturally more compelling" than images of nonhuman animals. Diamond's position, then, is that we ought to care about nonhuman animals as direct beneficiaries of concern, but that there is some sense of naturalness or givenness according to which we do (and perhaps ought) to care more about our fellow human beings than about our nonhuman sentient fellows. In this connection, Diamond is implicitly appealing to the same "basic sympathies" to which Blackburn appeals.

But what if it turns out that we have misunderstood these basic sympathies, and perhaps a great deal more pertaining to our embodied (affective) nature, due to our culture's millennia of adherence to a pointedly anthropocentric background ideal of living? Consider two very different ways of approaching "reality." One reduces reality to subject-object representation and subsequently assigns values to things. The other heeds Stein's advice to Lord Jim: "in the destructive element immerse!" The destructive element is precisely what becomes disclosed when we pursue the Schopenhauerian path of the will: a realm of sheer contingency, threatening and unpredictable but also populated with beings who exhibit clear signs of consciousness and share a common fate with us. The path of representation characterizes beings in terms of their calculability and manipulability, and only subsequently (if at all) does it take up the question of the appropriateness or permissibility of such manipulation.[12] Given that the interest in control is the motivating presupposition for the turn to subject-object representation, it hardly seems reasonable to suppose that we might articulate a truly robust nonhuman animal ethics from out of those presuppositions. Anyone who believes otherwise would do well to consider the vast and ever-growing number of nonhuman animals we kill and consume every year.

To embrace the call to immerse ourselves in the destructive element is to retreat from the confident proclamations (about both fact and value) proffered by the anthropocentric tradition and to acknowledge and try to come to terms in a new way with our fundamental vulnerability. There are signs of an awareness of this inherent frailty in the anthropocentric tradition itself; Susan Bordo, for example, notes the tremendous anxiety that motivates Cartesian metaphysics, citing it as a prime example of "a certain instability, a dark underside, to the bold rationalist vision."[13] That instability is a function of two facts: that reality resists our efforts to reduce it to objectivity, *and* that we human beings have not proved ourselves particularly skilled at confronting our own self-serving prejudices. It is not simply for the sake of nonhuman animals but for ourselves as well that we need to confront these prejudices in the spirit of openness and humility that I have urged from the start of this book. Needless to say, it will require unprecedented courage and honesty on the part of human beings.

A moment ago I noted that pathocentrism (the ascription of absolute primacy to emotion over reason) is beset with some serious difficulties. While it is true that our embodiment (our affective encounter with things) is absolutely essential for disclosing the sphere of meaning in which reason operates, it is equally true that we human beings suffer from a problematic relationship to our own embodiment that appears *not* to beset nonhuman animals: We seem to find it extremely easy to rationalize courses of action that have pernicious consequences for others (and perhaps for ourselves), and one form that this rationalization takes is to mischaracterize the value of goods disclosed by our affective attunements. This is what Descartes had in mind when he suggested that "the will's 'proper' weapons" against the passions are "firm and determinate judgments bearing upon the knowledge of good and evil." In doing so, Descartes sought to treat Eleatic reason as the sole arbiter of value (our sense of good and evil), and he saw the passions principally if not exclusively as threats to our well-being.[14] Descartes also saw certain passions, notably generosity and *caritas* (charity or love), as playing a crucial role in our social existence, although his characterization of embodiment in terms of objectivity prevented him from undertaking a full exploration of the affective dimension of our existence.[15] I cannot help but wonder whether the concrete content of the commitments that we classify under terms such as 'generosity' and 'caritas', both of which gesture toward a meaningful whole of which I am simply a subsidiary part, would begin to look very different if we were to effect a genuine shift from the comparatively closed perspective of anthropocentrism to the open vantage on reality afforded by our lived embodiment.

To move toward such a prospect would require us to employ both reason and affect. Our affective lives seem to be considerably more obscure to us than the standpoint of Eleatic detachment that holds out the promise of systematic control. If only because we have become so enmired in anthropocentric ways of thinking, our own lived embodiment has become more rather than less obscure to us as time has gone by. (*Grey's Anatomy* tells us nothing at all about *lived* embodiment, but only about the body-cum-object suitable for manipulation). Thus, while it may well not be "inherent" in human nature to misunderstand our own affective constitution, it certainly seems to be the case that history has taken us to such an impasse and that we need to find a way to get beyond it if we are to arrive at a truer sense of ourselves and of our relationship to nonhuman animals and the rest of the world. It is here that reason is vitally needed: to do exactly what Descartes said it needs to do, namely, to provide a "check" on exaggerated affective responses.

I will explore the prospects of reason to work in cooperation with our affective constitution in the balance of this chapter. What I would like to propose in the end is an ideal that draws inspiration from Tzvetan Todorov's call for "a well-tempered humanism." This may seem odd, given the profound historical associations between humanism and the endeavor to exercise dominion. I adhere to the language of humanism because I believe that the community of sentient beings comprises *individuals*, and because I believe that certain aspects of the traditional model of human agency are entirely correct even though those considerations do not in any way confer a superior moral status on human beings.

The endeavor to move past the anthropocentric mindset involves a rethinking of the notions of individuality and agency. In previous chapters I have shown an unremitting tendency in the Western tradition to proceed as if only those beings who possess linguistic rationality in the specifically human sense can possibly count as "individuals," where an individual is a being with a unique self-reflective personality as well as the capacity for detached rational contemplation (Cicero's two "personae" discussed in Chapter 1). In this connection, philosophers have often had to invoke some curious considerations in an effort to save children and adults with significant cognitive impairments from the verdict that they lack full and direct moral status. (This is an important part of the appeal of Diamond's reflections on the basis for inclusion in the moral community.) What major exponents of the tradition have resisted (and many have staunchly denied) is the notion that *a sentient being can count as an individual even if it lacks the capacity for conceptual abstraction and predication*. Korsgaard seems to have something like this in mind when she proposes, contra Kant, that we acknowledge the status of nonhuman animals as ends-in-themselves and thus as direct beneficiaries of moral concern.

And yet the term 'humanism' carries with it very strong historical associations with the ideal of human superiority, one excellent example of which is Pico Della Mirandola's *Oration in the Dignity of Man*, in which Pico asserts that human freedom (the chameleon-like capacity for self-determination) is such a precious endowment that even God and the angels envy us.[16] The reason I seek to retain the notion of humanism is that—and here I part company with a great many advocates of nonhuman animal rights—I believe there is something profoundly right in the tradition's ascription of a special kind of agency to human beings that other animals appear to lack. Rational capacity, as I hope to have shown by now, is a double-edged sword: It facilitates reflection on (and hence the prospect of control over) our prejudices and untoward inclinations, but it also has given rise to some epic distortions of reality. Recall Carruthers's blithe dismissal of nonhuman animal pain as not having any claim on our sympathy inasmuch as that pain is "non-conscious." Recall also my veterinarian's immediate response—that anyone who has spent five minutes in a veterinary emergency room knows that nonhuman animals are well aware of their pain. A key problem with which the Eleatic tradition has left us is how to establish "rationally" that this (and, more generally, the notion that nonhuman animals are "autotelic") is in fact the case; thinkers such as Carruthers (who is considerably more representative of contemporary sensibilities than you might imagine) demand definitive proof that these organic entities do indeed possess the capacities that people like me attribute to them, when in fact the terms of Eleatic proof procedures are much too limited to permit a demonstration of what ought to be apparent to us in our affective encounters with nonhuman animals—not what *is* apparent to us in light of our anthropocentric inheritance, but what *ought* to be.

This 'ought' is an indication of a capacity in human beings to *take responsibility* for their actions, which involves a complex set of rational capacities that I do believe we human beings possess, even if too often we fail to live up to their potential. When I hear people say (I've heard it many times) that if it is permissible for lions to kill gazelles then there is nothing wrong with humans killing nonhuman animals, I have to scratch my head. Few if any people seriously believe that non-humans such as lions have the capacity to reflect on their actions in such a manner as to be able to assess actions and events in terms of right versus wrong and per-missible versus impermissible. Consider two views discussed earlier in this book: Rowlands's view that nonhuman animals lack the reflective capacity possessed by moral agents, and Korsgaard's suggestion that nonhuman animals lack the rational capacity that would permit them to articulate reasons for action. I believe that Row-lands and Korsgaard are right to make each of these claims, but I do not see why that should entail that nonhuman animals are not individuals—any more than it should entail that human children are not individuals.

When I argue for a well-tempered humanism, I am arguing for a recognition that human beings possess the rational capacity to evaluate their actions critically and classify them in terms such as "morally permissible" and "morally impermissible." I see no reason why key humanistic ideals such as the dignity of the individual should not be retained, as they have great power to help us think through and regu-late our conduct in the social (i.e., the human) sphere as well as in our relations with the nonhuman world. Where we need to depart from traditional humanism is precisely in its pointedly anthropocentric commitment to the proposition that rea-son renders us morally superior to nonhuman animals and the rest of nature. Thus I will return to Karl Löwith's contrast, noted near the end of Chapter 1, between "world" and "human world" (where the latter is inscribed within the former) and Schopenhauer's parallel distinction between "cosmic" and "social" justice. In the end, the kind of well-tempered humanism I have in mind will involve an open rec-ognition that justice is not simply a matter of human artifice but extends to all of sentient life—a conclusion that, needless to say, exponents of the anthropocentric mindset are likely to reject out of hand.

2. The Power and Essential Limits of Feeling or Emotion

The traditional concern about the affective dimension of our being is relatively straightforward: It is clear that our emotions often "get the better of us" in the se that we lose rather than retain control over our intentions. For Aristotle, *epith* (desire or appetite) requires rational guidance. For Descartes we need "proper weapons" against most if not all of our passions. For Kant, inclination is the enemy of moral integrity. This anxiety about our affective nature leads Seneca to dismiss states such as anger as irrational disturbances that we ought to quell as much as possible. It does not appear to have occurred to thinkers such as Seneca that states such as anger, while often extremely dangerous, also possess distinctive *disclosive power* that can be capitalized upon if they are brought into proper interplay with our capacity for rational reflection.

Together with this general ascription of subordinate status to the affects, the anthropocentric tradition has gone further and either denied nonhuman animals emotional capacities altogether or profoundly circumscribed the range of nonhuman animal emotions to the comparatively more "primitive" sorts of emotional responses, such as fear. Exponents of the tradition consider it absurd to suppose that any sort of nonhuman animal could possibly experience a state such as grief (which they take to require linguistic rationality), in spite of growing evidence that a variety of nonhuman animals (in particular, crows and elephants) engage in conspicuous behavior that for all the world appears to reflect a sense of tragic loss.[17] To suppose that a being can experience a state such as grief is, on the traditional view, to attribute to that being capacities such as conceptual abstraction. After all, how can a being have a sense of enduring loss—the loss must be deep and enduring to evoke a sense of grief rather than, say, mere disappointment—if that being has no abstract sense of time as a whole, and thus cannot relate consciously to the more-than-immediate past and future?

I have noted already that the standard view in Western philosophy, even among some potential advocates of nonhuman animals such as Schopenhauer, is that nonhuman animals are bereft of linguistic-conceptual ability and therefore are effectively imprisoned in an eternal present. To suppose that a being with no ability to think about the future could experience a state such as grief is simply illogical on the traditional view. The same holds, *mutatis mutandis*, for affective states such as generosity, which on the view of thinkers such as Aquinas and Descartes requires the ability to step back in acts of abstraction and contemplate one's place in a cosmic whole greater than oneself.[18] In short, the proposition that some (perhaps many) nonhuman animals experience affective states such as grief and generosity is wildly at odds with the anthropocentric background ideal of living.

By now it is clear that the anthropocentric and non-anthropocentric ideals take pointedly different approaches to the question whether any nonhuman animals can experience states such as grief. The anthropocentric approach is to place the burden of proof on anyone who would aver that the answer to this question is 'yes', by demanding irrefutable proof—proof that, as I have argued throughout this book, is rendered a practical impossibility given the grounding ideological assumption of human superiority and prerogative central to the anthropocentric ideal. The tradition generally assumes that human beings are unique among sentient creatures in possessing an awareness of death. I noted earlier that Heidegger exemplifies this prejudice when he suggests that nonhuman animals merely "perish," whereas human Dasein is truly subject to (which is to say, we actively structure our lives in terms of our confrontation with) death. This should come as no surprise, given Heidegger's denigration of the nonhuman animal as "world-poor" in comparison with us "world-forming" human beings. To anyone who intimates that an active relationship to death is not exclusive to human beings, the anthropocentric response is, again, *prove it!*, with the proviso that the question whether or not certain considerations constitute 'proof' is to be determined entirely by criteria that are cut to the measure of the anthropocentric ideal.

The non-anthropocentric alternative is one of openness—to our own ignorance, our own deep-seated prejudices, and crucially to the possibility that we humans are not nearly as unique as we desperately need to think we are. This is the tremendous potential of the approach gestured toward by Grimm and Taureck: It holds the promise of redirecting our thinking about nonhuman animals away from self-serving Eleatic-anthropocentric distortions and toward the kind of openness that "lets beings be"—in this case, individual nonhuman sentient beings. "Letting beings be" does not simply mean leaving them alone or refraining from interfering with their lives; it crucially means *letting them be the beings they truly are* rather than forcing them to be the beings to which we have reduced them through objectification, and this demands a tremendous act of modesty on our part.

An important feature of the non-anthropocentric ideal is that *it remains agnostic in principle on the question of experiential capacities in nonhuman animals*. If we really find the modesty to open ourselves to the "Tier an sich," we may just find ourselves surprised at the vista of the extraordinary range of nonhuman animal achievements (virtually all of them relegated to the status of instinctive reflexes by the anthropocentric orientation) that we attain. I myself have on occasion suggested that ethologists such as Marc Bekoff attribute too much to nonhuman animals when they suggest that many nonhuman species exhibit a sense of justice. I have to consider the possibility (indeed, perhaps even the likelihood) that in rendering that judgment I was, if only unwittingly, employing precisely the anthropocentric criteria that I have just suggested are categorically ill-suited to the endeavor to rethink the experiential capacities—the lives—of nonhuman animals.

A closely related feature of the non-anthropocentric approach is that it proceeds from the proposition that the moral status of nonhuman animals does not depend on whether, for example, nonhuman animals have the same relationship to death that we have, nor on the question whether nonhuman animals are capable of affective states such as love or grief.[19] In this connection, consider two rather different interpretations of the following experiment. Researchers recorded the calls of young vervet monkeys and discovered that if these calls were later broadcast within hearing range of the mother, the mother's response differed depending on whether her child had died in the interval between recording and playback. Mothers who heard the calls after their young had died gradually turned off their distressed response, leading the researchers to conclude that vervets understand the difference between lost and deceased offspring.[20] From the anthropocentric standpoint, this conclusion is both highly speculative and conceptually underdetermined: what, after all, can we expect a vervet monkey to "understand" about death, if a creature of that kind is bereft of predicative language? At the extreme, there may well be no understanding at all, but simply a shift of attention from an old object of concern to new ones—as if maternity, caring, filial bonds, and a sense of community generally were simply beyond the grasp of a mere vervet. As if "maternity" in a vervet were simply the essentially blind carrying out of biological imperatives.[21]

The non-anthropocentric approach is considerably more open to the prospect that many capacities and virtues we find in ourselves are not our exclusive possession but may well be an essential part of what it means to be a sentient creature. Where the anthropocentric orientation implicitly proceeds on the assumption that conditions such as maternity cannot possibly be the same in human and nonhuman animals, the non-anthropocentric approach openly acknowledges the many evident affinities between maternity in humans and in a wide variety of nonhuman animals. In this connection, the non-anthropocentric approach finds inspiration in an unlikely place: the thought of the Stoics, who as I noted in Chapter 1 assert the categorical superiority of human beings and exclude nonhuman animals from the sphere of justice altogether.

The Stoic doctrine of *oikeiosis* is a doctrine of belonging or affiliation according to which many sentient creatures (not only humans) are endowed by nature with the ability to form communal bonds. That which is 'oikeion' pertains to (belongs to) the household; the notion of household membership serves as an image for the much broader notion of belonging to (or, as the Stoics put it, living in accordance with) nature. Kinship bonds, in other words, are natural, indeed to such an extent that they are not found exclusively in human relations but permeate the entire animal world. In this connection, 'nature' is best conceived not in the comparatively closed terms of the anthropocentric-Eleatic mindset, but rather in the open terms in which thinkers such as Porphyry and Schopenhauer endeavored to encounter the world. From the standpoint of subject-object representation, it is virtually impossible to demonstrate that, say, a mother hen experiences an active, conscious, loving relationship to her young; as I have observed, it is all too easy from that vantage point to relegate that mother's conduct to the dustbin of blind instinct. What the non-anthropocentric approach remains open to is the prospect that it is misleading in the extreme to characterize the consciousness of a nonhuman animal mother as simultaneously caring and blind. (That approach also remains keenly sensitive to the extent to which *human* behavior is instinctive.)

This, however, was not what the Stoic philosophers had in mind when they developed the doctrine of *oikeiosis*. As should be amply clear by now, the Stoics (who excluded nonhuman animals from the sphere of justice altogether) were no more open to human-nonhuman animal community than Aristotle, Descartes, or Kant. For all these thinkers, the sense that nonhuman animals have no place in the human social community entails that there can be no kind of community whatsoever with nonhuman animals. The Stoics proceed from an organic conception of the whole of reality rather than taking a Cartesian-Kantian self as the starting point. This enables them to recognize kinship relations in the natural world, not simply within the human social world. *Oikeiosis* occurs in stages that trace out increasingly inclusive circles of belonging, only to terminate in a circle of belonging that categorically excludes all nonhuman animals. In other words, rather than taking the doctrine of *oikeiosis* to what could have been its non-anthropocentric apex in a conception of cosmic (as opposed to merely human social) community, the Stoics tailor the final stage of the doctrine to eminently traditional, anthropocentric purposes.

For the Stoics, every sentient creature has a sense of being at one with its own embodiment.

> Immediately upon birth . . . a living creature feels an attachment for itself, and an impulse to preserve itself and to feel affection for its own constitution and for those things which tend to preserve that constitution; while on the other hand it feels an antipathy to destruction and to those things which appear to threaten destruction.

In providing this characterization of the initial stage of *oikeiosis*, Cato (as reported by Cicero) seems fairly clearly to have sentient creatures (not living beings generally) in mind, as is apparent from his suggestion that "self-affection" requires "self-consciousness" and further that "it is love of self which supplies the primary impulse to action."[22] Every sentient creature, not simply human beings, possesses a conscious awareness of its embodiment, an awareness that is not initially separate from its sense of caring for itself.[23]

Subsequent stages of *oikeiosis* take the form of broadening the scope of concern so as to include caring for others. Like the first stage, this second stage is characteristic of a great many nonhuman sentient creatures.

> Nature creates in parents an affection for their children. . . . It could not be consistent that nature should at once intend offspring to be born and make no provision for that offspring when born to be loved and cherished. Even in the lower animals nature's operation can be clearly discerned; when we observe the labour that they spend on bearing and rearing their young, we seem to be listening to the actual voice of nature.

This sense of caring pertains in the first instance to offspring and subsequently to broader affiliations, as when ants, bees, or storks "do certain actions for the sake of others besides themselves."[24]

Up to this point, the doctrine of *oikeiosis* views human and nonhuman sentient creatures as essentially the same: embodied, conscious, feeling, caring, social, purposive. It is not clear whether or to what extent this characterization of the

subjective lives of nonhuman animals is ultimately compatible with Seneca's characterization of them as essentially biological reaction devices; indeed, the tension
between these two ways of viewing nonhuman animals is one that I have noted in
a number of other thinkers, notably Kant (who struggles with the idea of gratitude
toward a "thing") and Schopenhauer (who embraces an organic standpoint that at
least in broad strokes resembles that of the Stoics but who nonetheless assigns a
subordinate moral status to nonhuman animals). What is important is that, notwithstanding its anthropocentric conclusion, the doctrine of *oikeiosis* lays the ground-
work and provides inspiration for *a reaffirmation of the primacy of embodiment
and caring in the lives of sentient creatures*, a reaffirmation that brings with it a new
sense of the place of human beings in the cosmic scheme.

In the thought of the Stoic philosophers themselves, this prospect is foreclosed
by the nature of the third and putatively highest stage of *oikeiosis*, a stage attainable
exclusively by human beings in virtue of our possession of linguistic rationality. At
this stage, human beings capitalize on the (supposed) fact that "with human beings
[the] bond of mutual aid is far more intimate" than among nonhuman animals, such
that we humans alone are "fitted to form unions, societies and states."[25] But rather
than this greater capacity to form bonds of mutual aid giving rise to a conception
of community that embraces all sentient creatures (all creatures who participate in
earlier or "lower" stages of *oikeiosis*), for the Stoics it leads to a highly *exclusionary* conception of universal human community:

> The outermost and largest circle [of *oikeiosis*], which encompasses all the rest,
> is that of the whole human race. . . . It is the task of the well-tempered man,
> in his proper treatment of each group, to draw the circles together somehow
> towards the centre.[26]

This is a theory of felt kinship, but it is a highly selective one: It proceeds from an
acknowledgment of some deep affinities between human and nonhuman animals,
only to abandon that acknowledgment once human rationality enters the picture.
Once it does enter the picture, the previously noted continuities and sense of shared
fate are quickly forgotten and the process of establishing essential distinctions
between human beings and nonhuman animals takes center stage.

Cicero's notion of two human "personae," which I presented in Chapter 1, finds
its place at this third stage of *oikeiosis*. Recall that those two *personae* are the
individual (my *persona* or role as Gary Steiner) and the universal (my *persona* as
a generic member of a polity, who can recognize the need to establish and honor
reciprocal entitlements and duties with my fellow citizens). Cicero argues that both
depend crucially on rational capacity, the Stoic conclusion being that nonhuman
animals can be neither concrete individuals nor members of a universal polity. This
in turn means that "no right exists as between man and beast," one straightforward
consequence being that "men can make use of beasts for their own purposes without injustice."[27]

This is the moment in the Eleatic mindset when the common-sense acknowledgment of human animality and our deep continuity with the rest of the sentient world becomes abruptly abandoned in the name of human exceptionalism, the proclamation of the uniqueness and categorical superiority of human beings over all other worldly creatures. As I have argued from the start, this abandonment is a product not of error or trenchant insight, but of *ideology*: There is an active motivation at work that guides the operation of reason at this third and putatively highest level of *oikeiosis*, although it may well outwardly appear that reason is really operating impartially here. The Stoics proceed from an open acknowledgment that nature has implanted a strong bond of filial affection in all sentient creatures, to the conclusion that nonhuman animals are essentially objects of use bereft of the kind of "real" emotional states and responses that human beings regularly experience.

This effort to establish a clear dividing line between human and nonhuman animals has been facilitated greatly by a feature of human existence to which I have already drawn attention: our sheer ignorance. Human beings have struggled for millennia with the difficulty of gaining insight into the lives of nonhuman animals. Our endowment of linguistic rationality is, as I have suggested, a double-edged sword: It enables us to make fine distinctions, identify stable causal laws, and do things like send rovers to Mars, but it also profoundly obscures our efforts to grasp the embodied life of a non-linguistic sentient being. Because we are capable of acts of detached reflection and can represent our affective states to ourselves as if they were determinate objects, we are accustomed to thinking of states such as love as ones that require sophisticated cognitive capacities like the ones we employ when we are thinking in the abstract about our emotional constitution. To anyone inured to thinking of love as involving *logos* and reflection, it is likely to seem absurd to attribute anything like love to a nonhuman animal. What does such a creature actually experience, if not "real" love? It must be some pale simulacrum of love, founded entirely on instinct.

From the most common-sensical of standpoints, this proposition strikes me as highly implausible. Love takes many forms, some of them involving extensive reflection and some of them comparatively immediate (unreflective). Human beings appear to experience love in a variety of forms, and it is far from clear that we have ever arrived at a definitive definition of what love is. In a great many instances, human beings appear to experience love without any apparent reflection at all—which would appear to place them in very close proximity to nonhuman animals. So it would seem that the mistake consists not in attributing affective states such as love to nonhuman animals, but rather in failing or refusing to do so.

Of course, the defender of the anthropocentric ethos will demand a clear demonstration that, for example, mother hens actually experience love for their offspring. Such a demand will presumably bring with it an expectation that the demonstration will include a clear and uncontroversial definition of love that can then be applied to human and nonhuman animals alike. For reasons I have attempted to sketch, such a demonstration is not in the cards—not because nonhuman animals are incapable of affective states such as love, but because of our own inability to work out such a demonstration, together with a prior ideological interest in there being no such demonstration. In Heideggerian language, the fact that we humans possess apophantic *logos* (the *logos* of explicit assertion) in addition to hermeneutic *logos* (the *logos* of engaged practice) prevents us from contemplating in any depth what life is like for a being who possesses hermeneutic but not apophantic *logos*. The anthropocentric assumption is essentially that beings lacking apophantic *logos* must necessarily lack hermeneutic *logos* as well.

The question of emotion in nonhuman animals is a deeply vexed one. I do not consider it necessary to delve deeper into it here, other than to stipulate that considerations of evolutionary continuity, physiological similarity, and common sense all point to the conclusion that many nonhuman animals have rich affective lives—lives whose broad outlines we can grasp, but whose fine-grained detail remains largely hidden from us due to our cognitive limitations and ideological conditioning. Heeding Grimm's and Taureck's call for modesty in assessing the nature and capacities of nonhuman animals, I do not think that the specific affective (or other experiential) capacities a given nonhuman animal possesses have any significance whatsoever for that creature's moral worth—any more than, say, one person with a greater capacity to love somehow "counts" more in the moral calculus than someone with a lesser capacity to love. Thus, as I have stated, I remain agnostic on the question of specific experiential capacities (both their presence and their character) in nonhuman animals. The foundational unit of measure for the kind of humanism I envision is the sentient individual, where being an individual does not depend in any essential way on the possession of human language or rationality. It simply depends on whether or not a being is, in Derrida's terminology, "autotelic." I am very curious about the various experiential capacities of nonhuman animals; but I do not need to wait for breaking news from the world of ethology (nor do I consider it necessary to catalogue in any detail the specific affective capacities of various nonhuman animals) to be committed to the proposition that there are extensive continuities between human and nonhuman animal experience, and that these continuities are sufficient to ground an ethic of felt kinship with nonhuman animals.[28]

3. Toward a Dialectical Conception of the Reason-Emotion Dichotomy

In Chapter 4 I noted Virginia Woolf's observation about the difficulty involved in the endeavor to "untwist" reason and emotion from one another, and her thought that both have their origin "deep in the darkness of ancestral memory." This preoccupation with origins is evident in the thought of many philosophers, notably in early twentieth-century thinkers such as Spengler and Heidegger. There has also been a good deal of resistance to this "archaic" form of thinking, in part due to the susceptibility of historical reflection to the problem of ideologically motivated (or perhaps just wildly speculative) mythologizing (e.g., was there ever really a "golden age" of the kind described by Hesiod and Ovid?) That is and will always be a danger that attends thought. A related danger is the tendency of anthropo-centric-Eleatic thinking and valuing to dismiss as mere myth virtually any and all ideas or claims that fail to satisfy the demands of proof procedures carried out through apophansis. That is the fate suffered by the sorts of claims that Schopen-hauer makes about the access that the will affords us to the inner nature of things, as well as by the suggestion that there is something like "the darkness of ancestral memory" out of which the divide between reason and emotion first emerged. It is in this connection that Stein counsels Lord Jim to "immerse" in "the destructive ele-ment," that inherently mysterious primordial element of existence (Schopenhauer's world will) out of which individuals became distilled into their fleeting moment of awareness.

In the endeavor to move past the monolithic anthropocentric orientation, an essential task is to find a way to vindicate appeals to truths about existence that fall outside the relatively narrow scope of proof as it pertains to apophantically-structured discourse. If thinkers such as Plutarch, Porphyry, and Schopenhauer are right—and I believe they are—then there are a great many truths about real-ity, some of them very straightforward and evident when contemplated apart from the influence of the anthropocentric mindset, that merit our acknowledgment. The shared vulnerability and mortality of all sentient creatures is one such truth.

To attribute the status of a "myth" to such a statement is not, however, necessar-ily to discredit it as the product of someone's wild lyrical imagination. In the end, it may constitute *a different kind of truth* than the truths forged by subject-object representation, for reasons I have already examined in depth. I remain open on the question whether "myth" is an appropriate characterization of this kind of truth. Here I would simply note that anyone who seeks to dismiss all myth as fanciful or irrelevant to the tasks of life would do well to consider Hans Blumenberg's analysis of myth as a way of coming to grips with "the absolutism of reality," where absolut-ism signifies for Blumenberg what *ananke* signified for Freud.[29] One consideration offered by Blumenberg that merits reflection in the present context is his suggestion that Descartes's appeal to the *cogito* is itself a mythic "fiction of the zero point" in the sense that it constitutes a denial or refusal of the absoluteness of reality by positing the rational self as an absolute origin ("zero point") of certainty.[30] From the vantage point of Cartesian detachment, what we look at when we contemplate nature is not any sort of "destructive element," certainly not one unfathomable for rational beings, but rather an ordering of determinate objects subject to universal forces and general laws.

Schopenhauer's appeal to the world will, Stein's call to immerse in the destruc-tive element, and Woolf's intimation of an origin that eludes specification in definitive apophantic terms are all candidates for rapid dismissal by exponents of anthropocentric-Eleatic thinking and valuing. And yet all point toward a level of experience beneath the superstructure of apophantic abstraction, a level at which the human condition shows up as having a great many more continuities with nonhuman sentient life than discontinuities. In previous chapters I have stressed the tradition's central focus on control, and I have presented considerations in support of the conclusion that we human beings are not alone in seeking to exercise control over "the absolutism of reality." *All* sentient creatures face this challenge, not simply human beings. The difference between the two as concerns the endeavor to exercise control, however, seems clear: the supplement of apophantic *logos*, which appears to be unique to human beings, facilitates not only a degree of material control unattained by any nonhuman animals, but also and crucially the devising of aspirations (such as that of *systematic* dominance over the entirety of the nonhuman) that find no counterpart in nonhuman animals. Philosophers sometimes suggest that nonhuman animals are incapable of good and evil, inasmuch as they have no concept of the good and are driven essentially by instinct; our conceptual ability, these thinkers aver, renders human beings capable of good and evil. Not surprisingly, defenders of human exceptionalism place their primary focus on the human capacity for good and the putative "fact" that this capacity renders us morally superior to nonhuman animals. From the very start of this book I have disputed this characterization of human beings, on the grounds that we have employed our complex cognitive powers much more for destructive and oppressive purposes than for salutary ones—and, even more fundamentally, on the grounds that there is no good reason to suppose that cognitive complexity or sophistication *per se* has any significance for assessing the moral worth of a sentient being. (If it *were* relevant, would that mean that human geniuses possess greater moral worth than human beings of ordinary intelligence?).

The acknowledgment that human beings are not alone in the endeavor to exercise control constitutes a step in the direction of working through the ideological underpinnings of anthropocentrism. As I have noted, one of Dewey's concerns was to probe "beneath" the level of apophantic detachment and rethink human experience at a more fundamental level. Where Woolf appeals to the depths of ancestral memory to locate this deeper level of experience, Dewey appeals to the notion of "quality" in the sense of an irreducible, global, felt sense of orientation in the world. What we encounter first of all is not an ordered set of objects, nor raw matter that stands in need of organization. Instead we encounter "a permeating qualitative unity" that is "felt rather than thought." Dewey cautions against reifying this felt encounter into "a" feeling, stressing that feeling at this primordial level discloses not discrete objects or states but rather "an unanalyzed whole."[31] Dewey seeks to explore the origin of distinctions such as reason-affect in "some prior unity" in the field of experience.[32] And like Heidegger, Dewey sees individual psychological moods and affective states as presupposing a global disclosure of things. For Dewey as for Heidegger and Merleau-Ponty, that global disclosure is a disclosure of *meaning*, not of discrete sense data or objects. The putatively "infallible spectatorship" of "a consciousness which is set on the outside over against the course of nature" could not even occur to us if the notion had not emerged from out of a prior "consummatory union of environment and organism."[33]

Dewey's conviction that our most fundamental encounter with things is in the mode of a felt disclosure of an entire context of meaning makes a contribution to the endeavor to broaden the range of full and direct moral concern so as to include all of sentient life. Dewey, for his own part, does not pursue this path but instead hews closely to the traditional prejudice that nonhuman animals have no share in meaning. In the case of human beings, "body-mind simply designates what actually takes place when a living body is implicated in situations of discourse, communication, and participation." Dewey makes it clear that the specific kinds of bodies capable of entering into discourse are exclusively human: On his view, not only are nonhuman animals incapable of "thinking," but they are utterly excluded from the sphere of meaning in virtue of their being *aloga*. And even though he acknowledges the capacity for feeling in nonhuman animals, Dewey characterizes feeling even in putatively higher nonhuman species as taking the form of "vague and massive uneasiness, comfort, vigor and exhaustion."[34]

In Chapter 4 I offered considerations in support of the proposition that to be sentient is *eo ipso* to be an active participant in meaning. Dewey is well aware that detached analysis of qualitative relations can distort the underlying realities, as when the tradition posited a strict metaphysical dualism of mind and body. He himself appears to have fallen prey to this danger when he acknowledges that "total quality operates with animals," only to cast serious doubt on the possibility that any nonhuman animals are capable of "symbolization and analysis."[35] To say that total quality operates in nonhuman animals is, given Dewey's conception of "quality," to acknowledge that sentient creatures encounter a whole environment invested with challenges and threats. To say that nonhuman animals are not capable of symbolization and analysis is to assume, a little too hastily, that reckoning consciously with one's environment must necessarily take the form of apophantic *logos*. It is evident that a great many nonhuman animals actively confront and work through challenges all the time, and moreover that they appear to have a considerably more harmonious relationship with the natural world than we humans seem capable of establishing. Again, the problem is not that nonhuman animals lack these capacities, but rather that we seem unable (and, if my central thesis is right, *unwilling*) to grasp what it means for a being lacking symbolic language to engage in any sort of "analysis" or scrutiny of the contingencies it faces.

I take this failing on the part of traditional thought to be a product of inadequate reflection on the dialectical relationship between reason and affect in human experience, a relationship that appears to spring from a deeper origin in experience. The tradition was right to proceed on the recognition that the supplement of reason has important ramifications for our affective lives. It gives rise to a variety of capacities and endeavors that seem foreclosed to nonhuman animals, including the setting of goals such as the systematic domination of the natural world, or the long-term pursuit of a vendetta. It also gives rise to the pursuit of moral virtue for its own sake, a pursuit that virtually the entire tradition denies to nonhuman animals. A central challenge in the endeavor to bring nonhuman animals into the moral community as direct and full beneficiaries is to disentangle the respective roles of reason and emotion in our own experience, so as to open ourselves to the essential continuity that we share with the experience of nonhuman animals. Even in human beings, reason never operates in a fully autonomous manner (not outside of fields such as mathematics and logic, at any rate) but is interwoven with affective dispositions and ideological commitments. The more we acknowledge this fact, the likelier we will be to be receptive to the notion that we are more "clever beasts who have to die" than "the titular lords of nature."

The reciprocal relationship between reason and affect in human beings is not well understood, and in any case there is no settled account of that relationship. I doubt that it will be well understood, if ever, unless and until we have subjected traditional commitments about reason and emotion to the kind of deep and sustained non-anthropocentric reflection that they vitally need but have not yet received. Thus my remarks here are admittedly and necessarily tentative.

Like a number of thinkers I have discussed in this book, Arne Vetlesen characterizes emotions as involving "a certain way of seeing," such that they are "instrumental in giving us a fundamental first access" to the relevance and urgency of concrete situations.[36] Vetlesen thus departs from Kant's "cognitivistic one-sidedness" (recall Diamond's "rational Martians") and opts for an approach that holds the promise of enriching our notion of what it means to feel real concern for the suffering of others. What is needed to feel such concern is not simply some detached objective judgment or idea about the rightness or wrongness of an action or event, but rather "already to have established an *emotional bond* between myself and the person I 'see' suffering."[37] The formation of such a bond involves a "sensuous-cognitive-emotional openness to the world" that Vetlesen associates with the Heideggerian notion of mood or disposition that I examined in Chapter 4.[38]

Like thinkers such as Diamond, Vetlesen proceeds on the assumption that our emotional bonds with our fellow human beings have an essential priority over our sense of emotional connection with anything or anyone nonhuman. Following a line of thinking advanced by Charles Taylor, Vetlesen confines his discussion to human moral agency. With regard to sentient nonhuman life, he states that nonhuman "animals qua objects of suffering . . . inspire compassion in us."[39] This call for compassion is more revealing than it might appear at first blush. By following Taylor's line of thinking, Vetlesen implicitly accepts the proposition that justice is for humans, whereas compassion is the most we might owe to nonhuman animals. Taylor repeats the traditional prejudice that only "persons," i.e., only those beings who are self-aware and capable of evaluation and choice, merit respect; Taylor explicitly appeals to "species" membership as a criterion for personhood, thereby categorically excluding the possibility that any nonhuman animal could be considered a moral "subject."[40] Thus Taylor makes an anthropocentrically motivated mistake common to almost the entire tradition: the assumption that respect (and, thereby, justice) is owed only to linguistic beings, i.e., that only moral agents can (or should) be considered to possess full moral worth.

Vetlesen, for his part, acknowledges that "an animal can be harmed, can be hurt, can suffer, and it can for that very reason be an object of unjust and immoral conduct."[41] But he says nothing about what sorts of justice relations we might have with nonhuman animals because, like so many thinkers, he attempts to formulate a conception of human moral community (including both compassion and justice) *prior* to any consideration of the place and status that nonhuman animals might have in such a community. On any such approach, I fear, nonhuman animals will always be treated as supplements or afterthoughts to our moral dealings rather than as central. It is perfectly understandable that we privilege our "local" affiliations (self, immediate family, other loved ones) over more global ones; and it is perfectly understandable that many of us privilege our fellow humans over any nonhuman animals, even those we love very deeply. The question that has received inadequate attention is whether the mere fact that we *do* privilege local affiliations and humans in this way entails that we *ought* to privilege them as we do.

Taylor's implicit response to this question is to place the same premium on language that all major exponents of anthropocentric reason have placed on it: He acknowledges that many nonhuman animals experience feelings and that feelings convey import, but he privileges specifically human forms of feeling on the grounds that language facilitates the formation of special forms of import unique to human beings.[42] Thus nonhuman animals, just like machines, are incapable of feeling shame; and if, say, baboons are capable of anything like dignity, it can only be "some analogue of human dignity" inasmuch as baboons are incapable of the linguistic facility that would permit them to identify, reflect on, and take a reasoned stand on their lives.[43] Where I part company with thinkers such as Taylor is in rejecting the capacity for shame or dignity as a marker of moral worth: I do not see why we should consider nonhuman animals to be morally inferior (less worthy of full and direct moral consideration) simply in virtue of their (supposedly) being incapable of a state such as shame or the specifically human shape that dignity takes. Moreover, it strikes me as revealing that we human beings, and not nonhuman animals, seem to encounter so very many occasions on which we *ought* to feel ashamed of ourselves.

Richard Posner dispenses with rational considerations altogether and asserts that our privileging of human beings over nonhuman animals is the product of "a moral intuition deeper than any reason."[44] In making this claim, Posner implicitly embraces the kind of pathocentrism that I consider to be dangerous for reasons that I have discussed several times in this book in connection with Kierkegaard and Schmitt. Simply asserting that you have a feeling or instinct with moral weight, and that there is simply no discussion about it, leads very quickly to totalitarian forms of thought and valuing. Consider the confrontation between two radically opposed ways of thinking about the moral status of nonhuman animals: Posner's and mine. I maintain that it is possible to retrieve a sense of felt kinship with nonhuman animals from out of the depths of ancestral memory. Posner maintains that the very idea of rights for nonhuman animals is "weird" and "insane." What are we to do in the face of such a pronounced disagreement? The terms of a total critique of reason leave us with no way to resolve such a disagreement other than sheer force or a parting of the ways (a permanent disruption of community). I take it as axiomatic that a robust ideal of community that seeks to honor autotelic individuals must not seek to resolve disputes with force, unless that force is the force of the better reason. And on the account I have been developing, I believe that these "better reasons" themselves are informed by our affective constitution.

The problem with which we are left in such a dispute is how to find common ground. I have attempted to lay some common ground by drawing inspiration from the very tradition that I believe has taken a very wrong road. Naturally there is no guarantee that any given set of considerations will automatically change the mind of a Carruthers, a Watson, or a Posner. In fact, the very thought that some specific set of considerations could automatically change the mind of a person deeply invested in rejecting them strikes me as a sign of the distorting power of Eleatic reason—as if there were some Kantian Transcendental Deduction of the notion that rights are not simply for human beings. To the extent that reason by its very nature depends on the guidance of our affective constitution, there can be no such deduction. What there can be is a process of negotiation—between people, and between the competing thoughts and feelings of each individual person—that could lead us collectively to the prospect of re-envisioning in non-anthropocentric form our tradition's most deeply held convictions about moral worth.

The very idea of moral accountability, that capacity that has been vaunted so emphatically in humans by the philosophical tradition, is an important part of the moral landscape, and we abandon it at our own grave peril. Moral accountability involves *giving reasons* for one's conduct that the human community can discuss and assess in common. That does not mean that reason is the origin of our moral commitments, but rather that reason performs the important function of helping us to distinguish between selfish or partial affiliations from those that genuinely merit being honored, as well as of facilitating the process of *cultivating* our affective dispositions rather than treating them as innate endowments incapable of development. In other words, simply to say that there are moral intuitions "deeper than any reason" is not *eo ipso* to say that reason plays no role in the evaluation of our intuitions.

I believe that reason, as I suggested early in Chapter 1, is essentially in dialogue with the emotions in the case of human beings. In all sentient creatures, meaningful lived experience is the basic unit of measure; in this respect, all sentient creatures are autotelic agents. Human beings appear to be alone in possessing the apophantic capacity for fine-grained analysis, and that renders us moral agents capable of contemplating and taking on explicitly articulated rights and obligations. Given the state of the world, replete as it is with such epic anthropogenic violence, I cannot follow the tradition in seeing this superstructure of moral agency as indicative of some categorical superiority on the part of human beings.

I am considerably more inclined to see it as analogous to what Freud said about consciousness—that just as for Freud, consciousness is simply the tip of the iceberg of mental life, our moral agency is simply the extreme manifestation of an entire set of processes and capacities that we share with all of sentient life. What lies beneath the tip of the iceberg is the vast sweep of our sense of orientation and significance in the world, a sense that we share with a great many sentient creatures and that we have fought hard to deny—just as Schopenhauer notes that individuals, once they have emerged from the world will for their brief instant of detached awareness, are overwhelmingly inclined to treat their temporary vantage point as if it were their very essence.

In previous chapters I have presented the basis for a rethinking of what lies beneath the layer of detached, linguistically-informed self-awareness. Schopenhauer's appeal to intuitive as opposed to discursive knowledge, Ortega's recognition that lived experience is prior to theorizing, Miller's identification of the act as the "noncognitive basis of cognition," Dewey's insight into quality, Heidegger's appeal to hermeneutic *logos*, and Merleau-Ponty's open acknowledgment that a great many sentient beings are full participants in meaning all point toward resources for a radical rethinking of what it means to be human, what it means more generally to be animal, and what it means for human and nonhuman animals to be members of the moral community. As Nietzsche intimates at one point, when you embark upon an open sea of discovery, you do not know what you are liable to find; there is something inherently *experimental* about the endeavor. Given how inadequately the ideal of moral community with nonhuman animals has been explored up to now, I do not believe we are anywhere near being in a position to make definitive pronouncements about exactly which sorts of entitlements nonhuman animals would enjoy in such a broadened community, just as we are nowhere near being in a position to make definitive pronouncements about the specific cognitive and other experiential capacities of nonhuman animals. But there is one grounding idea for such an ideal that provides us with a preliminary sense of direction and inspiration for further reflection.

That idea is the notion of justice, broadened so as to have implications for all of sentient life. The tradition has been more or less unremitting in proclaiming that justice is a relation among rational contractors (i.e., among human agents), and that at most what we owe nonhuman animals is compassion. Thinkers such as Martha Nussbaum, as I noted in Chapter 1, aver that nonhuman animals merit justice; but these thinkers conceive of the entitlements of nonhuman animals as "less than," not simply different from, those of human beings. (Recall that, for Nussbaum, not only is there nothing unjust about using horses in racing and dressage, but in effect we are benefiting them by using them in these ways.) I noted a moment ago that Vetlesen wavers between focusing on compassion and on justice with regard to nonhuman animals. Too often, it seems that thinkers are comfortable with the proposition that compassion is all that is required in the case of nonhuman animals, both on the grounds that nonhuman animals are incapable of entering into explicit reciprocal arrangements and on the grounds that a sense of compassion is adequate by itself to ensure that the entitlements of nonhuman animals are respected.

As I have argued in previous work, this call for compassion for nonhuman animals outwardly looks entirely worthwhile but in fact falls far short of what we actually owe to our fellow sentient creatures.[45] There is something a little strange in the supposition that we have to contemplate our relations with other human beings in an explicitly rational manner, but that when it comes to nonhuman animals all we have to do is essentially be nice. The vast majority of humanity has been anything but nice to nonhuman animals (leaving aside pets, of course) in the course of history, and yet I am willing to bet that quite a few people like to think of themselves as living a life of compassion toward nonhuman animals. The problem is that we do all sorts of things to nonhuman animals all the time, on an ever-wider scale, and we purport to do many of these things "compassionately." Why is it that industries that sell products such as meat and milk go to such great pains to inform the public that their cows are "happy"? Is using a cow as a milk delivery device for a handful of years and then sending her, well before she has arrived at her natural life expectancy, to the slaughterhouse "compassionate"? Is using nonhuman animals in invasive research "compassionate"?

This is a place where it is vitally important that reason come into dialogue with our affective dispositions. A simple example: I became vegetarian a little over a decade before I became vegan. I became vegetarian after reading a great deal about ethology and rights theory, my grounding insight being that sentient beings have a right to life and that killing them (with the exception of true life-and-death situations, which few people reading this book are likely ever to face) is categorically wrong. (This turn to thinking seriously about the entitlements of nonhuman animals was initiated, as I have mentioned already, by a certain shocking experience.) My thinking was that inasmuch as practices such as dairy and egg production do not immediately lead to the death of the nonhuman animals involved, there must be a moral difference between consuming flesh and consuming dairy or eggs. This process of reflection led, more or less immediately in my case, to a basic *change in the way I felt* about using nonhuman animals as sources of food. But my exclusive focus was on meat. As I continued to read and become more conversant with the nature of the dairy and egg industries, I became sensitive to the nature and extent of confinement and treatment, as well as of the fact that these individuals will end their lives in the same slaughterhouses as the nonhuman animals we raise for meat. It became apparent to me that the difference between meat on the one hand and dairy and eggs on the other was merely a superficial one. These rational insights brought about a change in the very feelings I experience deep in my gut, and I made the further move to veganism.

Again, there is no guarantee that another person would respond to the facts about dairy and egg production the way I did. There are naturally many factors at work in the formulation of a response, including matters of individual emotional psychology that are often opaque to us and the influence of several millennia of pervasive anthropocentric valuing that have shaped our affective responses as well as our thinking. If there is to be any undoing or working past this heritage of anthropocentric prejudice, it will take time and nobody can rightfully claim to have a clear vision of what life will look like. What I am suggesting as a preliminary impetus is the proposition that compassion is important, but it is not enough by itself. It has got to lead to operations of reason that inform and enrich our sense of compassion and what it requires of us. I take principles of justice to be assertions of explicit guidelines, always tentative and subject to further discussion and revision, that help us to see when our professions of compassion for others are genuine and when they are self-serving or otherwise deficient. The Stoic *oikeiosis* doctrine reminds us that we have deep proclivities to feel a sense of community with others, and further that these feelings can be developed into a full-fledged sense of justice. A non-anthropocentric ideal of living will be one that capitalizes on this Stoic insight without delimiting it to the strictly human.

What I am arguing for is both compassion *and* justice for nonhuman animals, not simply one or the other. Rawls was right to state that we need a theory of the natural order and the place of human beings in it, but he was wrong to follow Epicurus in maintaining that justice must be understood as a set of arrangements between beings who share the same (specifically human) cognitive endowments. Thinkers such as Rawls conceive of justice in narrowly human *social* terms, whereas what is needed is the sort of appeal that Schopenhauer makes to a notion of *cosmic* justice. Near the end of Chapter 1, I noted Karl Löwith's distinction between 'world' and 'human world', the latter being inscribed within the former rather than being considered to tower above it. Löwith's call for a "cosmo-politics" is a call, implicitly inspired by the Stoic doctrine of *oikeiosis*, for a reconceiving of the human social and political realm within a broader ideal of cosmic community that acknowledges our rootedness in nature and the prospect of obligations to the nonhuman.[46]

The Stoic *oikeiosis* doctrine is a doctrine of *felt kinship*: The Stoics acknowledge that human community originates in and has deep continuities with nature. But they effectively betray this insight when, at the third level of *oikeiosis*, they characterize human beings as essentially different from and superior to the rest of nature. As I note in *Animals and the Moral Community*, this commitment to human exceptionalism finds a parallel within the human social and political community, in the form of prejudices surrounding key features of embodied existence such as race, gender, ethnicity, and religious affiliation.[47] (Recall for example Kant's uninformed pronouncements about black Africans, noted in Chapter 2.) In both cases, there is a failure of justice when arbitrary or self-serving prejudice leads a human individual or society to proclaim its categorical superiority over others, be those other humans or sentient nonhumans. To make sense of the more expansive sense of justice and injustice that is at work here, it is necessary to move beyond the narrowly circumscribed scope of Epicurean-Rawlsian justice and embrace what I call "cosmic holism." I stated in *Animals and the Moral Community* that "the ideal of cosmic holism acknowledges human duties of justice toward animals on the grounds that animals are teleological centers of life and that many animals are sentient beings whose lives matter to them."[48] The ideal of cosmic holism is that of a moral community broadened so as to include all of sentient life, without regard to the question whether a given member of the community possesses linguistic rationality and is capable of taking on obligations. As Löwith recognizes, a sense of rootedness in and commitment to the cosmos rather than simply to the human political community demands the kind of *openness and generosity* that thinkers such as Grimm and Taureck require of us.

It is in this spirit that Schopenhauer appeals to an ideal of "heavenly justice," a sense of justice that originates in compassion and extends the Epicurean premium on non-harm to the entire sentient community rather than confining it to explicit contractual arrangements arrived at by human beings.[49] In situating the human community within the larger framework of sentient life generally, Schopenhauer departs radically from the tradition by locating the problem of injustice prior to the question of rights: The proper starting point is not the supposed entitlements of rational beings, as Kant would have it, but rather the antecedent problem of manifest injustice.[50] In this respect, Schopenhauer anticipates contemporary critiques of "ideal" as opposed to "non-ideal" political theory by arguing that we must start with concrete injustices rather than with detached abstractions about entitlements. At the same time, a robust conception of rights can and must be developed from out of the compassion that is our most authentic response to the injustices we observe; otherwise, we could easily find ourselves in the position of doing no more than simply noting for the record, as if in a detached bureaucratic manner, that some injustice has been visited on a sentient nonhuman.[51] The virtue of Schopenhauer's appeal to heavenly or cosmic justice is that it provides a rationale for attributing rights to sentient nonhuman life—as Schopenhauer maintains, not only are nonhuman animals members of the moral community, but they merit rights.[52]

A claim of this kind, at first blush, could easily appear to reflect a commitment to moral realism. And while that does appear to be the conception of morality that Schopenhauer embraces—he sees the rights of nonhuman animals not as a matter of human thought or choice, but rather of cosmic endowment—the considerations I offered in Chapter 3 about the specifically historical character of reason effectively foreclose the possibility of moral realism. The Protagorean aspiration to undistorted contact with reality presented by Heidegger must forever remain a regulative ideal. Thus it might seem impossible to articulate a compelling justification of the notion of nonhuman animal rights. What I have attempted to demonstrate in previous chapters is that the very idea of moral realism is based on a misguided reliance on the traditional assumption that reality is simply "out there," waiting to be discovered by us, and moreover that that reality "out there" has a specifically axiological character. Ortega's and Millers' reflections on the historically situated character of reason point toward a resolution of this dilemma: They remind us that there are no bare moral facts, but rather human commitments that are always already at work and always subject to ongoing contestation and revision. Where

Rorty would aver that this process has no center but is subject to sheer contingency, defenders of historical reason recognize that the process of rational reflection is one of refining our moral commitments in accordance with an ever-developing sense among the human community not only of what matters, but what *ought* to matter. The 'ought', rather than enjoying any kind of transcendental status, emerges immanently within the process of historical reflection and contestation.

The central concern of historical reason is not the past, but rather the present and future: What are the contingencies that we face now, and how are we to *act* on them? The open, non-anthropocentric standpoint for which I am arguing is one from which the epic injustices we regularly visit on nonhuman animals should be utterly undeniable and considered to be in urgent need of being redressed—even though we find ourselves unable to arrive at a clear sense of what the experience of various nonhuman animals is like, and even though we are so inured to an anthropocentric (contractualist) conception of justice that it is difficult to envision in any precise detail what justice for a non-contractor can ultimately mean. The common sense reactions of most people to this suggestion seem to accord with the verdict offered by the major exponents of the philosophical tradition. The typical reaction is to assume human superiority and take Varner's position that in the absence of definitive proof we should assume that nonhuman animals *lack* a given capacity. Recall, for example, Carruthers's claim that nonhuman animals are not really conscious of their pain—a claim that could have come straight out of Descartes, with his confident assertion that nonhuman animals are nothing more than biological machines with no inner subjective awareness whatsoever. And yet, as I have suggested from the start, there are resources in the anthropocentric tradition for making progress toward a genuinely non-anthropocentric way of seeing and valuing things beyond the human. Those resources draw at least as much on our affective constitution as on our linguistic capacities.

One such resource is Adam Smith's notion of the impartial spectator, which brings together the dimensions of thought and felt evaluation in a relatively seamless manner. Smith recognizes that human beings are endowed with "passive" feelings that reflect a selfish orientation, and he further recognizes the need for human beings to establish a sense of concern for others if we are to honor the ideal of a harmonious community. A robust sense of community requires the "moderation" of our passions, and this requires the operation of reason. "Reason, principle, conscience, the inhabitant of the breast, the man within, the great judge and arbiter of our conduct" is the faculty that permits each of us to see our character and conduct in the light of a more universal standpoint that honors an ideal of humanity.[53] That universal standpoint underwrites and reinforces our endeavor to "respect the sentiments and judgments of [one's] bretheren" in the face of our tendencies to privilege our own selfish interests.[54] The ideal of humanity that is at work here finds its proper expression in "the love of what is honourable and noble," where Smith invokes the wisdom of the Stoics in characterizing honor and nobility with exclusive reference to "the whole security and peace of human society."[55] The ideal of impartiality, in other words, extends no further than the human, and it is bound up with a commitment to the essential equality of all human individuals as well as to the proposition that our deepest desire is for "the love, the gratitude, the admiration of mankind."[56]

I believe that this appeal to the ideal of the impartial spectator is at work whenever a human being is struggling with the dilemma between serving their own individual interests and honoring the interests of the larger community. It is here that reason's essential role in the process of refining our moral sensibilities and judgments is most evident; Smith openly acknowledges the essential tension between selfish impulses and what I call "other-regarding" commitments. But unlike Hume, he does not treat reason as simply "the slave of the passions"; and unlike Kant, he does not treat our emotional constitution as fully subordinate to reason. Smith's account has the added benefit of avoiding anything like Hutcheson's appeal to an innate "moral sense"; as Smith recognizes, if there were such a sense, then it would be considerably easier than it actually is for each of us to acknowledge our own selfish tendencies.[57] Thus Smith's account leaves room for a process of maturation in our sentiments and judgments, anticipating Mill in suggesting that only "the most artificial and refined education . . . can correct the inequalities of our passive feelings."[58]

Smith's appeal to the impartial spectator reflects his awareness that in order to evaluate our own conduct and values in more than a local, biased manner we have to try to see ourselves from the standpoint of others. "Our sensibility to the feelings of others, so far from being inconsistent with the manhood of self-command, is the very principle upon which that manhood is founded." This sensibility constitutes "the most perfect virtue."[59] And while Smith recognizes that we can feel some sentiments toward nonhuman animals—like Kant after him, he seems untroubled by professions of gratitude toward nonhuman animals who have been used in service to humans—he asserts a clear demarcation between humans and nonhuman animals by excluding the latter (he calls them "wild beasts") from community with humans and hence from the sphere of justice.[60]

From the anthropocentric standpoint occupied by Smith, there is no need to take the sentiments of nonhuman animals into consideration when we make determinations about what constitutes proper versus improper treatment. Smith states that he understands the proclivity to feel gratitude toward a nonhuman animal who has performed service for us, but his remarks make it crystal clear that such feelings of gratitude are by no means incumbent upon us inasmuch as we owe nothing to nonhuman animals; this much is clear from Smith's categorical exclusion (not surprising, given his repeated invocations of the Stoics) of nonhuman animals from the sphere of justice. But what if we were to use our imagination and suppose, entirely counter to the closed thinking of anthropocentrism, that nonhuman animals *could* respond meaningfully to our treatment of them and *could* express their sentiments to us about that treatment? What sorts of feelings would they share with us? Would they endorse Julian Franklin's suggestion that, in general, we ought to take their use by human beings to be "virtually voluntary"?[61] Or would they express outrage and condemnation of the horrors we systematically visit upon them? A very typical anthropocentric response to questions of this kind is that nonhuman animals, being "wild beasts," essentially have no cognitive grasp of what is happening to them much of the time, and that in any case they are categorically incapable of experiencing states such as outrage. This sort of response owes a great debt to thinkers such as Seneca, who denies not only thought in nonhuman animals but the capacity for emotion as well. The shift to a non-anthropocentric standpoint for perceiving and valuing things in the world demands, as I have argued from the start, *a pathos of humility* that remains open—not necessarily to the proposition that nonhuman animals would definitely respond in a specific way (nor to the proposition that all would respond in the same way), but at least to the very real possibility (I consider it overwhelmingly likely) that they would not approve of our conduct, to say the least.

In a conception of *oikeiosis* denuded of its anthropocentric limitations, Smith's ideal of the impartial spectator would function at the third and most encompassing level, that at which rational agents contemplate their place in and relationship to a larger cosmic whole in which the basic unit of membership is not linguistic rationality but rather sentience. Concern for the sentiments of others is Smith's guiding principle in the endeavor to realize the ideal of harmonious community. Contemplated in the light of Heidegger's notion of worldhood (that of a global, implicit background of meaning disclosed to us first of all through our affective constitution) and Merleau-Ponty's acknowledgment that many nonhuman animals are active participants in meaning, the idea of being intimately related to others in virtue of sentiments that bind us together could pave the way toward an open acknowledgment that the affective constitutions of human and nonhuman animals have a great many commonalities—perhaps a great many more commonalities than differences.

I have stated a number of times that the moral status of a given sentient being should not be assessed on the basis of its supposed cognitive sophistication. In the early chapters of this book, I suggested that the tradition seized upon the criterion of cognitive sophistication not because it is actually relevant to considerations of moral status, but because that criterion held the greatest promise of vindicating human exceptionalism. Reliance on this criterion permitted major exponents of the tradition to deny *logos* in nonhuman animals; this in turn permitted them to deny many emotional states in nonhuman animals, thinkers such as Seneca denying emotional capacities in nonhuman animals altogether. Today it seems that most people are willing to acknowledge at least some "primitive" emotional reactions in nonhuman animals, fear being the least controversial of them. For many thinkers, the proposition that some nonhuman animals can feel putatively "higher" states such as pride, shame, remorse, or anger makes no sense whatsoever, given that no nonhuman animals appear to be capable of the kind of linguistically-informed self-awareness traditionally assumed to be essential for states such as pride or anger. At most, these thinkers believe, nonhuman animals have what Aristotle attributed to them: not capacities such as ingenuity (*synesis*), etc., but merely (presumably pale) "analogues" of such capacities. Ever since Aristotle offered this qualification, we have struggled with the problem of how to think (and feel) about the experiential capacities and lived experience of sentient beings who appear not to be participants in symbolic language. The default position has been to assume that nonhuman animals lack or have less of the relevant capacities than human beings have.

In Chapter 1, I noted Derrida's suggestion that Bentham "changes everything" when he bases membership in the moral community on sentience rather than on linguistic or rational capacity. I am not confident that any given figure in the philosophical tradition has really "changed everything" when it comes to our sensibilities about nonhuman animals; but if I were to choose a candidate for this assessment, it would be Darwin rather than Bentham. Darwin does more than any thinker in the tradition to express a real openness to the emotional lives of nonhuman animals, freely acknowledging capacities for reason, sympathy, love, and heroism in nonhuman animals and suggesting that even insects experience emotions.[62] Even though he ultimately attributes a superior status to human beings, calling "man . . . the very summit of the organic scale," he is entirely open to the notion that nonhuman animals possess "mental individuality."[63] Thus Darwin has no problem attributing "something very like a conscience" to dogs and presumably to other nonhuman animals possessing putatively "higher" cognitive capacities.[64]

Darwin's achievement is to dispense with the traditional assumption of a sharp divide between human beings and nonhuman animals, doing more than any other thinker to make a compelling case for the proposition that the differences between human beings and nonhuman animals are really matters of degree—even though Darwin ultimately retains the ancient commitment to human superiority. By acknowledging that *nonhuman animals are individuals*, Darwin makes a contribution to the effort to rethink the very idea of membership in the moral community. Why include some individuals but exclude others from the community? If, as Darwin argues, many nonhuman animals possess reason (including deliberative capacity), some can even make at least limited sense of human language, and a great many have emotional lives that overlap with our own, the answer seems to be that inclusion and exclusion are to be determined entirely on grounds of species membership. In other words, it is only by dint of anthropocentrism or speciesism, the arbitrary privileging of the human, that we rationalize the exclusion of nonhuman animals from the sphere of morality and justice.

In this connection, Cora Diamond's reflection on the predicament of the vulnerable girl is instructive: At the most elemental level, it is not considerations of cognitive sophistication that move us to consider a given individual a member of the moral community and thus a proper recipient of our concern; instead the motivating factor is a deeply felt sense of connection, and at this pre-apophantic (pre-predicative) level of engagement with the other there is not (yet) any consideration of the specific experiential capacities the other possesses. There is simply the sense of felt kinship that binds us together.

4. Toward a Well-Tempered Humanism

But the problem remains: What about the great many people who purport to have no intimation whatsoever of such a sense of felt kinship? I have offered considerations regarding the specifically historical character of reason, together with a dialectical picture of the way in which the interplay of reason and emotion can help to facilitate the process of emotional maturing. Stated bluntly, I believe that the state of our culture's emotional sophistication about sentient nonhuman life is as immature as our technological sophistication is advanced. As a culture, we have an exceedingly long way to go before (if ever) it becomes commonplace that nonhuman animals possess inherent worth and merit respect and perhaps even some kind of reverence.

What we face today is an antinomy between the proclamation of categorical human superiority and that of an essential parity among all sentient life. Each assertion outwardly appears justifiable through appeal to "rational" considerations. Kant's method for taking a stand on an antinomy is to appeal to a Transcendental Idea of Reason, as when he seeks to defend the idea of human freedom by appealing to "another causality" than the kind exhibited by efficient causes in nature.[65] Kant is clear that this appeal to the Transcendental Idea of freedom is not a proof; indeed, he is clear that proof pertains only to the realm of nature that has been constructed by the mind. Given the essential limits that I have sought to attribute to our rational capacity, no purely "rational" resolution of the conflict between anthropocentric and non-anthropocentric ideals of living seems possible. Instead our rational considerations must proceed in concert with our affective dispositions and responses, which as I argued in Chapter 4 afford us with our primary sense of orientation in the world.

Naturally our affective dispositions cannot be taken at face value as straightforward indications of what matters, and for two reasons. First, those dispositions are not all mutually compatible with one another; some are selfish or partial, while others are other-regarding. Rorty believes that this tension (which he characterizes as prevailing between our interest in autonomy and our interest in justice) is irreducible.[66] For my own part I cannot help but wonder whether the conception of autonomy that Rorty has in mind is one according to which the individual has no obligations to anyone or anything other than their own personal creative vision. I take that to be a conception of individuality that draws a bit too much inspiration from Nietzsche—as if respecting the needs and interests of others were a grievous infringement on personal dignity. It seems clear that there has always been (and almost certainly always will be) an essential tension between the selfish and other-regarding poles of our existence. The proper response is not to abandon the prospect of working toward a resolution of this tension, but rather to find a way to work through it productively.

This working through requires a process of maturation that affects both our reason and our affective constitution—as we become clearer about what ought to matter in the moral community (as opposed to accepting what is currently taken to matter), both our reasoning and our felt sense of obligation will evolve. Toward what end? The historical character of reason is such that we cannot, in Eleatic style, posit an ultimate "end" to history in advance. Instead we must work toward the progressive disclosure of that end as we struggle to overcome the influence of historical prejudice. As I will discuss in the concluding section of this chapter, this struggle must be directed at relations between human beings as well as at relations between human and nonhuman animals. The struggle takes the form of "critical finitude," which is a Millerian way of characterizing "the way autonomy emerges in time."[67]

In Chapter 3, I suggested that Ortega's and Miller's turn to historical reason constitutes a major advance over the Eleatic conception. But I also noted a certain lacuna in the historical approach: Notwithstanding Miller's identification of the body as "the primary functioning object," the historical approach pays inadequate attention to the role of our affective constitution in the uncovering, contestation, and revision of meaningful commitments. At the same time, there is a clear if implicit ideal of humanity at work in the thought of both Ortega and Miller, one that anticipates a potential unification of the two poles of commitment that Rorty considers irreconcilable. Conceived non-anthropocentrically, that ideal of humanity is one that takes seriously the question how a being who is not subject to our particular biases would evaluate the conduct and motivations of the human race. In this connection, Smith suggests that the truly virtuous individual imitates "the divine artist," i.e., the appeal to the standpoint of the impartial spectator is in essence an appeal to a standard of judgment that must forever remain a regulative ideal to humans due to our seemingly intractable selfishness.[68] As such, this standard is one that we must approach through our imagination as well as through our reason, and in this respect the process of "critical finitude" is one in which the dimension of judgment should be conceived on the model of Kantian aesthetic judgment rather than on the model of, say, scientific judgment. It is a process, in other words, in which the determining ground of judgment is subjective rather than objective; and yet for all that, the judgment is far from arbitrary.[69] As in aesthetic judgment, in the process of moral reflection we seek to find a harmony between concepts and imagination; on the view I am proposing, imagination here needs to be understood as a faculty of embodied (felt rather than thought) contemplation, although both concept and affect operate on the basis of global background commitments.

To this extent, we need to contemplate the standpoint of the divine artist both cognitively and affectively, and in both cases we will be appealing implicitly to some background ideal of living. In the early chapters of this book I have shown how an anthropocentrically-conceived divine being would evaluate our treatment of nonhuman animals: Such a being might well not sanction wanton cruelty but would have no scruple about human beings treating nonhuman animals in essentially the same exact ways we have been treating them for millennia. After all, that divine being assigned us the status of "lords of nature" in virtue of our rational-linguistic abilities, and assigned to nonhuman animals the status of living instrumentalities. But what if the "divine artist" we imagine is not as invested in the will to power as we are, and occupies a specifically non-anthropocentric standpoint? In that case, I cannot help but wonder whether the divine being in whom we seek a measure for the propriety of our conduct would respond very much in the way that I have intimated that nonhuman animals would respond were they capable—with shock, frustration, and bitter disappointment rather than pride.[70]

We cannot know in advance, or in the abstract, exactly how such a divine being would assess our conduct. All we can do is work toward a progressively clear sense of the ideals we should honor. But as I have argued from the start, this does not mean that the process of moral reflection should start from a Cartesian "zero point," nor indeed *can* it mean that. In principle, we are forever heirs to a tradition of thinking, valuing, and acting that is not of our devising; our task is not to reinvent meaning and value from scratch, but to take a stand on the meanings and values with which we find ourselves confronted. I have tried to show that the anthropocentric tradition is not one we should seek to reject in its entirety, but rather one in which we find vital resources for thinking past the anthropocentric limitations of that tradition. In that tradition, as I have tried to show, insights into the value of the individual (conceived non-anthropocentrically) can be distilled from out of a persistent historical treatment of human beings as the only "real" individuals.

In the *oikeiosis* doctrine, the Stoics proceed from an inclusive to an archly exclusive ideal of belonging. To move from an anthropocentric to a non-anthropocentric value orientation is to challenge the Stoics' characterization of the putatively "highest" and most encompassing circle of belonging. But that does not mean rejecting their insights altogether. The Stoics, like Epicurus, seize upon something very important when they place their central focus on the capacity of human beings to establish reciprocal rules and expectations for conduct. In modernity this insight into what Freud thought of as our capacity to "adjust [the] mutual relations of human beings" evolved into the ideal of democracy, which Rorty treats as an essentially contingent commitment (if you opt for hierarchy or cruelty, who's to say you're wrong?) but which thinkers such as Dewey and Miller see as a conspicuous achievement on the part of humanity.[71]

Dewey and Miller present complementary conceptions of the democratic ideal as an historical achievement that has emerged out of the process of critical finitude. The starting point for Miller is the recognition that

Kant left obscure the place of quite personal experience in his highly impersonal account of nature. . . . Idealism has been tolerably successful in relating the law to biography and to history. Its characteristic difficulty occurs in its omission of individuality and finitude, rather than in an exaggeration of the personal stream of consciousness.[72]

The process of critical finitude is one in which the lived experience of concrete individual human agents is the primary unit of measure. Our finitude, in Miller's language, is "constitutional" in the sense that it "is no abstraction. It is individual, not particular. It is actual, the here-and-now. It is the center of any orderly discourse, and this center is essential."[73] Not simply just another "particular" entity or feature that happens to show up in the field of experience, the individual is a special kind of being possessing freedom, reason, and the imperative to take a critical stand on the historical situation into which we have contingently been cast. "One can do anything one likes with knowledge of objects except one thing: one cannot by its means define the free man or the democratic society."[74] Miller's "actualism" is a philosophy of "the present active participle," i.e., concrete acts of individual historical agents are the motor that drives understanding, commitment, and history itself.

For Miller, this means that society, like philosophy itself, must be democratic. Miller embraces the ideal of the liberal state, arguing that

> the affirmation of liberalism is the unity of men in self-critical association. It is humanism, but self-disciplined humanism. It respects another's opinion where that opinion confesses its openness to dispute, its possible error, and its subordination to the kind of test available to all.[75]

This is a humanism that honors tolerance and mutual disagreement; this ideal demands "an education of the will and not of the intellect," inasmuch as it needs to "reconcile" the passions of different human beings.[76] Although Miller does not delve nearly as deeply into the role of affect or passion in human experience as he might have done, he does recognize that "politics is a science of the power of men over themselves and over others. Its meaning vanishes when human nature lacks self-assertion and the passionate egoism which is the spring of all control."[77] Miller's acknowledgment of egoism is his affirmation of the central role of the individual in society and history. His characterization of history proceeds from this affirmation as well as from the recognition that conflicts in society are "constitutional," which is to say that they are ineliminable in principle. The goal for a democratic society is to honor the ideal of equal and full participation of human agents in ongoing negotiations over law and right.

In a similar vein, Dewey identifies democracy as "an individualism of freedom, of responsibility, of initiative to and for the ethical ideal, not an individualism of lawlessness."[78] Dewey characterizes this notion of the individual in terms not simply of a given person's ability to do whatever they please, but in the more demanding terms of personality. "Personality is as universal as humanity. . . . It means that in every individual there lives an infinite and universal possibility; that of being a king and a priest." To be a king is to affirm that "every man is an absolute end in himself," while to be a priest is to recognize one's relationship and responsibility to the larger community.[79] Democracy, for Dewey as for Miller, is not simply another form of government. "Democracy, in a word, is a social, that is to say, an ethical conception. . . . Democracy is a form of government only because it is a form of moral and spiritual association" that, unlike forms of government such as aristocracy, truly honors the agency of the human individual.[80] Democracy is "a personal, individual way of life . . . controlled by a working faith in the possibilities of human nature" and the prospects for "every person to lead his own life free from coercion and imposition by others provided right conditions are supplied."[81]

Miller and Dewey offer us the outlines of an ethics of democracy informed by historical reason. They also shed some important light on the role of affect or passion in the formation and maintenance of our axiological commitments, Dewey emphasizing that quality is disclosed through feeling and Miller affirming the significance of the body as the primary functioning object and posing an ideal of "passionate egoism." Both recognize that passion does not simply disappear or lose its primacy once thought and language enter the scene. Rorty is right that there is no definitive "knock-down" argument that liberal democracy is the best form of government, but that is simply because our axiological commitments reach deeper than reason alone. This does not mean, however, that Posner is reasonable in claiming to possess "a moral intuition deeper than any reason" when he dismisses certain ideas as "weird" and "insane." Posner is at the opposite end of the spectrum from Diamond's rational Martian in asserting, in effect, that there simply is no arguing about matters such as nonhuman animal rights, and that is a very dangerous position to embrace. It is based on the assumption that there is no possibility of harmonizing reason and passion in the manner gestured toward by Miller. The ideal of human democratic community is one according to which none of us is entitled, in matters that affect others significantly, to take the blithe position that I have a special intuition and that there simply is no disputing or even discussing it. (Recall, in this connection, the misgivings I expressed in Chapter 3 about taking Kierkegaard's Abraham as a paragon of moral choice.) The democratic ideal is one that demands mutual recognition and the effort to find common ground. That ground, as I have argued throughout this book, is not an Eleatic given but is constantly evolving and subject to contestation.

Naturally "democracy" can be understood in different ways. The ideal sketched by Dewey and Miller requires ongoing engagement with others and ourselves—not only about ideas and propositions, but crucially about our affective responses. One obscure and challenging task that each individual faces in this connection is to *make a case* for the affective responses that we think others ought to share with us. A position such as Posner's on the question of nonhuman animal rights fails to satisfy this demand. In this respect, I believe that the conception of human choice that he presents is at odds with what a truly well-tempered humanism would entail.

I borrow the expression "well-tempered humanism" from Tzvetan Todorov, who employs it in an effort to rectify the internal contradictions of modern humanism. Todorov embraces the universalism of the humanist tradition, including its putative commitment to the equal inherent worth and empowerment of human individuals. He explores the prospects for living up to that ideal by diagnosing the persistent failures of our culture to live up to its own ideals. Todorov's central concern is the process of othering, whereby divisions are established within the human community (most particularly racial divisions) that serve to privilege certain groups and exclude or subordinate others. Thus Todorov addresses a problem within human relations that finds its parallel in relations between human and nonhuman animals—a betrayal, in the name of exercising dominance, of the very ideals that one purports to honor. As regards relations among human beings, Todorov concludes that "a well-tempered humanism could protect us from the misguided ways of the past and the present."[82] Those misguided ways have culminated in a betrayal of the ideal of justice, which Todorov characterizes as a concern for humanity as a whole.[83]

If we posit a well-tempered humanism as an ideal for humanity, then Todorov's reflections on matters such as racial injustice lead irretrievably to the conclusion that what our culture suffers from is an epic case of *ill*-tempered humanism—many of the right abstract ideas, but a failure to establish and sustain any truly passionate commitment to them. Where the tradition has sought to proceed with a pathos of confident ambition, the facts, as I have tried to show, tell a very different story about our guiding motivations. We tend to think of philosophy as originating in a love of knowledge (per Aristotle) or in doubt (per thinkers such as Descartes), but it is equally born in *fear*. The acknowledgment of this essential feature of our existence accomplishes two things: It militates against the anthropocentric claim that human beings are ultimately quite different from nonhuman animals, by admitting that fear is the great equalizer among sentient beings; and it places in a very revealing light the repeated proclamations of the tradition that its investigations into human and nonhuman animal nature are motivated by a pure, disinterested desire for knowledge.

In characterizing the prevailing sensibilities about humanism and human community as ill-tempered, Todorov offers a diagnosis of our culture as hypocritical and sick. Needless to say, many people who benefit from this hypocrisy have an interest in perpetuating it, and one of the most effective means for perpetuating it is to conceal or deny it—most of all, to oneself. Of course, not all individuals relate to problems such as racism in the same way, and not all members of a given subgroup of humanity (say, wealthy Caucasians) react in the same way. Some possess the humility to acknowledge the injustice, even though in doing so they contribute to efforts that could eventually lead to their own loss of privilege. Others take the "none so blind as those who will not see" approach, denying or minimizing the problem and typically denying that they themselves are participants in the injustice. The ideal of a well-tempered humanism is the regulative ideal of a culture that has reflected on its prior (avowed) commitments to the point that its members' guiding affects (not just the abstract values it purports to honor) are congenial to rather than hostile to "others."

A well-tempered humanism, one that brings the insights of thought into fruitful resonance with our emotional constitution, is a regulative ideal. It is not a once-and-for-all achievement, but rather must forever remain an aspirational goal that we pursue as we work through the successive "constitutional conflicts" (Miller) that are inherent in the human condition. Where thinkers such as Descartes and Kant proceed on the confident assumption that we can neutralize doubt (and, implicitly, fear) by employing linguistic rationality to exercise systematic control, thinkers such as Miller show states such as doubt and fear to be endemic to the human condition. The most fruitful response, it seems to me, is not to deny or brush past doubt and fear, nor is it to seek to extinguish both through an approach to control that amounts to massive overkill. (Indications of overkill, many of them naturally bound up with the profit motive, come to mind: the invention and use of pernicious chemicals such as DDT, the ever-increasing threat of nuclear annihilation, the overturning of Reconstruction in the U.S. south after the Civil War and all the Strange Fruit that came with it, the massive and ever-expanding nonhuman animal agriculture industry, etc.)

The body is the primary functioning object, which is to say that each individual body is the primary locus of agency. This is true notwithstanding Heidegger's and Merleau-Ponty's characterization of meaning as a global phenomenon rather than one that occurs Cartesian-style in isolated egos. Each individual becomes initiated into meaning by being born into a cultural context already fully invested with significance. Each individual must then endeavor to take a stand on that inherited meaning, using only their natural endowments and the context of meaning with which they find themselves confronted. (Even Heidegger, one of the progenitors of the now relatively ubiquitous idea that "the individual is socially constructed," argues from the start that Dasein "is in each case mine" and faces the task of becoming an authentic *individual*.[84]) What thinkers such as Dewey, Miller, and Todorov recognize is something that thinkers such as Heidegger patently fail to accept: that human equality is an ideal that is best honored by passionately promoting the liberal ideal of an open society populated with equally empowered individual agents engaged in ongoing rationally-informed contestations over meaning and value as well as in ongoing acts of self-determination.[85]

To honor an even deeper commitment to justice, we need to inscribe this ideal of liberal society within the larger cosmic framework to which Rawls gestured but that he never took the trouble to explore. This is what Löwith had in mind when he proposed an inscription of the human world within the larger framework of the cosmos. This is a step that is either buried deep in our cultural memory, or it is one that we have never undertaken. (Again: was there ever really a "golden age" of the kind posited by Hesiod and Ovid?) In either case, it is a step that is largely if completely unfamiliar to the vast majority of the human race—even, I fear, to a great many pet owners and ethological researchers who proclaim their abiding concern for and devotion to nonhuman animals. Indeed, even those of us who, through our actions, confirm the clearest openness to the alterity of the nonhuman have considerable difficulty envisioning the kind of moral community that includes nonhuman animals as full and direct beneficiaries of moral concern.

And yet that, as I see it, is the task that lies before us. The endeavor to move from an anthropocentric to a non-anthropocentric ideal of living demands an essential *openness and humility* in the face of the world that we have for too long taken to be what Heidegger called "a gigantic gasoline filling station." Efforts to acknowledge and accommodate the interests of nonhuman animals from within the anthropocentric ethos have not succeeded but have simply given rise to repeated self-serving rationalizations, virtually all of them based on the confident assumption that nonhuman animals are not only different from human beings but precisely *less than*. I believe that this verdict holds even for Donaldson and Kymlicka's wildly popular effort in *Zoopolis* to use citizenship theory to classify and sketch the parameters of human obligation toward (and prerogatives on the part of) various groupings (domestic, liminal, wild) of nonhuman animals: If only against their own best intention, Donaldson and Kymlicka operate within an essentially anthropocentric worldview, one according to which human beings are entitled to decide the fates (or, more precisely, the political status) of nonhuman animals. After all, what if some or perhaps many nonhuman animals have no desire to become subsumed under human administration, which is essentially what the *Zoopolis* doctrine calls for?[86] There is an imposition of force in such an endeavor—if not outright physical force, then at the very least conceptual and moral force.

What Grimm and Taureck call on us to do is to suspend judgment at least provisionally regarding the experiential capacities and moral status of nonhuman animals. Donaldson and Kymlicka assume a little too quickly that the robust pursuit of abolitionist animal rights (which would require the cessation of all uses of nonhuman animals to benefit human beings) would lead to complete apartheid between human and nonhuman animals.[87] Grimm and Taureck offer us an alternative way of proceeding: to acknowledge at the outset that we are embarking upon an open sea, and that we cannot reasonably claim to know exactly what things would look like if we were truly to *let nonhuman animals be*. It was in this spirit that I chose the following epigram for *Anthropocentrism and Its Discontents*, one that must strike anthropocentric common sense as absurd: "For the life of a man is of no greater importance to the universe than that of an oyster."[88] Unless and until we confess our own pride and ignorance, it will not be possible to entertain such a suggestion even as a programmatic prelude to a more fine-grained consideration of the actual lives and struggles of nonhuman animals—and only in the course of such a deeper consideration will we find ourselves in a position to take an open and honest stand on questions such as that of pet ownership, which strike so many people as benign and eminently defensible in spite of the fact that it involves the imposition of extensive control over nonhuman animals.

* * *

The ideal of a well-tempered humanism, denuded of its anthropocentric prejudice, holds the promise of expanding the scope of moral concern so as to embrace and honor *all* individuals, not just those relative few endowed with human linguistic rationality. In antiquity, the Stoics expressed an essential insight into our caring nature when they seized upon maternal instinct as the bridge between self-regarding and other-regarding tendencies. But they betrayed this insight when they based inclusion in the moral community on rational rather than affective capacity. In modernity, Kant devoted a great deal of attention to the significance of moral love in human life. But he staunchly and implausibly insisted that that love is simply an adjunct to a sense of moral commitment, which he considered to be arrived at through purely rational means. The only way we will ever get into the right moral relationship to ourselves, nonhuman animals, and perhaps the rest of nature as well is to reflect in an open and honest way about our own fear and arrogance, our own limitations, and the troubling extent to which *we have failed* to live up to an ideal of moral purity to which we have purported to be committed for millennia. Only then can we begin to contemplate what true justice, and not simply compassion, for nonhuman animals might actually look and feel like. We cannot know with any certainty where this open sea of exploration will lead us. Perhaps we will find ourselves surprised to learn that an 'is'—the reality of suffering in sentient life, much of it caused by us—*can* lead to an 'ought': the imperative to render loving care and aid to vulnerable others who most need it, regardless of their species membership.

Notes

1 Richard A. Watson, "Self-Consciousness and the Rights of Nonhuman Animals and Nature," p. 9.
2 Paul W. Taylor, "The Ethics of Respect for Nature," pp. 105–6.
3 *Dialectic of Enlightenment*, pp. 163, 156. Horkheimer and Adorno's central concern is the ideology of capitalism; much in their analysis holds, *mutatis mutandis*, for the problem of ideological distortion evident in the anthropocentric background ideal of living.
4 Richard A. Posner, "Animal Rights: Legal, Philosophical, and Pragmatic Perspectives," p. 65.
5 Jacques Derrida, "But as for Me, Who Am I (following)?," p. 94. See also my remarks in *Animals and the Limits of Postmodernism*, 112–25.
6 One of the many recent thinkers to place stress on this fact is Dewey. See *Experience and Nature*, Ch. 8, p. 235.
7 See my remarks in *Anthropocentrism and Its Discontents*, pp. 62, 119.
8 *Experience and Nature*, p. 179.
9 Dewey, on the other hand, resists this equation of human and nonhuman animals as participants in meaning and interpretation, consigning the latter to the black box of instinct. See *Experience and Nature*, pp. 138–40.
10 Simon Blackburn, *Ruling Passions*, p. 131. See also p. 133: Deliberation is an "accounting" of affects.
11 *Ruling Passions*, pp. 306, 304.
12 More precisely, Schopenhauer believes that the kind of representations that conform to the Principle of Sufficient Reason (Kant's forms of space, time, and causality) are directed at manipulation, whereas the representations that are produced through will or embodiment pertain to phenomena (such as sheer force) that cannot ultimately be accounted for through appeal to the Principle of Sufficient Reason. See *The World as Will and Representation*, vol. 1, sec. 7 and 26.
13 Susan Bordo, "The Cartesian Masculinization of Thought," p. 247. See also p. 251: Cartesian inwardness reflects "deep epistemological alienation." Whether Bordo is right to see in Descartes's approach a retreat "from the organic female universe of the Middle

Ages and the Renaissance" (pp. 248–9) is considerably more controversial and raises many questions. Have human beings ever actually inhabited such a universe? (Is this the way medieval and Renaissance Europeans experienced their relationship with nature?) Does the nature of human consciousness permit this kind of total integration in nature? Does it remain fruitful today to think about reality in strict gender binaries such as masculine and feminine?

14 In *Descartes as a Moral Thinker*, I argue that Descartes appeals to an extra-rational source of meaning only implicitly and against his own intention.

15 Descartes devoted his last book, *The Passions of the Soul*, to an analysis and inventory of the passions; but he proceeds entirely on the basis of the subject-object model of representation, thereby failing to appreciate (or perhaps actively evading) the prospects for a radically different mode of access to the real. On Descartes's ideas about generosity and *caritas*, see my remarks in *Descartes as a Moral Thinker*, pp. 164–77.

16 *Oration on the Dignity of Man* (1486), p. 112.

17 There is a large and rapidly growing body of literature on this question. See for example Margo DeMello, ed., *Mourning Animals: Rituals and Practices Surrounding Animal Death*.

18 On Descartes's notion of *générosité* and Aquinas's notion of *magnanimatas*, see my remarks in *Descartes as a Moral Thinker*, pp. 186–97.

19 Gary Francione rejects what he calls "similar minds theory," the view that the moral status of a given nonhuman animal depends on how similar its mental capabilities are to those of typical human beings. See Gary L. Francione, *Animals as Persons*, pp. 124–5.

20 Colin Allen and Marc D. Hauser, "Concept Attribution in Nonhuman Animals."

21 As I note in *Anthropocentrism and Its Discontents* (see pp. 72–6), Aristotle takes a comparable approach in the zoological texts when he attributes states such as ingenuity (*synesis*) to nonhuman animals as a way of accounting for their sophisticated adaptive powers, only to qualify these attributions by suggesting that nonhuman animals simply behave "as if" they possessed these sophisticated capacities.

22 *On Ends*, book 3, sec. 5, pp. 233–5.

23 It is with implicit reliance on this sense of self-belonging that Bentham observes that of all the rules that society needs to establish in order to do justice to utilitarian ideals, the one kind of rule that is *not* needed is one that would require prudence on the part of individuals. Bentham's reasoning is that we are endowed by nature with self-interest, such that rules requiring prudence would be entirely superfluous. See *An Introduction to the Principles of Morals and Legislation*, Ch. 17, sec. 1.

24 *On Ends*, book 3, sec. 19, pp. 283–5. Cf. Charles Darwin, *The Expression of the Emotions in Man and Animals*, p. 82: "No emotion is stronger than maternal love."

25 *On Ends*, book 3, sec. 19, p. 285.

26 Hierocles (Stobaeus 4.671.7–7.673.11), A.A. Long and D.N. Sedley, *The Hellenistic Philosophers*, 57G. This is the ideal of community that Diogenes of Sinope had when he declared himself "*kosmopolites*"—a citizen of the world. See *Lives of the Eminent Philosophers*, vol. 2, pp. 64–5.

27 *On Ends*, book 3, sec. 20, p. 287.

28 It is worth noting that the discipline of ethology itself is a double-edged sword: It serves the desire to know that Aristotle identified as universal to human beings, but it does so at the expense of a lot of confinement and control over the lives of nonhuman animals. In light of this latter consideration and the economics of ethological research, it does not strike me as far-fetched to characterize the field of animal behavior research in its current form as the Ethological-Industrial Complex.

29 Hans Blumenberg, *Work on Myth*, p. 266.

30 *Work on Myth*, p. 377; on the notion of the "zero point," see also Hans Blumenberg, *The Legitimacy of the Modern Age*, p. 379.

31 John Dewey, "Qualitative Thought," pp. 248–9.

32 *Experience and Nature*, p. 116; see also John Dewey, "The Ethics of Democracy," p. 238.

33 *Experience and Nature*, p. 259.

34 *Experience and Nature*, pp. 217, 215, 198, 197.

35 "Qualitative Thought," p. 260n4.

36 Arne Johan Vetlesen, *Perception, Empathy, and Judgment*, p. 166.

37 *Perception, Empathy, and Judgment*, pp. 158–9.

38 *Perception, Empathy, and Judgment*, pp. 162, 172–3.

39 *Perception, Empathy, and Judgment*, pp. 169, 189.

40 Charles Taylor, "The Concept of a Person," p. 103; Charles Taylor, "Self-Interpreting Animals," 58 (where Taylor states that nonhuman animals, "idiots," and infants are excluded from the sphere of moral subjects).

41 *Perception, Empathy, and Judgment*, p. 169.

42 "Self-Interpreting Animals," pp. 51, 62, 73.

43 "The Concept of a Person," p. 110; "Self-Interpreting Animals," pp. 72–3.

44 "Animal Rights: Legal, Philosophical, and Pragmatic Perspectives," p. 65.

45 I present my ideas about cosmic justice and felt kinship in greater depth in Chapters 5 and 6 of *Animals and the Moral Community*.

46 See *Welt und Menschenwelt*, p. 303: "The shared life of human beings in a polis cannot be in order if that life is not established in accordance with the cosmos. As one whose life is humanly ordered, one who is magnanimous is in tune with the cosmos. In this respect, Greek philosophers and the Eastern way think very similarly, i.e., cosmo-politically in the literal sense."

47 *Animals and the Moral Community*, p. 138.

48 *Animals and the Moral Community*, p. 143.

49 *On the Basis of Morality*, sec. 17, p. 152.

50 *On the Basis of Morality*, sec. 17, p. 152.

51 See *Contingency, Irony, and Solidarity*, p. 177. Here Rorty does not explicitly commit himself to the view that we need do no more than observe injustices perpetrated on non-human animals; I believe it is implicit in his effort here to exclude nonhuman animals from moral community with humans, which he seeks to accomplish by asserting a distinction in kind between pain in humans and nonhuman animals. Rorty leaves room for compassion for nonhuman animals, but he seems categorically uninterested in the notion of justice for them.

52 *On the Basis of Morality*, sec. 19, pp. 175, 180.

53 *The Theory of Moral Sentiments*, Part 3, Ch. 3, sec. 4, p. 158; Part 7, sec. 2, Ch. 3, art. 21, p. 361.

54 *The Theory of Moral Sentiments*, Part 3, Ch. 2, sec. 31, p. 149.

55 *The Theory of Moral Sentiments*, Part 3, Ch. 3, sec. 6, p. 159.

56 *The Theory of Moral Sentiments*, Part 2, Ch. 2, sec. 1, pp. 97–8; Part 3, Ch. 4, sec. 7, p. 185.

57 *The Theory of Moral Sentiments*, Part 7, sec. 3, Ch. 2.

58 *The Theory of Moral Sentiments*, Part 3, Ch. 3, sec. 7, p. 160. In John Stuart Mill, *The Subjection of Women*, Mill argues for equal educational opportunity as vital to the endeavor to redress gender inequalities in society.

59 *The Theory of Moral Sentiments*, Part 3, Ch. 3, sec. 34 and 35, p. 176.

60 *The Theory of Moral Sentiments*, Part 2, sec. 3, Ch. 1, art. 3, pp. 111–2 (nonhuman animals are "less improper objects of gratitude and resentment than inanimated objects" but "are still far from being complete and perfect objects, either of gratitude or resentment"); Part 2, sec. 2, Ch. 2, art. 4, pp. 101–2.

61 Julian Franklin, *Animal Rights and Moral Philosophy*, pp. 62–3.

62 Charles Darwin, *The Descent of Man*, Ch. 3–4, passim; *The Expression of the Emotions in Man and Animals*, p. 347 (emotion in insects).

63   *The Descent of Man*, Ch. 21, p. 920; cf. Ch. 3, p. 460, where Darwin states that there is "no great improbability" that "higher animals" possess "self-consciousness."

64   *The Descent of Man*, Ch. 3, p. 476. See also my remarks on Darwin in *Anthropocentrism and Its Discontents*, pp. 190–7.

65   *Critique of Pure Reason*, p. 409 (at A444/B472).

66   *Contingency, Irony, and Solidarity*, p. xv.

67   Gary Stahl, "Making the Moral World," p. 120.

68   *The Theory of Moral Sentiments*, Part 6, sec. 3, p. 292.

69   *Critique of the Power of Judgment*, "Introduction," sec. 4, p. 67; sec. 8, p. 99; *The Philosophy of History with Reflections and Aphorisms*, p. 145. See also "Making the Moral World," p. 120.

70   One might even imagine such a divine being responding with anger, notwithstanding Seneca's assurance that anger is utterly irrational—Seneca's implication being that no self-respecting god would ever be caught in a state such as anger. For an alternative perspective on the potential *legitimacy* of anger, see Amia Srinivasan, "The Aptness of Anger."

71   Sigmund Freud, *Civilization and Its Discontents*, p. 63 (translation altered).

72   John William Miller, "Idealism and Freedom," p. 253. Recall again Diamond's "rational Martians."

73   Miller, "The Owl," p. 224.

74   *The Paradox of Cause and Other Essays*, pp. 100–1. See also Miller, *In Defense of the Psychological*, p. 154.

75   John William Miller, "Tolerance and Its Paradoxes," p. 285.

76   "Tolerance and Its Paradoxes," pp. 283–4.

77   *The Paradox of Cause and Other Essays*, p. 101.

78   "The Ethics of Democracy," p. 244.

79   "The Ethics of Democracy," pp. 245–6. Here Dewey implicitly endorses Cicero's notion of two "personae" unique to human beings examined in Chapter 1.

80   "The Ethics of Democracy," pp. 238, 240.

81   John Dewey, "Creative Democracy—The Task Before Us," pp. 226–7. Cf. "The Ethics of Democracy," p. 246: One such condition is "a democracy of wealth," although Dewey acknowledges that we cannot know at present what a democracy of wealth would look like.

82   Tzvetan Todorov, *Nous et les autres*, p. 523.

83   *Nous et les autres*, p. 514.

84   *Being and Time*, Division One, Ch. 1, par. 9, p. 67; Division Two, Ch. 2, par. 53, p. 310.

85   On Heidegger's betrayal of the liberal ideal, see Gary Steiner, "The Perils of a Total Critique of Reason."

86   I develop this critique at length in Gary Steiner, "Tiere als Personen, aber nicht als Staatsbürger" [Animals as Persons, but Not as Citizens].

87   *Zoopolis*, pp. 7, 49, 62, 178, 180.

88   David Hume, "Of Suicide," p. 583.

BIBLIOGRAPHY

Ackerman, Felicia Nimue. "Fried Chicken Heaven." Letter to the Editor. *New York Times*. May 26, 2019. https://www.nytimes.com/2019/05.26/opinion/letters/vegetarianism.html.

Allen, Colin and Marc D. Hauser. "Concept Attribution in Nonhuman Animals: Theoretical and Methodological Problems in Ascribing Complex Mental Processes." *Philosophy of Science* vol. 58 (1991): 221–40.

Aquinas, Saint Thomas. *Basic Writings of Saint Thomas Aquinas*, vol. 2. Ed. Anton C. Pegis. Indianapolis: Hackett, 1997.

———. "Summa Contra Gentiles." *Basic Writings of Saint Thomas Aquinas*, vol. 2, pp. 3–224.

Aristotle. *History of Animals, Books VII–X*. Trans. D.M. Balme. Cambridge/London: Harvard University Press, 1991.

———. *Complete Works of Aristotle*, vol. 2. Ed. Jonathan Barnes. Princeton: Princeton University Press, 1995.

———. "Eudemian Ethics." *Complete Works of Aristotle*, vol. 2, pp. 1922–81.

———. "Metaphysics." *Complete Works of Aristotle*, vol. 2, pp. 1552–1728.

———. "Nicomachean Ethics." *Complete Works of Aristotle*, vol. 2, pp. 1729–1867.

———. "On the Soul." *Complete Works of Aristotle*, vol. 1, pp. 641–92.

———. "Physics." *Complete Works of Aristotle*, vol. 1, pp. 315–446.

———. "Politics." *Complete Works of Aristotle*, vol. 2, pp. 1986–2129.

Augustine, St. *Confessions*. Trans. R-W. Pine-Coffin. London: Penguin, 1961.

———. *The Catholic and Manichaean Ways of Life*. Trans. Donald A. Gallagher and Idella J. Gallagher. Fathers of the Church, vol. 56. Washington, DC: Catholic University of America Press, 1966.

Balcombe, Jonathan. *What a Fish Knows: The Inner Lives of Our Underwater Cousins*. New York: Scientific American/Farrar, Straus, and Giroux, 2017.

Bekoff, Marc and Jessica Pierce. *Wild Justice: The Moral Lives of Animals*. Chicago: University of Chicago, 2010.

Bentham, Jeremy. *An Introduction to the Principles of Morals and Legislation*. New York: Hafner/Macmillan, 1948.

Berkeley, George. "The Principles of Human Knowledge." *Philosophical Writings*. Ed. Desmond M. Clark. Cambridge: Cambridge University Press, 2008, pp. 67–149.

Biemel, Walter. *Philosophische Analysen zur Kunst der Gegenwart.* The Hague: Martinus Nijhoff, 1968.

Blackburn, Simon. *Ruling Passions: A Theory of Practical Reasoning.* Oxford: Clarendon Press, 1998.

Blumenberg, Hans. *The Legitimacy of the Modern Age.* Trans. Robert M. Wallace. Cambridge: M.I.T. Press, 1983.

———. *Work on Myth.* Trans. Robert M. Wallace. Cambridge: M.I.T. Press, 1985.

Boethius. *The Consolation of Philosophy.* Trans. David R. Slavitt. Cambridge: Harvard University Press, 2008.

Bordo, Susan. "The Cartesian Masculinization of Thought." *Sex and Scientific Inquiry.* Ed. Susan Gubar and Jean F. O'Barr. Chicago: University of Chicago, 1987, pp. 247–64.

Candland, Douglas Keith. *Feral Children and Clever Animals: Reflections on Human Nature.* New York: Oxford University Press, 1993.

Carruthers, Peter. *The Animals Issue: Moral Theory in Practice.* Cambridge: Cambridge University Press, 1992.

Cicero. *On Duties/De officiis.* Trans. Walter Miller. Cambridge: Harvard University Press, 2001.

———. *On Ends/De finibus bonorum et malorum.* Trans. H. Rackham. Cambridge, MA: Harvard University Press, 1999.

Coetzee, J.M. *The Lives of Animals.* Princeton: Princeton University Press, 2001.

Conrad, Joseph. *Lord Jim.* London: Penguin, 2007.

Curley, Edwin M. *Descartes Against the Skeptics.* Cambridge: Harvard University Press, 1978.

Darwin, Charles. "The Descent of Man." *The Origin of Species by Means of Natural Selection or the Preservation of Favored Races in the Struggle for Life and The Descent of Man and Selection in Relation to Sex.* New York: Modern Library, n.d.

———. *The Expression of the Emotions in Man and Animals,* Definitive ed. New York: Oxford University Press, 1998.

Debord, Guy. *Society of the Spectacle.* Detroit: Black and Red, 1983.

DeMello, Margo, ed. *Mourning Animals: Rituals and Practices Surrounding Animal Death.* East Lansing: Michigan State University Press, 2016.

Derrida, Jacques. *The Animal That Therefore I Am.* Ed. Marie-Louise Mallet. New York: Fordham University Press, 2008.

———. "But as for Me, Who Am I (following)?" *The Animal That Therefore I Am,* p. 52–118.

———. "Hospitality, Justice and Responsibility: A Dialogue with Jacques Derrida." *Questioning Ethics: Contemporary Debates in Philosophy.* Ed. Richard Kearney and Mark Dooley. London/New York: Routledge, 1999, pp. 65–83.

———. *La bête et le souverain 2002–2003,* vol. 2. Paris: Éditions Galilée, 2010.

———. *Of Grammatology.* Trans. Gayatri Chakravorty Spivak. Baltimore/London: The Johns Hopkins University Press, 1976.

Descartes, René. "Author's Replies to the Second Set of Objections." *The Philosophical Writings of Descartes,* vol. 2, pp. 93–120.

———. "Dedicatory Letter to the Sorbonne." *The Philosophical Writings of Descartes,* vol. 2, pp. 3–6.

———. "Discourse on Method." *The Philosophical Writings of Descartes,* vol. 1, pp. 111–51.

———. "The Passions of the Soul." *The Philosophical Writings of Descartes,* vol. 1, pp. 325–404.

———. *The Philosophical Writings of Descartes,* vol. 1. Trans. John Cottingham et al. Cambridge: Cambridge University Press, 1985.

————. *The Philosophical Writings of Descartes*, vol. 2. Trans. John Cottingham et al. Cambridge: Cambridge University Press, 1984.

————. *The Principles of Philosophy. The Philosophical Writings of Descartes*, vol. 1, pp. 190–292.

Dewey, John. "Creative Democracy—The Task Before Us." *The Later Works, 1925–1953. Vol. 14: 1939–41: Essays, Reviews, and Miscellany*. Ed. Jo Ann Boydston. Carmbondale: Southern Illinois University Press, 1988, pp. 224–34.

————. "The Ethics of Democracy." *The Early Works, 1882–1898. Vol. 1: 1882–1888: Early Essays and Leibniz's New Essays*. Ed. Jo Ann Boydston. Carbondale: Southern Illinois University Press, 2008, pp. 227–49.

————. *Experience and Nature*. Carbondale: Southern Illinois University Press, 2008.

————. "Moral Theory and Practice." *International Journal of Ethics* vol. 1, no. 2 (January 1891): 186–203.

————. "Qualitative Thought." *Later Works, 1925–53. Vol. 5: 1929–30: Essays, The Sources of a Science of Education, Individualism, Old and New, and Construction and Criticism*. Ed. Jo Ann Boydston. Carbondale: Southern Illinois University Press, 2008, pp. 243–62.

Diamond, Cora. "The Difficulty of Reality and the Difficulty of Philosophy." *Philosophy and Animal Life*. Ed. Stanley Cavell et al. New York: Columbia University Press, 2008, pp. 43–89.

————. "Eating Meat and Eating People." *Animal Rights: Current Debates and New Directions*. Ed. Cass R. Sunstein and Martha C. Nussbaum. Oxford: Oxford University Press, 2004, pp. 93–107.

————. "The Importance of Being Human." *Royal Institute of Philosophy Supplement* vol. 29 (1991): 35–62.

————. "Injustice and Animals." *Slow Cures and Bad Philosophers: Essays on Wittgenstein, Medicine, and Bioethics*. Ed. Carl Elliott. Durham: Duke University Press, 2001, pp. 118–48.

Dierauer, Urs. *Tier und Mensch im Denken der Antike. Studien zur Tierpsychologie, Anthropologie und Ethik*. Amsterdam: Grüner, 1977.

Diogenes Laertius. *Lives of the Eminent Philosophers*, vol. 2. Trans. R.D. Hicks. Cambridge, MA: Harvard University Press, 2000.

Donaldson, Sue and Will Kymlicka. *Zoopolis: A Political Theory of Animal Rights*. Oxford: Oxford University Press, 2011.

Emerson, Ralph Waldo. *The Selected Writings of Ralph Waldo Emerson*. Ed. Brooks Atkinson. New York: Modern Library, 1992.

————. "Self-Reliance." *The Essential Writings of Ralph Waldo Emerson*, Ed. Brooks Atkinson. New York: Modern Library, 2000, pp. 132–53.

Epictetus. *The Discourses as Reported by Arrian Books I-II*. Trans. W.A. Oldfather. Cambridge, MA: Harvard University Press, 2000.

Eze, Emmanuel Chukwudi. "The Color of Reason: The Idea of 'Race' in Kant's Anthropology." *Postcolonial African Philosophy: A Critical Reader*. Ed. Emmanuel Chukwudi Eze. Cambridge, MA: Blackwell, 1997, pp. 103–40.

Fell III, Joseph P. *Emotion in the Thought of Sartre*. New York: Columbia University Press, 1965.

Fontenay, Elisabeth de. "Pourquoi les animaux n'auraient-ils droit à un droit des animaux?" *Le Debat* vol. 109 (2000): 138–55.

Forst, Rainer. *Das Recht auf Rechtfertigung. Elemente einer konstruktivistischen Theorie der Gerechtigkeit*. Frankfurt: Suhrkamp, 2007.

Francione, Gary L. *Animals as Persons: Essays on the Abolition of Animal Exploitation.* New York: Columbia University Press, 2008.

———. *Animals, Property, and the Law.* Philadelphia: Temple University Press, 1995.

Frankfurt, Harry R. *Demons, Dreamers, and Madmen: The Defense of Reason in Descartes's Meditations.* Indianapolis: Bobbs-Merrill, 1970.

Franklin, Julian. *Animal Rights and Moral Philosophy.* New York: Columbia University Press, 2005.

Freud, Sigmund. *Civilization and Its Discontents.* Trans. James Strachey. New York: Norton, 2010.

Friedländer, Paul. *Plato: An Introduction.* Princeton: Bollingen/Princeton, 1973.

Grimm, Herwig. "Das *Tier an sich*? Auf der Suche nach dem Menschen in der Tierethik." *Tiere: Der Mensch und seine Natur.* Ed. Konrad Paul Liessmann. Vienna: Paul Zsolnay Verlag, 2013, pp. 33–70.

Gueroult, Martial. *Descartes's Philosophy Interpreted According to the Order of Reasons,* vol. 2. Ed. Roger Ariew. Minneapolis: University of Minnesota Press, 1985.

Hargrove, Eugene C., ed. *The Animal Rights/Environmental Ethics Debate: The Environmental Perspective.* Albany: SUNY Press, 1992.

Hegel, G.W.F. *Introduction to the Philosophy of History.* Trans. Leo Rausch. Indianapolis: Hackett, 1988.

———. *The Philosophy of Right.* Trans. T.M. Knox. London: Oxford University Press, 1967.

———. *The Science of Logic.* Trans. George Di Giovanni. Cambridge: Cambridge University Press, 2010.

Heidegger, Martin. "The Age of the World Picture." *The Question Concerning Technology and Other Essays,* pp. 115–54.

———. *The Basic Problems of Phenomenology.* Trans. Albert Hofstadter. Bloomington: Indiana University Press, 1982.

———. *Being and Time.* Trans. John Macquarrie and Edward Robinson. New York: Harper and Row, 1962.

———. *The Fundamental Concepts of Metaphysics: World, Finitude, Solitude.* Trans. William McNeill and Nicholas Walker. Bloomington: Indiana University Press, 1995.

———. "Letter on 'Humanism'." *Pathmarks,* pp. 239–76.

———. *Logic: The Question of Truth.* Trans. Thomas Sheehan. Bloomington: Indiana University Press, 2010.

———. *Nietzsche. Vol. 4: Nihilism.* Trans. Frank A. Capuzzi. San Francisco: Harper and Row, 1982.

———. *1. Nietzsches Metaphysik. 2. Einleitung in die Philosophie. Denken und Dichten. Gesamtausgabe,* vol. 50. Frankfurt: Klostermann, 1990.

———. "On the Essence of Truth." *Pathmarks,* pp. 136–54.

———. *Pathmarks.* Ed. William McNeill. Cambridge: Cambridge University Press, 1998.

———. "Plato's Doctrine of Truth." *Pathmarks,* pp. 155–82.

———. *Poetry, Language, Thought.* Trans. Albert Hofstadter. New York: Perennial Classics, 2001.

———. "Postscript to 'What Is Metaphysics'?" *Pathmarks,* pp. 231–38.

———. *The Question Concerning Technology and Other Essays.* Trans. William Lovitt. New York: Harper Torchbooks, 1997.

———. "The Question Concerning Technology." *The Question Concerning Technology and Other Essays,* pp. 3–35.

————. "The Rectorate 1933/34: Facts and Thoughts." *Martin Heidegger and National Socialism*. Ed. Gunther Neske and Emil Kettering. New York: Paragon House, 1990, pp. 15–32.

————. "The Thing." *Poetry, Language, Thought*, pp. 161–184.

————. *Vorträge. Teil 2: 1935 bis 1967*, Gesamtausgabe, vol. 80.2. Frankfurt: Klostermann, 2020.

————. "What Are Poets for?" *Poetry, Language, Thought*, pp. 87–140.

————. "What Is Metaphysics?" *Pathmarks*, pp. 82–96.

Helm, Bennett W. "The Significance of Emotions." *American Philosophical Quarterly* vol. 31, no. 4 (1994): 319–31.

————. *Emotional Reason: Deliberation, Motivation, and the Nature of Value*. Cambridge: Cambridge University Press, 2001.

Herder, Johann Gottfried von. "Treatise on the Origin of Language." *Philosophical Writings*. Trans. and Ed. Michael N. Forster. Cambridge: Cambridge University Press, 2002, pp. 65–166.

Hobbes, Thomas. *Leviathan*. Ed. Richard Tuck. Cambridge: Cambridge University Press, 2013.

Horkheimer, Max and Theodor W. Adorno. *Dialectic of Enlightenment: Philosophical Fragments*. Ed. Gunzelin Schmid Noerr. Stanford: Stanford University Press, 2002.

Hume, David. "Of Suicide." *Essays Moral, Political, and Literary*, Revised ed. Ed. Eugene F. Miller. Indianapolis: Liberty Fund, 1987, pp. 577–89.

————. *A Treatise of Human Nature*, 2nd ed. Ed. L.A. Selby-Bigge. Oxford: Clarendon, 1981.

Husserl, Edmund. *The Crisis of the European Sciences and Transcendental Phenomenology: An Introduction to Phenomenological Philosophy*. Trans. David Carr. Evanston: Northwestern University Press, 1970.

————. *Ideen zu einer reinen Phänomenologie und phänomenologischen Philosophie*, book 2. Ed. Marly Biemel. The Hague: Martinus Nijhoff, 1952.

Huth, Martin. "Reflexionen zu einer Ethik des vulnerablen Leibes." *Zeitschrift für praktische Philosophie* vol. 3, no. 1 (2016): 273–304.

Jacobs, Andrew. "Is Dairy Farming Cruel to Cows?" *New York Times*. December 29, 2020. https://www.nytimes.com/2020/12/29/science/dairy-farming-cows-milk.html.

James, William. "Pragmatism." *Writings 1902–1910*. New York: Library of America, 1987, pp. 479–624.

————. *The Principles of Psychology*, vol. 2. New York: Dover, 2014.

Jay, Martin. "Sartre, Merleau-Ponty, and the Search for a New Ontology of Sight." *Modernity and the Hegemony of Vision*, pp. 143–85.

Jonas, Hans. *The Gnostic Religion: The Message of the Alien God and the Beginnings of Christianity*. Boston: Beacon Press, 1991.

Kant, Immanuel. "An Answer to the Question: What Is Enlightenment?" *Political Writings*, pp. 54–60.

————. *Anthropology, History, and Education*. Ed. Günter Zöller and Robert P. Louden. Cambridge: Cambridge University Press, 2009.

————. "Anthropology from a Pragmatic Point of View." *Anthropology, History, and Education*, pp. 227–429.

————. "Conjectural Beginning of Human History." *Anthropology, History, and Education*, pp. 160–75.

————. *Critique of the Power of Judgment*. Ed. Paul Guyer. Cambridge: Cambridge University Press, 2008.

————. *Critique of Practical Reason*, 3rd ed. Trans. Lewis White Beck. Upper Saddle River: Library of Liberal Arts/Prentice Hall, 1993.

————. *Critique of Pure Reason*. Trans. Norman Kemp Smith. New York: Humanities Press, 1950.

————. *Grounding for the Metaphysics of Morals*. Trans. James W. Ellington. Indianapolis: Hackett, 1981.

————. *Lectures on Ethics*. Ed. Peter Heath and J.B. Schneewind. Cambridge: Cambridge University Press, 1997.

————. *The Metaphysics of Morals*. Ed. Mary Gregor. Cambridge: Cambridge University Press, 1996.

————. "Of the Different Races of Human Beings." *Anthropology, History, and Education*, pp. 82–97.

————. "Perpetual Peace: A Philosophical Sketch." *Political Writings*, pp. 93–130.

————. *Political Writings*, 2nd ed. Ed. Hans Reiss. Cambridge: Cambridge University Press, 1991.

————. "What Is Orientation in Thinking?" *Political Writings*, pp. 237–49.

Kierkegaard, Søren. *Fear and Trembling/Repetition*. Ed. and Trans. Howard V. Hong and Edna H. Hong. Princeton: Princeton University Press, 1983.

Kohak, Erazim. *The Embers and the Stars: A Philosophical Inquiry into the Moral Sense of Nature*. Chicago: University of Chicago Press, 1984.

Korsgaard, Christine M. *The Constitution of Agency: Essays on Practical Reason and Moral Psychology*. Oxford: Oxford University Press, 2008.

————. *Fellow Creatures: Our Obligations to the Other Animals*. Oxford: Oxford University Press, 2018.

Kramer, Matthew H. *Moral Realism as a Moral Doctrine*. West Sussex: Wiley-Blackwell, 2009.

Kuhn, Thomas. *The Structure of Scientific Revolutions*, 2nd ed. Chicago: Unversity of Chicago Press, 1970.

Lestel, Dominique. *Eat This Book: A Carnivore's Manifesto*. Trans. Gary Steiner. New York: Columbia University Press, 2016.

Levin, David Michael. *Modernity and the Hegemony of Vision*. Berkeley: University of California Press, 1993.

Lipsius, Justus. *De constantia/Von der Standhaftigkeit*, Latin-German. Trans. Florian Numann. Mainz: Dieterisch'sche Verlagsbuchhandlung, 1998.

————. *On Constancy*. Ed. John Sellars. Liverpool: Liverpool University Press, 2006.

Long, A.A. and D.N. Sedley, eds. *The Hellenistic Philosophers*, vol. 2. Cambridge: Cambridge University Press, 1990.

Löwith, Karl. "Der Weltbegriff der neuzeitlichen Philosophie." *Sitzunsberichte der Heidelberger Akademie der Wissenschaften, Philosophisch-historischer Klasse* vol. 44 (1960): 7–23.

————. *Meaning in History: The Theological Implications of the Philosophy of History*. Chicago: University of Chicago/Phoenix Books, 1949.

————. "The Occasional Decisionism of Martin Heidegger." *Martin Heidegger and European Nihilism*. Trans. Gary Steiner. New York: Columbia University Press, 1995, pp. 137–72.

————. "Welt und Menschenwelt." *Mensch und Menschenwelt: Beiträge zur Anthropologie*. Stuttgart: Metzler, 1981, pp. 295–328.

Luther, Martin. *Open Letter to the Christian Nobility of the German Nation Concerning the Reform of the Christian Estate*, 1520.

MacKinnon, Michael. "Animals, Economics, and Culture in the Athenian Agora: Comparative Zooarchaeological Investigations." *Hesperia: The Journal of the American School of Classical Studies at Athens* vol. 83, no. 2 (April-June 2014): 189–255.

Marcuse, Herbert. *One-Dimensional Man: Studies in the Ideology of Advanced Industrial Society*. Boston: Beacon Press, 1964.

Merleau-Ponty, Maurice. *Nature: Course Notes from the College de France*. Ed. Dominique Séglard. Evanston: Northwestern University Press, 2003.

———. *Phenomenology of Perception*. Trans. Colin Smith. London: Routledge & Kegan Paul, 1978.

———. "The Primacy of Perception and Its Philosophical Consequences." *The Primacy of Perception and Other Essays on Phenomenological Psychology, the Philosophy of Art, History, and Politics*. Ed. James M. Edie. Evanston: Northwestern University Press, 1964, pp. 12–42.

———. *The Visible and the Invisible*. Ed. Claude Lefort. Evanston: Northwestern University Press, 1968.

Mill, John Stuart. *On Liberty and Other Essays*. Ed. John Gray. Oxford: Oxford University Press, 1998.

———. *The Subjection of Women*, 1869.

———. "Utilitarianism." *On Liberty and Other Essays*, pp. 131–204.

Miller, John William. *In Defense of the Psychological*. New York: Norton, 1983.

———. *The Definition of the Thing with Some Notes on Language*. New York: Norton, 1980.

———. "Idealism and Freedom." *The Task of Criticism: Essays on Philosophy, History, and Community*, pp. 249–60.

———. *The Midworld of Symbols and Functioning Objects*. New York: Norton, 1982.

———. "The Owl." *The Task of Criticism: Essays on Philosophy, History, and Community*, pp. 221–32.

———. *The Paradox of Cause and Other Essays*. New York: Norton, 1978.

———. *The Philosophy of History with Reflections and Aphorisms*. New York: Norton, 1981.

———. "The Quality of Philosophy Is Not Strained." *The Task of Criticism: Essays on Philosophy, History, and Community*, pp. 47–52.

———. *The Task of Criticism: Essays on Philosophy, History, and Community*. Ed. Joseph P. Fell, Vincent Colapietro and Michael J. McGandy. New York: Norton, 2005.

———. "Tolerance and Its Paradoxes." *The Task of Criticism: Essays on Philosophy, History, and Community*, pp. 279–88.

Montaigne, Michel de. "Apology for Raymond Seybond." *The Complete Works*, pp. 386–556.

———. *The Complete Works*. Trans. Donald M. Frame. New York: Everyman's Library, 2003.

Morgan, C. Lloyd. *Animal Behavior*. London: Edward Arnold, 1900.

Nagel, Thomas. *The Possibility of Altruism*. Princeton: Princeton University Press, 1979.

———. "What is it Like to be a Bat?" *Mortal Questions*. Cambridge: Cambridge University Press, 1979, pp. 165–80.

———. "What We Owe a Rabbit." *New York Review of Books*. March 21, 2019. https://www.nybooks.com/articles/2019/03/21/christine-korsgaard-what-we-owe-a-rabbit/

Nibert, David. *Animal Oppression and Human Violence: Domesecration, Capitalism, and Global Conflict*. New York: Columbia University Press, 2013.

Nietzsche, Friedrich. *Basic Writings of Nietzsche*. Trans. Walter Kaufmann. New York: Modern Library, 1992.

———. "Beyond Good and Evil." *Basic Writings of Nietzsche*, pp. 179–436.

——. *The Gay Science*. Trans. Walter Kaufmann. New York: Vintage, 1974.

——. "On the Genealogy of Morals." *Basic Writings of Nietzsche*, pp. 437–600.

——. "On Truth and Lies in a Nonmoral Sense." *Philosophy and Truth: Selections from Nietzsche's Notebooks of the Early 1870's*. Ed. and Trans. Daniel Breazeale. Atlantic Highlands, NJ: Humanities Press International, 1992, pp. 79–97.

——. "Twilight of the Idols." *The Portable Nietzsche*. Ed. and Trans. Walter Kaufmann. New York: Penguin, 1976, pp. 465–563.

——. *The Will to Power*. Trans. Walter Kaufmann and R.J. Hollingdale. New York: Vintage Books, 1968.

Nozick, Robert. *Anarchy, State, and Utopia*. New York: Basic Books, 2013.

Nussbaum, Martha Craven. *Aristotle's De Motu Animalium*. Princeton: Princeton University Press, 1978.

——. *Frontiers of Justice: Disability, Nationality, Species Membership*. Cambridge: Belknap/Harvard University Press, 2006.

——. *Justice for Animals: Our Collective Responsibility*. New York: Simon and Schuster, 2023.

——. *Upheavals of Thought: The Intelligence of Emotions*. Cambridge: Cambridge University Press, 2001.

Ortega y Gasset, José. *Historical Reason*. Trans. Philip W. Silver. New York: Norton, 1984.

——. "History as a System." *History as a System and Other Essays Toward a Philosophy of History*, pp. 165–236.

——. *History as a System and Other Essays Toward a Philosophy of History*. New York: Norton, 1961.

——. "Man as Technician." *History as a System and Other Essays Toward a Philosophy of History*, pp. 87–164.

Ovid. *Metamorphoses Books 9–15*. Trans. Frank Justus Miller. Cambridge: Harvard University Press, 1977.

Patterson, Charles. *Eternal Treblinka: Our Treatment of Animals and the Holocaust*. New York: Lantern Books, 2002.

Pico Della Mirandola. *Oration on the Dignity of Man 1486*. Ed. Francesco Borghesi et al. Cambridge: Cambridge University Press, 2012.

Porphyry. *On Abstinence from Killing Animals*. Trans. Gillian Clark. Ithaca: Cornell University Press, 2000.

Posner, Richard A. "Animal Rights: Legal, Philosophical, and Pragmatic Perspectives." *Animal Rights: Current Debates and New Directions*. Ed. Cass R. Sunstein and Martha C. Nussbaum. Oxford: Oxford University Press, 2004, pp. 51–77.

Proust, Marcel. *Swann's Way. In Search of Lost Time*, vol. 1. Trans. C.K. Scott Montcrieff et al. New York: Modern Library, 2003.

Rawls, John. *Justice as Fairness: A Restatement*. Ed. Erin Kelly. Cambridge: Belknap/Harvard University Press, 2001.

——. *Political Liberalism*, Expanded ed. New York: Columbia University Press, 2005.

——. *A Theory of Justice*, Revised ed. Cambridge: Harvard/Belknap, 2000.

Regan, Tom. *The Case for Animal Rights*. Berkeley: University of California Press, 1983.

Rorty, Richard. *Contingency, Irony, and Solidarity*. Cambridge: Cambridge University Press, 2009.

Rowlands, Mark. *Can Animals Be Moral?* Oxford: Oxford University Press, 2012.

Santayana, George. *The Life of Reason*. Amherst, NY: Prometheus Books, 1998.

Sartre, Jean-Paul. *Existentialism and Human Emotions*. Secaucus: Carol Publishing Group, 1998.

Scanlon, T.M. *Being Realistic About Reasons*. Oxford: Oxford University Press, 2014.

Scarry, Elaine. *The Body in Pain: The Making and Unmaking of the World*. New York: Oxford University Press, 1987.

Scheler, Max. *Die Stellung des Menschen im Kosmos*. Munich: Nymphenburger Verlagshandlung, 1949.

———. *Wesen und Formen der Sympathie*. Bern/Munich: Francke Verlag, 1973.

Schmitt, Carl. *The Concept of the Political*. Trans. George Schwab. Chicago: University of Chicago Press, 1996.

———. *Political Theology: Four Chapters on the Concept of Sovereignty*. Trans. George Schwab. Chicago: University of Chicago Press, 2005.

Schopenhauer, Arthur. *The Fourfold Root of the Principle of Sufficient Reason*. Trans. E.F.J. Payne. La Salle, IL: Open Court, 1974.

———. *On the Basis of Morality*. Trans. E.F.J. Payne. Providence: Berghahn Books, 1995.

———. "On the Doctrine of the Indestructibility of our True Nature by Death." *Parerga and Paralipomena*, vol. 2, pp. 267–82.

———. "On Religion." *Parerga and Paralipomena*, vol. 2, pp. 324–94.

———. *Parerga and Paralipomena*, vol. 2. Trans. E.F.J. Payne. Oxford: Clarendon Press, 2000.

———. *The World as Will and Representation*, vol. 1. Trans. E.F.J. Payne. Indian Hills, CO: Falcon's Wing Press, 1958.

Seneca. *Epistles 1–65*. Trans. Richard M. Gummere. Cambridge, MA: Harvard University Press, 1917.

———. *Epistles 66–92*. Trans. Richard M. Gummere. Cambridge, MA: Harvard University Press, 1920.

———. *Epistles 93–124*. Trans. Richard M. Gummere. Cambridge, MA: Harvard University Press, 1925.

———. *Moral Essays*, vol. 1. Trans. John W. Basore. Cambridge, MA: Harvard University Press, 1928.

———. *Moral Essays*, vol. 2. Trans. John W. Basore. Cambridge, MA: Harvard University Press, 1932.

———. "On Anger." *Moral Essays*, vol. 1, pp. 106–355.

———. "On Firmness." *Moral Essays*, vol. 1, pp. 48–105.

———. "On the Happy Life." *Moral Essays*, vol. 2, 98–179.

Shafer-Landau, Russ. *Moral Realism: A Defence*. Oxford: Clarendon, 2003.

Singer, Peter. *Animal Liberation*, Updated ed. New York: Harper Perennial, 2009.

———. *Practical Ethics*, 3rd ed. Cambridge: Cambridge University Press, 2011.

Smith, Adam. *An Inquiry into the Nature and Causes of the Wealth of Nations*, vol. 2. Ed. R.H. Campbell and A.S. Skinner. Oxford: Oxford University Press, 1976.

———. *The Theory of Moral Sentiments*. Cambridge: Cambridge University Press, 2002.

Sorabji, Richard. *Animal Minds and Human Morals: The Origins of the Western Debate*. Ithaca: Cornell University Press, 1993.

Srinivasan, Amia. "The Aptness of Anger." *The Journal of Political Philosophy* vol. 26, no. 2 (2018): 123–44.

Stahl, Gary. "Making the Moral World." *The Philosophy of John William Miller*, Bucknell review. Ed. Joseph P. Fell. Lewisburg: Bucknell University Press, 1990, pp. 111–22.

Steiner, Gary. *Animals and the Limits of Postmodernism*. New York: Columbia University Press, 2013.

———. *Animals and the Moral Community: Mental Life, Moral Status, and Kinship*. New York: Columbia University Press, 2008.

————. *Anthropocentrism and Its Discontents: The Moral Status of Animals in the History of Western Philosophy*. Pittsburgh: University of Pittsburgh Press, 2005.

————. *Descartes as a Moral Thinker: Christianity, Technology, Nihilism*. Amherst, NY: Prometheus/Humanity Books, 2004.

————. "Kathy Rudy's Feel-Good Ethics." *Humanimalia* vol. 3, no. 2 (Spring 2012): 130–5.

————. "The Moral Schizophrenia of Catholicism." *The Ark* vol. 240 (Autumn 2018): 40–8.

————. "The Perils of a Total Critique of Reason: Rethinking Heidegger's Influence." *Philosophy Today* vol. 47 (2003): 93–111.

————. "Review of John J. Callanan and Lucy Allais, eds., *Kant and Animals*." *Journal of the History of Philosophy* vol. 59, no. 3 (July 2021): 517–9.

————. Review of Mark Rowlands, *Can Animals Be Moral? The Philosophers' Magazine* no. 65 (2d quarter 2014): 115–8.

————. "Tiere als Personen, aber nicht als Staatsbürger." *TIERethik* vol. 14, no. 1 (2017): 14–39.

————. "Toward a Nonanthropocentric Cosmopolitanism." *Animals and the Limits of Postmodernism*, pp. 167–194.

Taureck, Bernhard. *Manifest des veganen Humanismus*. Paderborn: Wilhelm Fink, 2015.

Taylor, Charles. "The Concept of a Person." *Human Agency and Language*, Philosophical Papers 1. Cambridge: Cambridge University Press, 2005, pp. 97–114.

————. *Hegel*. Cambridge: Cambridge University Press, 1975.

————. "Self-Interpreting Animals." *Human Agency and Language*, Philosophical Papers 1, pp. 45–76.

Taylor, Paul W. "The Ethics of Respect for Nature." *The Animal Rights/Environmental Ethics Debate*, pp. 95–120.

Todorov, Tzvetan. *Nous et les autres. La réflexion française sur la diversité humaine*. Paris: Éditions du Seuil, 1989.

Tugendhat, Ernst. *Der Wahrheitsbegriff bei Husserl und Heidegger*, 2nd ed. Berlin: Walter De Gruyter, 1970.

Varner, Gary E. *Personhood, Ethics, and Animal Cognition: Situating Animals in Hare's Two-Level Utilitarianism*. Oxford: Oxford University Press, 2012.

Vetlesen, Arne Johan. *Perception, Empathy, and Judgment: An Inquiry into the Preconditions of Moral Performance*. University Park: Penn State Press, 1994.

Warren, Mary Anne. "The Rights of the Nonhuman World." *The Animal Rights/Environmental Ethics Debate*, pp. 185–210.

Watson, Richard A. "Self-Consciousness and the Rights of Nonhuman Animals and Nature." *The Animal Rights/Environmental Ethics Debate*, pp. 1–36.

Williams, Bernard. *Descartes: The Project of Pure Enquiry*. Harmondsworth: Penguin, 1978.

————. "Internal and External Reasons." *Moral Luck: Philosophical Papers 1973–1980*. Cambridge: Cambridge University Press, 1983, pp. 101–13.

Woolf, Virginia. *Three Guineas*. Orlando: Harvest/Harcourt, 1966.

Yancy, George. "Judith Butler: When Killing Women Isn't a Crime: The Global Struggle to Stop Violence Against Women and the Struggle for Human Equality are Linked." *New York Times*. July 10, 2019. https://www.nytimes.com/2019/07/10/opinion/judith-butlergender html.

INDEX

Printed in the United States
by Baker & Taylor Publisher Services

Printed in the United States
by Baker & Taylor Publisher Services